T0281595

Waves and Mean Flows

Interactions between waves and mean flows play a crucial role in understanding the long-term aspects of atmospheric and oceanographic modelling. Indeed, our ability to predict climate change hinges on our ability to model waves accurately.

This book gives a modern account of the nonlinear interactions between waves and mean flows, such as shear flows and vortices. A detailed account of the theory of linear dispersive waves in moving media is followed by a thorough introduction to classical wave–mean interaction theory. The author then extends the scope of the classical theory and lifts its restriction to zonally symmetric mean flows. It can be used as a fundamental reference, a course text, or by geophysicists and physicists needing a first introduction.

This Second Edition includes new material, including a section on Langmuir circulations and the Craik–Leibovich instability. The author has also added exercises to aid students' learning.

OLIVER BÜHLER is a Professor of Mathematics and Atmosphere Ocean Science at the Courant Institute of Mathematical Sciences, New York University.

Established in 1952, the *Cambridge Monographs on Mechanics* series has maintained a reputation for the publication of outstanding monographs covering such areas as wave propagation, fluid dynamics, theoretical geophysics, combustion, and the mechanics of solids. The books are written for a wide audience and balance mathematical analysis with physical interpretation and experimental data where appropriate.

A complete list of books in the series can be found at www.cambridge.org/mathematics.

RECENT TITLES IN THIS SERIES

Turbulence, Coherent Structures, Dynamical Systems and Symmetry (Second Edition)
PHILIP HOLMES, JOHN L. LUMLEY, GAHL BERKOOZ & CLARENCE W. ROWLEY

Elastic Waves at High Frequencies
JOHN G. HARRIS

Gravity–Capillary Free-Surface Flows
JEAN-MARC VANDEN-BROECK

Lagrangian Fluid Dynamics
ANDREW F. BENNETT

Plasticity
S. NEMAT-NASSER

Reciprocity in Elastodynamics
J. D. ACHENBACH

Waves and Mean Flows

SECOND EDITION

OLIVER BÜHLER
Courant Institute of Mathematical Sciences
New York University

CAMBRIDGE
UNIVERSITY PRESS

CAMBRIDGE
UNIVERSITY PRESS

University Printing House, Cambridge CB2 8BS, United Kingdom

One Liberty Plaza, 20th Floor, New York, NY 10006, USA

477 Williamstown Road, Port Melbourne, VIC 3207, Australia

314-321, 3rd Floor, Plot 3, Splendor Forum, Jasola District Centre, New Delhi - 110025, India

103 Penang Road, #05-06/07, Visioncrest Commercial, Singapore 238467

Cambridge University Press is part of the University of Cambridge.

It furthers the University's mission by disseminating knowledge in the pursuit of education, learning and research at the highest international levels of excellence.

www.cambridge.org
Information on this title: www.cambridge.org/9781107669666

© Oliver Bühler 2009, 2014

This publication is in copyright. Subject to statutory exception and to the provisions of relevant collective licensing agreements, no reproduction of any part may take place without the written permission of Cambridge University Press.

First published 2009
Second Edition 2014

A catalogue record for this publication is available from the British Library

ISBN 978-1-107-66966-6 Paperback

Cambridge University Press has no responsibility for the persistence or accuracy of URLs for external or third-party internet websites referred to in this publication, and does not guarantee that any content on such websites is, or will remain, accurate or appropriate.

```
I will arise and go now, for always night and day
I hear lake water lapping with low sounds by the shore;
While I stand on the roadway, or on the pavements grey,
I hear it in the deep heart's core.
```

W. B. Yeats 'The Lake Isle Of Innisfree'

Contents

Preface

For the revised edition

The happy occasion of the revised paperback printing made it possible to add a section on Langmuir circulations and the Craik–Leibovich instability to chapter 11. These are important and fundamental topics that ought to have been included already in the first edition. This new material also prompted significant changes in section 13.4 on the vorticity generated by breaking surface-gravity waves, which hopefully make this crucial topic more transparent. In addition, there are smaller changes such as high-lighting the amazing curl–curvature formula for wave ray tracing in a weak vortical mean flow in §4.4.3, as well as numerous small fixes and some additional references. A small number of exercises has also been added to various chapters, which hopefully will aid the educational aspects of this book.

I am exceptionally grateful to Michael McIntyre for his very detailed reading of the first edition and for his support in preparing this revised edition. Thanks are also due to Rick Salmon and William Young for their insightful suggestions and to David Tranah for his continued support at Cambridge University Press.

Finally, I would like to dedicate this edition to the memory of my father by the last words of Mahler's *Lied der Erde: "Ewig, ewig"*.

New York, March 2013.

The aim of this book

This book is on waves and on their interactions with mean flows such as shear flows or vortices. Such interactions are generally a two-way street, with the waves being affected by the mean flow whilst the mean flow itself responds to the presence of the waves. For instance, readily observed examples of waves affected by mean flows are surface waves propagating on a sheared river current, or ripples that are refracted by a bath-tub vortex. Mean flows that are responding to waves are slightly less easily observed, here examples are given by the classic phenomenon of acoustic streaming, by longshore currents driven by breaking waves on beaches, and, as it turns out, also by many other flows in the atmosphere and ocean. Not surprisingly, *wave–mean interaction theory* is a very important topic in geophysical fluid dynamics (GFD).

For instance, the wave-induced transport of mass, momentum, and angular momentum plays a crucial role in the global-scale circulation of the atmosphere and the ocean, and in such complex multi-scale phenomena as the stratospheric ozone hole. However, many waves that contribute significantly to this transport are much too small in scale to be resolvable by even the most powerful present-day supercomputers, which implies that these small-scale waves, and their interactions with the large-scale mean flow, must be "parametrized" in numerical models, i.e., they must be put-in by hand based on a combination of theory and observational data. Our ability to predict the climate by large-scale numerical simulation therefore hinges delicately on our ability to model unresolvable wave processes in a consistent and accurate manner.

Arguably, the theoretical advances in wave–mean interaction theory that have been achieved in GFD in order to answer this need have far outpaced those in other fluid-dynamical fields, such as plasma physics, or condensed-matter physics including superfluids. For instance, the intricate interplay between the so-called *pseudomomentum* of the waves and the momentum and the vorticity of the mean flow has been fully worked out in GFD, but not yet in these other fields. Hopefully, by providing a comprehensive account of the tools and key results of wave–mean interaction theory in GFD, the present book will be equally useful for readers interested in GFD and also for readers who might want to apply such a theory in other fields.

The types of waves that are studied in this book include sound waves, surface and internal gravity waves, waves modified by Coriolis forces, and also Rossby waves, which are a peculiar form of large-scale vorticity waves that owe their principal restoring mechanism to the spherical curvature of the rapidly rotating Earth. To the best of my ability, I have tried to focus as closely as possible on the *fundamental principles of fluid dynamics* in the hope that this approach will best elucidate the essential workings of the diverse interaction effects that can occur.

Specifically, at its core and throughout all its chapters, this book aims to make obvious and to exploit the intimate connections between three key concepts in fluid dynamics:

1. *Kelvin's circulation theorem*,
2. the material invariance of *potential vorticity* in ideal fluid flow, and
3. the dynamics of the waves' *pseudomomentum vector* field.

As we shall see, putting the focus on the fundamental fluid dynamics becomes particularly important if one considers interactions outside the setting of spatially symmetric mean flows, which is the classical setting that is considered in detail in the current textbooks on GFD. The present book gives a detailed discussion of this classical setting, but it also moves significantly beyond it, with nontrivial implications for the "parametrization" problem as well as, for instance, the ocean-beach problem.

With regard to the mathematical methods of wave–mean interaction theory, I have tried to combine freely the standard fixed-position, Eulerian methods with the less standard particle-following, Lagrangian methods. Each kind of method has its merits and drawbacks in different situations, and I think it is very advantageous to know how to use both kinds. After all, the interplay between Eulerian and Lagrangian viewpoints lies at the very heart of understanding in fluid dynamics.

Before moving on to the specifics of the book content, I would like to make just one comment on terminology, namely on the use of the phrase "wave–mean interactions". It is far from obvious that this is the most logical term. Indeed, from a physical point of view it might be more logical to speak of interactions between "waves" and "vortices", so we should prefer the term "wave–vortex interactions". On the other hand, from a mathematical point of view one might prefer to juxtapose "disturbance fields" and "mean fields", as these can be defined rigorously in terms of a suitable averaging operator. After all, mean fields can be wave-like and some disturbance fields (such as Rossby waves) can be vortex-like. Thus perhaps one should prefer the term "disturbance–mean interactions". In the light of these valid objections, we will use the cross-bred term "wave–mean interactions" precisely because it reminds us that our physical and mathematical categories never fully catch all the facets of the slippery reality we seek to understand—but we try.

Contents outline

The book is structured such that it can be read cover-to-cover by a general reader with little previous knowledge of the subject matter, although one might have to assume a considerable enthusiasm for fluid dynamics in this case. Alternatively, I have tried to keep the parts and the chapters within the parts as self-contained as possible, with clear references to results established elsewhere, so for the specialist it should be easy to pull out particular sections of interest. Some of the chapter materials were first tested in several graduate classes at the Courant Institute and at various summer schools, but the book as it stands is all new.

Each chapter ends with a very brief discussion of some references, which are very far from being comprehensive. Basically, I have only endeavoured to refer to some original sources and to particular review articles or textbooks that I have read myself. Today, it is very easy to find and to obtain a multitude of references from the web once one knows what to look for.

Part I covers the linear theory of dispersive waves in moving media, which is a necessary basis for the study of interaction theory. This also includes a brief summary of fluid dynamics in order to keep the book self-contained. Particular attention is given to group-velocity concepts, geometric wave theory, WKB techniques, and ray tracing, because these are the essential theoretical tools for the study of small-scale waves propagating on a large-scale flow. The conservation laws for wave activity measures such as *wave action* and *pseudomomentum* are derived, and it is explained how these conservation laws are related to the continuous symmetries of the basic state on which the waves are propagating. Caustics, where ray tracing fails and predicts infinite wave amplitudes, are ubiquitous in the atmosphere and ocean, and their treatment within linear theory is discussed in some detail.

Part II presents the classic wave–mean interaction theory that has been developed mostly in atmospheric science. This theory is based on *simple geometry*, which assumes spatially periodic flows and uses zonal averaging, i.e., averaging over the periodic coordinate such as longitude in the atmosphere. Moreover, the basic state on which the waves are propagating is assumed to be zonally symmetric, which greatly simplifies the situation.

In simple geometry the focus is on the zonal mean flow and on the corresponding zonal component of the pseudomomentum of the waves. The central result in this area is the *dissipative pseudomomentum rule*, which relates the dissipation rate of pseudomomentum due to viscous effects or wave breaking to an effective mean force felt by the mean flow. The most fundamental derivation of this result is based on Kelvin's circulation theorem and on the potential vorticity of the mean flow.

Examples discussed in this part include internal waves forced by flow over undulating topography and the interplay between large-scale Rossby waves and two-dimensional turbulence. The latter problem is relevant not only on Earth, but also on other planets such as Jupiter.

Part II also includes a detailed introduction to the so-called *generalized Lagrangian-mean theory*, which is based on particle-following averages and which formally allows extending all of the earlier results to finite wave amplitude. In particular, a detailed account of Kelvin's circulation theorem for the Lagrangian-mean flow can be given, which makes obvious the essential role of the pseudomomentum vector in this connection. Also, the previous discussion of conserved wave activities is extended here to apply to fully nonlinear, finite-amplitude waves and to the symmetries of the Lagrangian-mean flow.[1]

Part III discusses wave–mean interactions outside simple geometry, which is a profound change in the physical and mathematical situation. This makes the theory at once more complicated, but also much more relevant to the study of small-scale waves propagating generally on a slowly varying basic flow. Such a setting is natural in the ocean, for instance, because with rare exceptions it is not possible to average oceanic flows zonally, as the continents get in the way. It is also needed in the "parametrization" problem.

The crucial role of long-range mean pressure effects is then discussed in detail and with several examples. Such pressure effects are absent in simple geometry, and this marks the essential difference between the two situations. It is argued that at this stage wave–mean interaction theory must restrict itself to the dynamics of vorticity rather than of momentum, essentially in order to control the non-local effects due to the mean pressure field. With some further assumptions it is then possible to formulate a conservation law for a suitably defined *impulse* of the mean flow plus the pseudomomentum of the waves, and this law replaces the classic pseudomomentum rule of simple geometry.

A study of wave-driven nearshore circulations on beaches is then used to illustrate the new situation, which here involves mean-flow vortices driven by the waves. This is followed by a number of idealized scenarios that show how wavepackets can interact with isolated vortices at long range via refraction effects. These interactions are intrinsically non-local in nature. Finally, the peculiar kinship between wavepackets with low intrinsic group velocities and vortex couples will be discussed, which is relevant to packets of internal gravity waves undergoing a scale cascade to smaller wavelengths. This provides the most intriguing example of the interplay between the three fundamental concepts listed above.

Acknowledgements

I owe thanks to many people and institutions who, in different ways, enabled me to write this book. Foremost, these include Michael McIntyre and the

[1] Practical evaluations of finite-amplitude wave activity measures are usually computed differently, namely using the elegant methods of Hamiltonian fluid mechanics. Excellent accounts of these methods exist in the literature, but here we work with Lagrangian-mean methods because the Hamiltonian methods cannot be applied with equal ease to the topic of wave–mean interactions, where the Lagrangian-mean theory excels.

Dept of Applied Mathematics and Theoretical Physics in Cambridge (UK), and Andrew Majda and the Courant Institute of Mathematical Sciences at New York University. I would also like to acknowledge the kind hospitality of Rupert Klein and his group, as well as that of the Zuse Zentrum at the Freie Universität Berlin (Germany), during my 2007-08 sabbatical year, when the bulk of this book was written. David Tranah and the staff at Cambridge University Press kept the faith over long dry spells. My research has been supported financially by the National Science Foundation. Some of the illustrations were created using an author licence for Matlab.

I have been fortunate to have had many excellent teachers of fluid dynamics over the years: Heinz Schade, Alfons Michalke, Ken Powell, Phil Roe, Michael McIntyre, Peter Haynes, and David Crighton. The Woods Hole summer study in GFD has provided a home-away-from-home for many scientists, and I am thankful for its staff, especially Lou Howard, Joe Keller, and Ed Spiegel, for their inspiration and their teaching in GFD.

Once more I would like to thank Michael McIntyre, who taught me almost all I know about the topic of this book, and who kindly shared with me his draft of an earlier book project in the same direction; Michael also commented somewhat extensively on chapter 9.

Finally, as ever, all would have been for naught had it not been for SCK, now SCB, and our corollaries JBB and AKB.

New York, October 2008.

PART ONE

FLUID DYNAMICS AND WAVES

1
Elements of fluid dynamics

It is convenient to start with a brief summary of fluid dynamics fundamentals in order to establish the mathematical notation and the physical concepts that will be used throughout this book. We will first look at the kinematics of fluid flow, especially at how to capture the evolution of material elements such as material points or lines.

This is followed by a description of perfect fluid dynamics, which is the natural point of departure for the study of flows at very high Reynolds numbers in the atmosphere and the ocean. In these flows the direct influence of viscous forces is confined to boundary layers and to sparse pockets of three-dimensional turbulence within the fluid.

The culmination of perfect fluid dynamics is Kelvin's circulation theorem and the various links of this theorem to vorticity dynamics. Indeed, as we go on it will become increasingly clear that the circulation theorem is also the key result in wave–mean interaction theory.

1.1 Flow kinematics

In continuum fluid mechanics the molecular structure of the fluid is ignored and the description of the physical state of the fluid is accomplished by specifying a finite number of flow fields as functions of position $\boldsymbol{x}$ and time t, say. How many fields are needed depends on the complexity of the fluid under consideration, but all fluid flows require a working mass and momentum budget, which leads to the definitions of the density and velocity fields.

1.1.1 Mass, momentum and velocity

The *mass density* per unit volume $\rho(\boldsymbol{x}, t) \geq 0$ is defined such that

$$M = \int_{\mathcal{D}} \rho \, dV \tag{1.1}$$

is the fluid mass contained in any region $\mathcal{D}$, including the limiting case as $\mathcal{D}$ shrinks to an infinitesimal region. The essential usage of ρ as an integrand suggests that $\rho(\boldsymbol{x}, t)$ need not be very smooth; indeed, it certainly makes sense to allow for *discontinuity surfaces*, say at the interface between two fluids such as the interface between water and air, where the density jumps by a factor of a thousand, or at internal jumps such as compressible shocks in gas dynamics. Some fluids are mixtures that consist of several fluid species sharing the same volume and then each species has its own mass density. Examples occur in plasma physics, in superfluids, or even in the more familiar situation of rain drops or ice particles embedded in moist air. However, we will not consider mixtures in this book, so there will only be one fluid mass density.

The *momentum density* defines the fluid velocity vector $\boldsymbol{u}(\boldsymbol{x}, t)$ such that

$$\int_{\mathcal{D}} \rho \boldsymbol{u} \, dV \tag{1.2}$$

is the fluid momentum contained in any region $\mathcal{D}$, again including the infinitesimal limit. In other words, the fluid velocity at a point $\boldsymbol{x}$ is defined as the ratio between momentum and mass as $\mathcal{D}$ shrinks to zero volume around $\boldsymbol{x}$. This is a basic statement of physics.

Another basic statement of physics is that mass is conserved, which means that M in (1.1) should be constant if the region $\mathcal{D}$ moves with the fluid. More precisely, if we consider a region $\mathcal{D}$ whose points move with velocity $\boldsymbol{v}$, say, then it follows from calculus that

$$\frac{dM}{dt} = \int_{\mathcal{D}} \frac{\partial \rho}{\partial t} \, dV + \oint_{\partial \mathcal{D}} \rho \boldsymbol{v} \cdot \boldsymbol{dA} = \int_{\mathcal{D}} \left(\frac{\partial \rho}{\partial t} + \boldsymbol{\nabla} \cdot (\rho \boldsymbol{v}) \right) dV \tag{1.3}$$

where $\boldsymbol{dA} = \boldsymbol{n} dA$ is the outward-pointing area element at the boundary $\partial \mathcal{D}$ and the second form has been obtained by using the divergence theorem. Now, the physical statement of *mass conservation* implies that (1.3) is zero if $\boldsymbol{v} = \boldsymbol{u}$, in which case $\mathcal{D}$ is called a material volume and (1.3) is called the *continuity equation* in integral form. For this to be true for any choice of $\mathcal{D}$ it must be the case that

$$\frac{\partial \rho}{\partial t} + \boldsymbol{\nabla} \cdot (\rho \boldsymbol{u}) = 0. \tag{1.4}$$

This is the *continuity equation* in differential form. It is worth noting the double role of $\rho \boldsymbol{u}$, which is both the momentum vector density and also the mass flux vector.

1.1.2 Material trajectories and derivatives

The integral curves of $\boldsymbol{u}$ are called *material trajectories* and they are given by the vector-valued functions $\boldsymbol{X}(t)$ that solve the initial-value problem

$$t \in [0, T]: \quad \frac{d\boldsymbol{X}}{dt} = \boldsymbol{u}(\boldsymbol{X}(t), t) \quad \text{and} \quad \boldsymbol{X}(0) = \boldsymbol{X}_0 \tag{1.5}$$

with some initial position $\boldsymbol{X}_0$. Thus, $\boldsymbol{X}(t)$ is the trajectory during the time interval $[0, T]$ of the *fluid particle*[1] that was at position $\boldsymbol{X}_0$ at time $t = 0$. It is a working assumption in fluid dynamics that (1.5) is well posed for some finite time of interest $T > 0$, which implies that during this time interval material trajectories are unique and do not cross. Thus a fluid particle is uniquely identified, or labelled, by its initial position $\boldsymbol{X}_0$. Now, using $\boldsymbol{X}(t)$ we can evaluate any field $\phi(\boldsymbol{x}, t)$ along the trajectory $\boldsymbol{x} = \boldsymbol{X}(t)$ to obtain the time evolution of ϕ as observed by following the particle. Specifically, the chain rule yields

$$\frac{d}{dt}\phi(\boldsymbol{X}(t), t) = \left(\frac{\partial}{\partial t} + \boldsymbol{u}(\boldsymbol{X}(t), t) \cdot \boldsymbol{\nabla}\right) \phi(\boldsymbol{X}(t), t) \tag{1.6}$$

for the rate of change of ϕ along the material trajectory. This shows how the velocity field $\boldsymbol{u}$ defines a directional derivative following the fluid motion, which is called the *material derivative*. Obviously, the material derivative can also be expressed without reference to $\boldsymbol{X}(t)$ by

$$\frac{D\phi}{Dt} \equiv \frac{\partial \phi}{\partial t} + (\boldsymbol{u} \cdot \boldsymbol{\nabla})\phi. \tag{1.7}$$

This gives the rate of change per unit time of ϕ as observed by following the fluid particle that is occupying the position $\boldsymbol{x}$ at time t. This is the most important mathematical definition in fluid dynamics.

For smooth flows the continuity equation (1.4) can be re-written in the equivalent form

$$\frac{D\rho}{Dt} + \rho \boldsymbol{\nabla} \cdot \boldsymbol{u} = 0, \tag{1.8}$$

which shows explicitly how the density following a particle evolves according to the divergence of the velocity field. In the special case of *incompressible*

[1] The concept of a 'fluid particle' here must not be confused with a fluid molecule or any other form of atomic particle, which is necessarily part of the molecular structure that is ignored in a continuum theory. Instead, the fluid particle here is a small region $\mathcal{D}$ surrounding a point $\boldsymbol{x}$ such that the variations of $\boldsymbol{u}$ across $\mathcal{D}$ are negligible. Although $\mathcal{D}$ is small in this sense it still contains very many atomic particles. This is a well-defined concept in smooth flows although some care is needed in the presence of discontinuity surfaces.

flow $D\rho/Dt = 0$ and we have the pair

$$\frac{D\rho}{Dt} = 0 \quad \text{and} \quad \boldsymbol{\nabla} \cdot \boldsymbol{u} = 0. \tag{1.9}$$

Incompressibility does not imply that ρ is spatially uniform, only that ρ is constant along material trajectories. This is an example of *material invariance*, which takes the form (1.9a) for scalar fields.

1.1.3 Lagrangian and Eulerian variables

In fluid kinematics there is a distinction between *Eulerian variables* that are defined as functions of fixed position $\boldsymbol{x}$ and *Lagrangian variables* that are defined relative to fixed fluid particles labelled by their initial positions $\boldsymbol{X}_0$, say. For example, we can think of the mass density as a Eulerian variable described by $\rho(\boldsymbol{x}, t)$ or as a Lagrangian variable described by a function of the initial particle positions $\rho(\boldsymbol{X}_0, t)$. Clearly, this choice affects the form of the governing equations, for instance the material derivative takes the alternative forms

$$\frac{D\rho}{Dt} = \left(\frac{\partial}{\partial t} + (\boldsymbol{u} \cdot \boldsymbol{\nabla})\right) \rho(\boldsymbol{x}, t) = \frac{\partial}{\partial t}\rho(\boldsymbol{X}_0, t). \tag{1.10}$$

Both descriptions are mathematically equivalent, but in practice (especially in numerical computations) the use of Eulerian variables dominates. This is because realistic fluid flows involve large, chaotic particle trajectories that would render a mathematical description or numerical computation based on exactly resolved trajectories infeasible after a very short time. This is obvious from everyday experiences such as stirring milk in coffee, for example. Nevertheless, Lagrangian variables have their conceptual merits because they greatly simplify the description of material invariants. In addition, some dynamical laws such as Kelvin's circulation theorem can only be exploited fully using Lagrangian concepts.

We will take the view that fluid dynamics is naturally a hybrid theory in which both Eulerian and Lagrangian variables play a useful role in understanding and computation. This will also be our approach to wave–mean interaction theory.

1.1.4 Evolution of material elements

The material derivative (1.7) extends naturally to infinitesimal material elements such as points, lines, areas and volumes. The material derivative of a point is

$$\frac{D\boldsymbol{x}}{Dt} = 0 + (\boldsymbol{u}\cdot\boldsymbol{\nabla})\boldsymbol{x} = \boldsymbol{u}. \tag{1.11}$$

Thus the material rate of change of position is given by the velocity. Two neighbouring material points $\boldsymbol{x}$ and $\boldsymbol{x}+\boldsymbol{dx}$ move with the respective velocities $\boldsymbol{u}$ and $\boldsymbol{u}+\boldsymbol{du}$ and therefore a material line element $\boldsymbol{dx}$ evolves according to

$$\frac{D}{Dt}(\boldsymbol{dx}) = \boldsymbol{du} = (\boldsymbol{dx}\cdot\boldsymbol{\nabla})\boldsymbol{u}. \tag{1.12}$$

The evolution of the volume content of a material region follows from (1.3) with $\rho = 1$ and $\boldsymbol{v} = \boldsymbol{u}$ as

$$\frac{d}{dt}\int_{\mathcal{D}} dV = \int_{\mathcal{D}} \boldsymbol{\nabla}\cdot\boldsymbol{u}\,dV \quad\Rightarrow\quad \frac{D}{Dt}(dV) = \boldsymbol{\nabla}\cdot\boldsymbol{u}\,dV. \tag{1.13}$$

This shows that positive divergence corresponds to volume expansion. Combining (1.13) and (1.8) we obtain the continuity equation in the form

$$\frac{D}{Dt}(\rho\,dV) = 0, \tag{1.14}$$

i.e., the mass of a fluid particle is constant. This yields a useful general formula for the rate of change of an integral over a material volume $\mathcal{D}$, namely

$$\frac{d}{dt}\int_{\mathcal{D}} \rho\phi\,dV = \int_{\mathcal{D}} \frac{D}{Dt}(\rho\phi\,dV) = \int_{\mathcal{D}} \rho\frac{D\phi}{Dt}\,dV \tag{1.15}$$

for any function ϕ. For example, this formula applies to the centre of mass of a material volume:

$$\frac{d}{dt}\left(\frac{1}{M}\int_{\mathcal{D}} \rho\boldsymbol{x}\,dV\right) = \frac{1}{M}\int_{\mathcal{D}} \rho\boldsymbol{u}\,dV. \tag{1.16}$$

Thus the centre of mass moves with the total momentum.

Finally, the evolution of a material area element $\boldsymbol{dA}$ follows most easily from $dV = \boldsymbol{dx}\cdot\boldsymbol{dA}$ and therefore

$$\frac{D}{Dt}(\boldsymbol{dx}\cdot\boldsymbol{dA}) = \frac{D\,\boldsymbol{dx}}{Dt}\cdot\boldsymbol{dA} + \boldsymbol{dx}\cdot\frac{D\,\boldsymbol{dA}}{Dt} = \boldsymbol{\nabla}\cdot\boldsymbol{u}\,dV. \tag{1.17}$$

Substituting from (1.12) and demanding that the remaining equality holds for arbitrary $\boldsymbol{dx}$ then implies

$$\frac{D}{Dt}(\boldsymbol{dA}) = (\boldsymbol{\nabla}\cdot\boldsymbol{u})\,\boldsymbol{dA} - (\boldsymbol{\nabla}\boldsymbol{u})\cdot\boldsymbol{dA}, \tag{1.18}$$

where in the last term $\boldsymbol{dA}$ contracts with $\boldsymbol{u}$ and not with $\boldsymbol{\nabla}$. For example, in the case of isotropic expansion with $\boldsymbol{u} = \boldsymbol{x}$ the right-hand side is $(n-1)\,\boldsymbol{dA}$, where n is the number of spatial dimensions. This situation is familiar from inflating a balloon. In non-isotropic flow fields the last term captures the tilting of $\boldsymbol{dA}$ by the strain field.

For completeness, we note that the gradient of a material invariant ϕ evolves according to

$$\frac{D\phi}{Dt} = 0 \quad \Rightarrow \quad \frac{D}{Dt}(\nabla\phi) = -(\nabla\boldsymbol{u}) \cdot \nabla\phi, \tag{1.19}$$

where again $\nabla\phi$ contracts with $\boldsymbol{u}$ and not with ∇. For incompressible flows this is identical to the equation governing material area elements in (1.18). Experience shows that it is easy to get the signs wrong between (1.12), (1.18) and (1.19).

1.2 Perfect fluid dynamics

The equation of motion in a fluid follows from Newton's law, which demands that the time rate of change of momentum of a material body $\mathcal{D}$, say, is equal to the net force applied to the body. In a *perfect fluid* there are no viscous forces and this leads to Euler's equation. To close the set of equations we require constitutive relations, which for an ideal gas brings in the entropy. In perfect fluid flow there is no diabatic heating and therefore specific entropy is a material invariant. This closes the set of equations for perfect fluid flow.

1.2.1 Euler's equation

In a perfect fluid the only forces acting on the fluid are due to a potential per unit mass Φ and a pressure p. By definition, the net force on $\mathcal{D}$ is then

$$\int_{\mathcal{D}} -\rho\nabla\Phi\, dV - \oint_{\partial\mathcal{D}} p\, \boldsymbol{dA}. \tag{1.20}$$

For example, in the case of the standard gravitational potential $\Phi = gz$ with altitude z and constant gravity g the net potential force is downward and equals the weight of the fluid in $\mathcal{D}$. There is a qualitative difference between these two kinds of forces: the potential force acts throughout the volume of $\mathcal{D}$ whilst the pressure force acts only on its boundary $\partial\mathcal{D}$. Newton's law in integral form is

$$\frac{d}{dt}\int_{\mathcal{D}} \rho\boldsymbol{u}\, dV = \int_{\mathcal{D}} (-\rho\nabla\Phi - \nabla p)\, dV \tag{1.21}$$

and its differential form is *Euler's equation* for a perfect fluid:

$$\frac{D\boldsymbol{u}}{Dt} + \frac{\nabla p}{\rho} = -\nabla\Phi. \tag{1.22}$$

The division by ρ is convenient for vorticity dynamics. The appropriate boundary conditions for $\boldsymbol{u}$ at a fixed wall are that the normal component of $\boldsymbol{u}$ vanishes at the boundary. However, the tangential velocity components are not constrained in a perfect fluid.

1.2.2 Constitutive relations

We consider Φ as given and then in n spatial dimensions the continuity and Euler equations provide $1 + n$ equations for the $2 + n$ fields $(\rho, p, \boldsymbol{u})$. Thus we need a relation that links ρ and p, which is usually derived from thermodynamics under the assumption of local thermodynamic equilibrium. However, in the case of incompressible flow this additional equation is simply $\nabla \cdot \boldsymbol{u} = 0$. In fact, this case is special because taking the divergence of (1.22) then leads to an elliptic equation for the pressure. This is a diagnostic equation, i.e., an equation without a time derivative, which makes clear that for incompressible flow the pressure is not an independent field, but can be computed, albeit non-locally, from the other flow fields.

Another important class of fluid models is that a of *barotropic fluid*, in which there is a non-degenerate functional relationship between density and pressure. This simple fluid model allows for compressible effects, and we will use it frequently in this book. The assumed functional relationship implies that the pressure force is irrotational in a barotropic fluid, i.e.,

$$\nabla \times \left(\frac{\nabla p}{\rho}\right) = \nabla \left(\frac{1}{\rho}\right) \times \nabla p = 0. \tag{1.23}$$

For fluids such as an *ideal gas* there is a functional relationship between density, pressure, and one more state variable s, which is the entropy density per unit mass. In a perfect fluid there is no diabatic heating due to irreversible processes or radiation effects and therefore entropy is materially invariant in smooth flow regions:

$$\frac{Ds}{Dt} = 0. \tag{1.24}$$

However, the entropy increases if fluid particles pass through a shock, as is well known in gas dynamics. The barotropic fluid can be viewed as a special case of an ideal gas in which s is spatially uniform; this is also called a *homentropic fluid*. So, in perfect smooth flow the continuity and Euler equations together with (1.24) and the functional relationship between (ρ, p, s) form a complete system for $3+n$ flow fields. This will be sufficient for the purposes of this book, but we note that more complex fluid models may require further thermodynamics fields, such as a humidity variable for moist air, or they may include deviations from the thermodynamic relationships derived under the assumption of local thermodynamic equilibrium.

Finally, even for an ideal gas it is often useful to define some further thermodynamic flow fields, such as the temperature T, which are then linked to (ρ, p) via additional relations such as the ideal gas law and so on. For us it proves convenient to introduce the *internal energy* per unit mass, which is a function $\epsilon(\rho, s)$ defined by the first law of thermodynamics:

$$d\epsilon = T\,ds - p\,d\left(\frac{1}{\rho}\right) = T\,ds + \frac{p}{\rho^2}\,d\rho \quad \Leftrightarrow \quad \frac{\partial \epsilon}{\partial \rho} = \frac{p}{\rho^2} \quad \text{and} \quad \frac{\partial \epsilon}{\partial s} = T. \tag{1.25}$$

This internal energy quantifies the change of elastic energy due to compression or dilation of fluid particles, which will be needed to formulate the energy conservation law in §1.3. Another useful field is the *enthalpy* per unit mass defined by

$$P = \epsilon + \frac{p}{\rho}. \tag{1.26}$$

The enthalpy is useful because its differential is

$$dP = d\epsilon + \frac{dp}{\rho} - \frac{p}{\rho^2}\,d\rho = \frac{dp}{\rho} + T\,ds, \tag{1.27}$$

which implies that Euler's equation can be re-written as

$$\frac{D\boldsymbol{u}}{Dt} + \boldsymbol{\nabla}\Phi = -\frac{\boldsymbol{\nabla} p}{\rho} = -\boldsymbol{\nabla} P + T\boldsymbol{\nabla} s. \tag{1.28}$$

For barotropic flows the terms involving the entropy s in (1.27) and (1.28) are absent and thus in this case the pressure force is irrotational.

1.2.3 The polytropic fluid model

The polytropic fluid model is a special case of a barotropic fluid with a power law dependence of pressure on density, i.e., $p \propto \rho^\gamma$ for some constant γ. For example, in the case of homentropic flow of an ideal gas γ would be the usual ratio of specific heats. Also, this model includes the standard shallow-water equations if $\gamma = 2$ and ρ is identified with the layer depth as a function of horizontal coordinates (cf. §1.6 below). We will use this model frequently in this book and therefore we summarize its mechanical structure here.

We write the pressure as

$$p(\rho) = \frac{\rho_0 c_0^2}{\gamma}\left(\frac{\rho}{\rho_0}\right)^\gamma \tag{1.29}$$

where ρ_0 is a reference density and c_0 is the linear sound speed for perturbations around $\rho = \rho_0$. This is consistent with the nonlinear definition of the sound speed as $c^2 = \partial p/\partial \rho$ where the derivative is taken at constant entropy s. The advantage of writing the polytropic pressure in the form (1.29) is that it keeps separate the effects of varying γ and of varying the linear sound speed, which is instructive in applications.

The corresponding specific internal energy $\epsilon(\rho)$ is

$$\epsilon(\rho) = \frac{c_0^2}{\gamma(\gamma-1)}\left(\frac{\rho}{\rho_0}\right)^{\gamma-1} \tag{1.30}$$

and the specific enthalpy is

$$P = \frac{c_0^2}{\gamma-1}\left(\frac{\rho}{\rho_0}\right)^{\gamma-1} = \gamma\epsilon. \tag{1.31}$$

The governing momentum equation is then given by (1.28) without the $\boldsymbol{\nabla} s$ term.

1.3 Conservation laws and energy

The continuity equation (1.4) is an example of a *local conservation law*, which expresses the conservation of mass with the scalar density ρ and the flux vector $\rho\boldsymbol{u}$. In this special case the flux is purely advective, i.e., the flux equals the density times the velocity. Using (1.4), Euler's equation (1.22) can also be written in conservation form for the momentum, namely

$$\frac{\partial(\rho\boldsymbol{u})}{\partial t} + \boldsymbol{\nabla}\cdot(\rho\boldsymbol{u}\boldsymbol{u} + p\boldsymbol{\delta}) = -\rho\boldsymbol{\nabla}\Phi \tag{1.32}$$

where $\boldsymbol{\delta}$ is the unit tensor. Thus, in the absence of the potential force (1.32) is a conservation law for the momentum vector with density $\rho\boldsymbol{u}$ and flux tensor $\rho\boldsymbol{u}\boldsymbol{u} + p\boldsymbol{\delta}$. In this case the advective momentum flux $\rho\boldsymbol{u}\boldsymbol{u}$ is augmented by the non-advective momentum flux $p\boldsymbol{\delta}$ due to the pressure. The useful identity

$$\rho\frac{D\phi}{Dt} = \frac{\partial(\rho\phi)}{\partial t} + \boldsymbol{\nabla}\cdot(\rho\phi\boldsymbol{u}) \tag{1.33}$$

makes clear that any local conservation law with non-advective flux $\boldsymbol{F}$, say, can be written in the two equivalent forms

$$\frac{D\phi}{Dt} + \frac{1}{\rho}\boldsymbol{\nabla}\cdot\boldsymbol{F} = 0 \quad \text{or} \quad \frac{\partial(\rho\phi)}{\partial t} + \boldsymbol{\nabla}\cdot(\rho\phi\boldsymbol{u} + \boldsymbol{F}) = 0. \tag{1.34}$$

For $\phi = 1$ and $\phi = \boldsymbol{u}$ this yields the two forms of the mass and momentum conservation laws, respectively.

Perfect fluid motion also conserves energy and the corresponding conservation law is derived by contracting Euler's equation with $\boldsymbol{u}$, which yields

$$\frac{\partial E}{\partial t} + \boldsymbol{\nabla} \cdot (E\boldsymbol{u} + p\boldsymbol{u}) = -\rho \frac{\partial \Phi}{\partial t} \tag{1.35}$$

where the energy density per unit volume E is given by

$$E = \rho \left(\frac{|\boldsymbol{u}|^2}{2} + \Phi + \epsilon(\rho, s) \right). \tag{1.36}$$

For time-independent potentials this is a conservation law for the sum of mechanical, potential and internal energy. For example, in case of $\Phi = gz$ the total potential energy is proportional to the altitude of the centre of mass of the fluid body. The non-advective energy flux is the 'pressure work' vector $p\boldsymbol{u}$. For incompressible flows the internal energy term (which stems from the term $-p\boldsymbol{\nabla} \cdot \boldsymbol{u}$ in the manipulations) does not appear.

One can ask why to bother with the energy conservation law if the system is already completely defined by the mass and momentum equations. The answer is that conservation of total energy is a property of smooth solutions that is known a priori, i.e., it is known to be a property of the solution, even if the actual solution is not known. It lies at the very heart of applied mathematics to discover properties of solutions to differential equations without actually solving the equations. Energy conservation is an important tool in this quest and the same is true for the conservation of circulation to be considered next.

1.4 Circulation and vorticity

The circulation and vorticity are vitally important concepts in theoretical fluid dynamics. This is because they allow a second perspective on fluid dynamics that is complementary to the standard, momentum-based view. The vorticity view greatly aids an intuitive understanding of many complex fluid-dynamical phenomena such as nonlinear vortex dynamics, which are almost incomprehensible from a momentum-based point of view. Vorticity concepts are also vital to wave–mean interaction theory, as will become apparent throughout this book.

1.4.1 Circulation theorem

The circulation Γ is defined by the contour integral

$$\Gamma = \oint_C \boldsymbol{u} \cdot d\boldsymbol{x} \tag{1.37}$$

where C is a closed material loop, i.e., a closed contour that moves with the flow. The evolution of Γ follows from taking the material derivative of the entire integrand in (1.37), which gives

$$\frac{d\Gamma}{dt} = \oint_C \frac{D\boldsymbol{u}}{Dt} \cdot d\boldsymbol{x} + \oint_C \boldsymbol{u} \cdot \frac{D}{Dt}(d\boldsymbol{x}). \tag{1.38}$$

Substitution for the second term from (1.12) yields

$$\oint_C \boldsymbol{u} \cdot d\boldsymbol{u} = \oint_C d\left(\frac{1}{2}|\boldsymbol{u}|^2\right) = 0. \tag{1.39}$$

This is a perfect differential and because $\boldsymbol{u}$ is single-valued the integral around a closed loop vanishes. This is a kinematic fact and unrelated to Euler's equation. For the first term in (1.38) we substitute from (1.22) and obtain

$$\frac{d\Gamma}{dt} = \oint_C \boldsymbol{\nabla}\Phi \cdot d\boldsymbol{x} - \oint_C \frac{\boldsymbol{\nabla}p}{\rho} \cdot d\boldsymbol{x} = \oint_C \left(d\Phi - \frac{dp}{\rho}\right). \tag{1.40}$$

If the potential Φ is single-valued then the first term is again zero.[2] Moreover, if ρ is a constant then the second term will also be zero (because p is single-valued) and we obtain the celebrated result known as *Kelvin's circulation theorem*:

$$\frac{d\Gamma}{dt} = 0. \tag{1.41}$$

This is the most interesting formula in fluid dynamics. Clearly, the circulation theorem remains valid for a barotropic fluid because in this case $dp/\rho = dP$ is still a perfect differential. On the other hand, the circulation theorem fails in the ideal gas case because $dP - T\,ds$ is not a perfect differential in general.

However, a most important special case arises if the material loop C lies on an *isentrope*, i.e., on a surface of constant entropy $s = s_0$, say. In this situation $ds = 0$ on C and therefore the integral again vanishes. Moreover, because s is a material invariant we know that the material loop C will continue to lie on the same isentrope $s = s_0$ even though the shape of this surface will move around in space. This version of the circulation theorem is very important for stratified flows, in which there is a strong vertical gradient of s.

An analogous construction holds for an incompressible flow with inhomogeneous density described by (1.9). Here the isentrope must be replaced by

[2] This may not be the case in multiply-connected domains, such as an annulus in two dimensions, for instance. In this case it is feasible to have a multi-valued potential, e.g. with Φ proportional to the azimuthal angle, and such a potential will indeed change the circulation of a loop that encircles the annulus.

a so-called isopycnal surface $\rho = \rho_0$. In an incompressible flow an isopycnal surface is again materially invariant and the circulation theorem therefore holds for isopycnal material loops. We can combine the isentropic and isopycnal cases by speaking of *stratification surfaces*, which by definition are materially invariant surfaces on which there is a functional relationship between ρ and p.

Another application of the circulation theorem concerns periodic domains. For instance, if the flow is periodic in x with finite period length L, say, then any material contour $\mathcal{C}$ that traverses the periodic domain and has identifiable end points at $x = 0$ and $x = L$ qualifies as a closed material loop for the purpose of Kelvin's circulation theorem. This is because the periodicity of the physical variables is sufficient to allow the proof to go through. In stratified flow the contour must also lie on a stratification surface.

1.4.2 Vorticity and potential vorticity

The *vorticity* $\boldsymbol{\nabla} \times \boldsymbol{u}$ is the curl of the velocity field and it carries the same information as the skew-symmetric part of the velocity gradient $\boldsymbol{\nabla u}$. There is a close relation between circulation and vorticity because Stokes's theorem implies that for smooth flows

$$\Gamma = \oint_{\mathcal{C}} \boldsymbol{u} \cdot \boldsymbol{dx} = \int_{\mathcal{A}} (\boldsymbol{\nabla} \times \boldsymbol{u}) \cdot \boldsymbol{dA} \tag{1.42}$$

holds for any surface $\mathcal{A}$ lying in the fluid such that $\mathcal{C} = \partial\mathcal{A}$, i.e., for any surface in the fluid whose boundary is given by the loop $\mathcal{C}$. Strictly speaking, $\mathcal{A}$ need not be a material surface (although its boundary is a material loop by assumption), but it is usually convenient to consider that case. Thus the circulation theorem can be seen to make a statement about the co-evolution of the vorticity vector and of the material surfaces pierced by it. Indeed, an infinitesimal version of the circulation theorem is

$$\frac{D}{Dt}((\boldsymbol{\nabla} \times \boldsymbol{u}) \cdot \boldsymbol{dA}) = 0 \tag{1.43}$$

for material surface elements $\boldsymbol{dA}$, which lie in surfaces of constant entropy as necessary such that $\boldsymbol{dA}$ and $\boldsymbol{\nabla} s$ are parallel. Now, the curl of Euler's equation yields

$$\frac{D}{Dt}(\boldsymbol{\nabla} \times \boldsymbol{u}) + (\boldsymbol{\nabla} \times \boldsymbol{u})\boldsymbol{\nabla} \cdot \boldsymbol{u} - (\boldsymbol{\nabla} \times \boldsymbol{u}) \cdot (\boldsymbol{\nabla u}) = +\frac{\boldsymbol{\nabla}\rho \times \boldsymbol{\nabla} p}{\rho^2}. \tag{1.44}$$

Note that the potential force never contributes to the vorticity evolution. Division by ρ eliminates the second term because of (1.8):

$$\frac{D}{Dt}\left(\frac{\nabla \times \boldsymbol{u}}{\rho}\right) - \left(\frac{\nabla \times \boldsymbol{u}}{\rho}\right) \cdot (\nabla \boldsymbol{u}) = +\frac{\nabla \rho \times \nabla p}{\rho^3}. \tag{1.45}$$

The operator on the left-hand side is the same as in (1.12), which gave the evolution of a material line element $d\boldsymbol{x}$. The term on the right-hand side is zero if the fluid is barotropic and in this case the scaled vorticity vector field evolves analogously to a bundle of material lines, i.e., the vector field is dragged along by the flow. This behaviour is vividly described by saying the scaled vorticity vector field is 'frozen' into the fluid.[3] This underlies the persistence in time of regions of *irrotational* flow, i.e., of material regions in which the vorticity is zero. Such material regions remain void of vorticity if they were so originally. This is in contrast to other properties such as momentum, which can change due to travelling pressure waves relative the fluid.

If the fluid is not barotropic then the right-hand side of (1.44) is non-zero and it provides the so-called *baroclinic* source for vorticity. This vorticity source is very important for stratified fluids under gravity and it plays a fundamental role for internal gravity waves. In particular, it allows the propagation of vorticity into material regions of irrotational flow.

Now, for a fluid of the ideal gas type the pressure is a function of ρ and s and hence the cross product in the baroclinic term will be proportional to $\nabla \rho \times \nabla s$, which in particular implies that it will be orthogonal to ∇s. If (1.45) is now contracted with ∇s then the right-hand side vanishes, and pulling ∇s into the material derivative then also eliminates the stretching term on the left because of cancellations due to (1.19) as applied to the material invariant s. Thus we are left with the unexpected exact result that if

$$q = \frac{(\nabla \times \boldsymbol{u}) \cdot \nabla s}{\rho} = \frac{\nabla \cdot (s \nabla \times \boldsymbol{u})}{\rho} \quad \text{then} \quad \frac{Dq}{Dt} = 0. \tag{1.46}$$

The materially invariant scalar q is called the *potential vorticity*, or Rossby–Ertel potential vorticity in the present case of an ideal gas. The second form is based on the identity $\nabla \cdot \nabla \times \boldsymbol{u} = 0$ and makes it obvious that ρq is the divergence of a vector. Again, an analogous definition holds for the case of incompressible inhomogeneous flow with $\nabla \rho$ replacing ∇s in (1.46).

The material invariance of the potential vorticity (PV) in (1.46) is important because it makes explicit that a certain part of the vorticity field remains essentially frozen into the fluid, with the attendant restrictions on

3 In other words, using the language of differential geometry, $(\nabla \times \boldsymbol{u})/\rho$ is 'Lie-dragged' by the flow.

the propagation of the PV-structure relative to the fluid. Intuitively, the scalar ρq quantifies the cloud of points formed by the local intersections of vorticity lines with a given stratification surface and as such it acts as a materially invariant density for the conserved circulation (1.42) within stratification surfaces. More specifically, if we consider a simple cylinder-shaped control volume $\mathcal{D}$ that intersects the stratification surface $s = s_0$, say, then the volume integral of $\rho q \delta(s - s_0)$ over $\mathcal{D}$ is

$$\int_{\mathcal{D}} \rho q \, \delta(s - s_0) \, dV = \int\limits_{\mathcal{D} \cap s = s_0} \frac{(\boldsymbol{\nabla} \times \boldsymbol{u}) \cdot \boldsymbol{\nabla} s}{|\boldsymbol{\nabla} s|} \, dA = \int\limits_{\mathcal{D} \cap s = s_0} (\boldsymbol{\nabla} \times \boldsymbol{u}) \cdot \boldsymbol{dA}, \quad (1.47)$$

which equals $\Gamma(\mathcal{C})$ where $\mathcal{C}$ is the boundary of the intersection of $\mathcal{D}$ and the isentrope. This uses $\boldsymbol{dA} = dA \boldsymbol{\nabla} s / |\boldsymbol{\nabla} s|$ and the properties of the delta function.

Finally, we note that the single differential equation (1.46) trivially implies an infinite number of conserved integrals, e.g.,

$$\frac{d}{dt} \int_{\mathcal{D}} f(q) \rho \, dV = 0 \quad (1.48)$$

for any smooth function $f(\cdot)$, say. A commonly used example is the conservation of so-called *enstrophy*, which by definition corresponds to $f(q) = q^2/2$. Still, it is clear that (1.46) is a qualitatively different and much more powerful statement than integral conservation of enstrophy. This is of some importance for understanding the numerical simulation of fluid flows with numerical models in which a discrete version of enstrophy is exactly conserved even though the models does not satisfy a discrete version of (1.46). Near the grid scale the numerical conservation of discrete enstrophy must then be achieved by non-physical processes.

1.5 Rotating frames of reference

Euler's equation in the form (1.22) is valid in an inertial frame of reference, but in a non-inertial frame fictitious forces must be added to account for the accelerations due to the frame motion. Specifically, the gravitating Earth spins with angular frequency $\Omega > 0$ around its pole-to-pole axis and if the rotation vector along this axis is denoted by $\boldsymbol{\Omega}$ then the momentum equations relative to the spinning Earth must be augmented by suitable Coriolis and *centrifugal forces* based on $\boldsymbol{\Omega}$. The latter derive from the centrifugal potential $-\frac{1}{2}\Omega^2 b^2$, where b denotes the distance from the rotation axis. However, this potential can be combined with the gravitational potential to form the so-called *geopotential*, and it is tacitly assumed that the background state of rest is already compatible with this geopotential. Physically, this corresponds to an appropriate mass distribution at rest that is in balance with the geopotential; this accounts for the slightly ellipsoidal shape of the Earth. The main point is that we do not need to consider centrifugal forces explicitly.

On the other hand, the *Coriolis forces* must be considered explicitly and for this a term $\boldsymbol{f} \times \boldsymbol{u}$ must be added to Euler's equation:

$$\frac{D\boldsymbol{u}}{Dt} + \boldsymbol{f} \times \boldsymbol{u} + \frac{\boldsymbol{\nabla} p}{\rho} = -\boldsymbol{\nabla}\Phi \quad \text{where} \quad \boldsymbol{f} \equiv 2\boldsymbol{\Omega} \tag{1.49}$$

is the *Coriolis vector*. The *Coriolis force* $-\boldsymbol{f} \times \boldsymbol{u}$ is always perpendicular to $\boldsymbol{u}$ and it always seeks to rotate $\boldsymbol{u}$ around the rotation axis $\boldsymbol{\Omega}$, but in a sense opposite to that of the frame rotation itself. It makes no contribution to the energy budget of the flow but it can profoundly influence the dynamics of the fluid motion, both wavelike and vortical.

The derivations for circulation and vorticity remain valid provided the relative velocity $\boldsymbol{u}$ (relative to the rotating frame, that is) is replaced in (1.37) and (1.42) by the *absolute velocity* $\boldsymbol{u} + \frac{1}{2}\boldsymbol{f} \times \boldsymbol{r}$, where $\boldsymbol{r}$ is measured from a fixed point on the rotation axis such as the centre of the Earth:

$$\Gamma = \oint_C \left(\boldsymbol{u} + \frac{1}{2}\boldsymbol{f} \times \boldsymbol{r}\right) \cdot d\boldsymbol{x} = \int_A (\boldsymbol{\nabla} \times \boldsymbol{u} + \boldsymbol{f}) \cdot d\boldsymbol{A}. \tag{1.50}$$

The corresponding *absolute vorticity* $\boldsymbol{\nabla} \times \boldsymbol{u} + \boldsymbol{f}$ is the vorticity as observed in an inertial frame fixed with the heavens. The physical conditions of validity for the circulation theorem remain unchanged. For instance, in a rotating frame the definition of PV in (1.46) is simply modified to

$$q = \frac{(\boldsymbol{\nabla} \times \boldsymbol{u} + \boldsymbol{f}) \cdot \boldsymbol{\nabla} s}{\rho}. \tag{1.51}$$

One detail is that in a periodic flow configuration the absolute circulation is *not* conserved anymore on contours that traverse the periodic domain; this can be repaired by re-defining the circulation as described in §10.4.1.

In a rotating frame the integral conservation of momentum is replaced by the conservation of so-called *absolute momentum*, whose density per unit mass is $\boldsymbol{u} + \boldsymbol{f} \times \boldsymbol{r}$. This follows from the identity

$$\boldsymbol{f} \times \boldsymbol{u} = \frac{D}{Dt}(\boldsymbol{f} \times \boldsymbol{r}). \tag{1.52}$$

Interestingly, this conserved momentum is *not* the absolute momentum as observed in an inertial frame, which would involve $\boldsymbol{\Omega} = \frac{1}{2}\boldsymbol{f}$ instead of $\boldsymbol{f}$. The reason for this discrepancy can be traced back to the earlier tacit assumption about the pressure field, which was assumed to balance the centrifugal force in a state of rest relative to the rotating frame.

1.6 Shallow-water system

The shallow-water system occupies a special place in fluid dynamics because it is the simplest two-dimensional model that has the generic properties of interest. including non-trivial mass, momentum and potential-vorticity budgets. Moreover, the shallow-water equations can be viewed either as describing a shallow layer of incompressible homogeneous fluid, or as describing the two-dimensional compressible flow of a polytropic fluid with $\gamma = 2$. The former makes the shallow-water system relevant to geophysical applications whilst the latter makes it relevant to gas dynamics and acoustics. The link to acoustics also offers a perspective on weak solutions of the shallow-water equations, which contain nonlinear wave breaking through shock formation. Alternatively, in the shallow-water view, these waves are surface gravity waves and the shocks are hydrostatic bores, as discussed more fully in §13.4.

In the geophysical interpretation the shallow-water equations describe the flow of a vertically thin but horizontally wide layer of a homogeneous and incompressible fluid bounded above by a free upper surface and below by a solid impermeable boundary. The gravitational force is strong, i.e., the free boundary is tightly controlled to be nearly flat by gravity. By construction, the aspect ratio of the three-dimensional flow is very small, with horizontal scales greatly exceeding the vertical scale. Specifically, if the typical layer depth is H then $kH \ll 1$ where k is a typical horizontal wavenumber of the flow field. In the shallow-water regime this small parameter is exploited asymptotically and it can be shown that the leading-order flow consists of horizontal velocities $\boldsymbol{u} = (u, v)$ that are constant across the layer and a much smaller vertical velocity w that varies linearly across the layer. Moreover, the pressure field inside the layer is determined by the *hydrostatic relation* $\partial p/\partial z = -\rho g$ together with the boundary condition $p = p_0$ at the free surface, where p_0 is the constant ambient pressure, say. This implies a *horizontal* pressure gradient that is again constant across the layer depth.

This leads to an effectively two-dimensional fluid system based on horizontal coordinates $\boldsymbol{x} = (x, y)$ such that only the horizontal velocity $\boldsymbol{u}(\boldsymbol{x}, t)$ and the layer depth $h(\boldsymbol{x}, t)$ appear explicitly. The notation is two-dimensional from now on such that $\boldsymbol{\nabla}$ denotes the two-dimensional gradient operator, for instance. The fluid layer depth is denoted by $h(\boldsymbol{x}, t)$ and the *bottom to-*

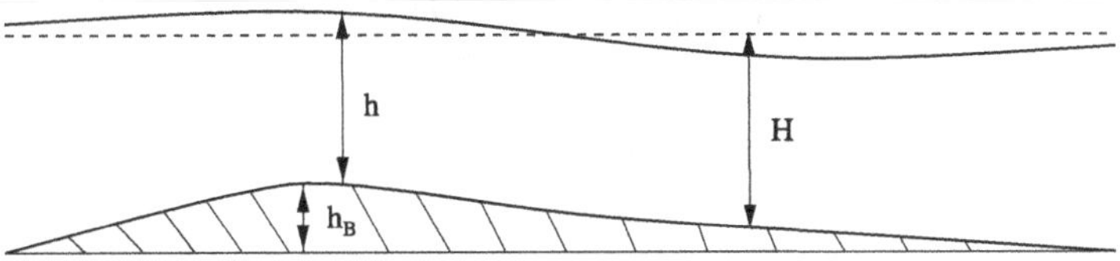

Figure 1.1 Shallow water layer with depth h, topography h_B and still water depth H.

pography elevation is denoted by $h_B(\boldsymbol{x})$ such that $h + h_B$ is the elevation of the free surface (see Figure 1.1). The *still water depth* $H(\boldsymbol{x}) > 0$ corresponds to the depth configuration with a flat surface, i.e., $\boldsymbol{\nabla} H = -\boldsymbol{\nabla} h_B$ where $\boldsymbol{\nabla}$ is the horizontal gradient. The depth h evolves according to the standard continuity equation

$$\frac{\partial h}{\partial t} + \boldsymbol{\nabla} \cdot (h\boldsymbol{u}) = 0 \quad \text{or} \quad \frac{Dh}{Dt} + h\boldsymbol{\nabla} \cdot \boldsymbol{u} = 0. \tag{1.53}$$

The hydrostatic pressure field implies a horizontal pressure gradient $\boldsymbol{\nabla} p = \rho g \boldsymbol{\nabla}(h + h_B)$, which is constant across the layer depth. Euler's equation takes the equivalent forms

$$\frac{D\boldsymbol{u}}{Dt} + g\boldsymbol{\nabla}(h + h_B) = 0 \quad \text{or} \quad \frac{D\boldsymbol{u}}{Dt} + g\boldsymbol{\nabla}(h - H) = 0. \tag{1.54}$$

The shallow-water PV follows from (1.53) and the curl of (1.54) as

$$q = \frac{\boldsymbol{\nabla} \times \boldsymbol{u}}{h} \quad \text{such that} \quad \frac{Dq}{Dt} = 0. \tag{1.55}$$

where here and throughout this book we write $\boldsymbol{\nabla} \times \boldsymbol{u} = v_x - u_y$ for the pseudoscalar that is the vertical component of vorticity in two dimensions. $Dq/Dt = 0$ corresponds to (1.43) applied to the free surface.

In the gas-dynamical interpretation the shallow-water equations (1.53) and (1.54) are viewed as the equations governing the two-dimensional flow of a hypothetical polytropic fluid with $\gamma = 2$. The equations take the standard form if we make the identifications

$$\rho = h, \quad p = g\frac{h^2}{2}, \quad \text{and} \quad \Phi = g h_B. \tag{1.56}$$

In this *gas-dynamical analogy* the effective two-dimensional pressure is proportional to the layer-integrated hydrostatic pressure field. This analogy is useful because it makes it obvious that shallow-water dynamics will contain all the features of two-dimensional gas dynamics, including sound-like non-dispersive waves, nonlinear wave steepening, and shock formation.

1.6.1 Available potential energy

Energy is conserved in the generic form (1.36) with $\epsilon(h) = gh/2$, which is most easily derived from the gas-dynamical analogy where $d\epsilon/dh = p/h^2$. This yields

$$h\frac{D}{Dt}\left(\frac{|\boldsymbol{u}|^2}{2} + g\frac{h}{2} + gh_B\right) + \boldsymbol{\nabla}\cdot\left(g\frac{h^2}{2}\boldsymbol{u}\right) = 0. \qquad (1.57)$$

However, this energy definition is not in the most convenient form because it does not make obvious that the gravitational energy should be minimal if the free surface is flat, i.e., when $h = H$ everywhere. This follows because mass conservation implies a finite lower limit for the total gravitational energy

$$\int_{\mathcal{D}}\left(g\frac{h}{2} + gh_B\right) h\,dx\,dy \quad \text{subject to} \quad \int_{\mathcal{D}} h\,dx\,dy = \text{const}, \qquad (1.58)$$

which is achieved if $h + h_B$ is constant. Physically, a raised surface at some location must correspond to a proportionally lowered surface somewhere else to conserve mass, and one can then mentally shave off the raised fluid portions to fill in the lowered regions in order to produce a flat surface. This procedure clearly lowers the centre of gravity of the fluid layer and therefore it lowers the total gravitational energy.

The minimal energy corresponding to a flat surface can be subtracted from (1.58) to yield the *available potential energy* of the layer. Here 'available' means available in principle for conversion into kinetic energy. This leads to the differential conservation law

$$\frac{\partial E}{\partial t} + \boldsymbol{\nabla}\cdot\left(E\boldsymbol{u} + gH\Delta h\boldsymbol{u} + g\frac{\Delta h^2}{2}\boldsymbol{u}\right) = 0 \qquad (1.59)$$

where

$$E = h\frac{|\boldsymbol{u}|^2}{2} + g\frac{\Delta h^2}{2} \quad \text{and} \quad \Delta h = h - H \qquad (1.60)$$

is the disturbance of the free surface elevation away from the still water depth H. In general, the concept of available potential energy, and especially its version for continuously stratified flows, is essential for a meaningful discussion of energy budgets in atmosphere ocean fluid dynamics.

1.7 Notes on the literature

Detailed descriptions of fundamental fluid dynamics can be found in the standard textbooks Batchelor (1967), Landau and Lifshitz (1982) or Whitham (1974); the latter is also a central reference for wave dynamics.

More specialized descriptions of geophysical fluid dynamics (GFD) appropriate for atmosphere and ocean flows are available in the textbooks Salmon (1998) and Vallis (2006).

2
Linear waves

Linear wave theory has a special place in applied mathematics. For example, the powerful concepts of linear wave theory, such as dispersion, group velocity or wave action conservation, are fundamental for describing the behaviour of solutions to many commonly occurring partial differential equations (PDEs). Also, whilst it is certainly not true that every linear wave problem has an explicit general solution, it is true that every linear problem can be approached by using *linear thinking*, i.e., by building up more complex solutions out of superpositions of simpler solutions. In some cases, this procedure can be carried to its logical conclusion and the complete general solution to a problem can be formulated as a sum over special solutions. For example, this works for PDEs with constant coefficients in a periodic domain, for which the general solution can be written as a sum of plane waves described mathematically by a Fourier series.

But even in cases where there is no explicit general solution, the possibility to develop special solutions using asymptotic methods and the ability to combine several simple solutions to form a more complex solution always deepens our understanding of the underlying problem, and such an improved understanding could then be used to aid a numerical simulation for situations of particular interest, for example. Thus time spent studying linear wave theory is time well spent.

We are particularly interested in the behaviour of small-scale waves propagating on an inhomogeneous basic state, because this is the natural setting for unresolved waves interacting with a resolved mean flow in a numerical model. Now, linear wave theory in fluid dynamics is an elementary textbook topic provided the medium is at rest in its basic state. However, if the medium is moving, as in the important case of waves on a shear flow, completely new phenomena such as critical layers may occur. We will start with the rest case and work our way up to the moving case.

2.1 Linear dynamics

In this section we consider the linear shallow-water equations without background rotation, i.e., without Coriolis forces. Coriolis effects, which are important and fundamental to GFD, are considered in §4.2.1.

The linear shallow-water equations relative to a state of rest are obtained by considering small deviations from a basic state $h = H$ and $\boldsymbol{u} = 0$. Thus we write

$$h = H(\boldsymbol{x}) + h'(\boldsymbol{x}, t) + O(a^2) \quad \text{and} \quad \boldsymbol{u} = \boldsymbol{u}'(\boldsymbol{x}, t) + O(a^2) \tag{2.1}$$

where h' and $\boldsymbol{u}'$ are $O(a)$ with $a \ll 1$ a small non-dimensional parameter measuring the amplitude of the flow deviations. The linearized equations are obtained by substituting (2.1) into the shallow-water equations and collecting terms linear in the small parameter a. This neglects terms $O(a^2)$ or higher, but these are small compared to the retained $O(a)$ terms in the asymptotic limit $a \to 0$. The result is

$$h'_t + \boldsymbol{\nabla} \cdot (H\boldsymbol{u}') = 0 \quad \text{and} \quad \boldsymbol{u}'_t + g\boldsymbol{\nabla} h' = 0. \tag{2.2}$$

Note that all effects due to sloping topography are contained in the single occurrence of $H(\boldsymbol{x})$ in the linear continuity equation. The linear PV is defined by $q = q' + O(a^2)$ such that

$$q' = \frac{\boldsymbol{\nabla} \times \boldsymbol{u}'}{H} \quad \text{and} \quad q'_t = 0. \tag{2.3}$$

Of course, because $H(\boldsymbol{x})$ does not depend on time normal vorticity can equally be used for PV in the present case.

The energy equation for the linear flow follows either from the exact equations (1.59) and (1.60) or it can be derived by contracting the linear equations (2.2) with $(h', H\boldsymbol{u}')$; the latter route is the more general approach in wave theory. Either way, the result is

$$E_t + \boldsymbol{\nabla} \cdot \boldsymbol{F} = 0 \quad \text{with} \quad E = H\frac{|\boldsymbol{u}'|^2}{2} + g\frac{h'^2}{2} \quad \text{and} \quad \boldsymbol{F} = gHh'\boldsymbol{u}'. \tag{2.4}$$

Here we used that $\Delta h = h'$ at leading order. This equation governs the evolution of the *disturbance energy*, which clearly involves a small-amplitude version of the available potential energy.

It is noteworthy that E and all the terms in (2.4) are $O(a^2)$ even though they depend only on the linear, $O(a)$ fields. The available potential energy is an example of an $O(a^2)$ *wave property*, which is defined as a quantity that is $O(a^2)$ for small deviations and that can be consistently evaluated at that order using only the linear, $O(a)$ solution. In contrast, energy in the form

(1.57) would not lead to a conserved wave property, because terms involving the unknown $O(a^2)$ corrections to h would enter the energy equation at this order. Of course, there is no inconsistency here: further analysis shows that the additional $O(a^2)$ terms in h can be computed from the governing equations and actually vanish because of the continuity equation evaluated at $O(a^2)$. This complication is another reason why the use of available potential energy is preferred: it neatly sidesteps this problem and leads directly to the disturbance energy.

2.1.1 Particle displacements and the virial theorem

It proves convenient to introduce linear particle displacements as an additional vector field for the description of linear waves in fluids. This is because the Lagrangian information carried by this field is helpful for visualizing the linear motion of fluid particles, and it also allows a more direct application of the standard methods of classical mechanics such as the *virial theorem*, which describes the partition of energy between its various forms.

We define the particle displacement vector field $\boldsymbol{\xi}'(\boldsymbol{x}, t) = (\xi', \eta')$ via

$$\boldsymbol{\xi}'_t = \boldsymbol{u}'. \tag{2.5}$$

This equation defines the linear particle displacements up to an initial condition, which can be chosen based on the problem at hand. For example, for time-periodic wave problems a zero-mean property for $\boldsymbol{\xi}'$ would be natural. Overall, in linear theory the particle displacement field $\boldsymbol{\xi}'(\boldsymbol{x}, t)$ is defined such that $\boldsymbol{x} + a\boldsymbol{\xi}'(\boldsymbol{x}, t)$ at time t is the actual position of the particle whose rest position is $\boldsymbol{x}$.

One immediate use of (2.5) is that it allows time-integrating the continuity equation in (2.2) as

$$h' + \boldsymbol{\nabla} \cdot (H\boldsymbol{\xi}') = 0. \tag{2.6}$$

The virial theorem is now derived by contracting the momentum equation in (2.2) with $H\boldsymbol{\xi}'$, which after short manipulation using (2.5) yields

$$\frac{\partial^2}{\partial t^2}\left(\frac{H|\boldsymbol{\xi}'|^2}{2}\right) + \boldsymbol{\nabla} \cdot (gh'H\boldsymbol{\xi}') = H|\boldsymbol{u}'|^2 - gh'^2. \tag{2.7}$$

In a bounded (or periodic) domain the second term vanishes after integrating over the domain. Moreover, under the assumption that $\boldsymbol{\xi}' \cdot \boldsymbol{u}$ is finite, it also follows that the time-average of the domain-integrated left-hand side is zero for large averaging time, and therefore the same must be true for the time-averaged domain-integrated difference between kinetic and potential energy on the right-hand side. This is the most general equipartition statement for linear waves in shallow water. Notably, equipartition of energy does not hold for all wave problems. For instance, it fails in shallow water if Coriolis forces are added.

2.1.2 Vortical and wave modes

The linear shallow-water dynamics features a mixture of vortical and wave-like modes in a manner that is typical for many GFD systems. The vortical or *balanced mode* is a steady solution to the equations, which implies the structure

$$g\nabla h'_b = 0 \quad \text{and} \quad \nabla\cdot(H\boldsymbol{u}'_b) = 0 \tag{2.8}$$

for the balanced flow fields, say. Thus the free surface is flat and the mass flux is non-divergent. In general, the term 'balance' in GFD describes flows in which the pressure fields is in approximate balance with the gravitational and/or Coriolis forces. In the present case the balanced pressure field is zero because there are no Coriolis forces. This steady balanced flow has non-zero disturbance energy density E, but zero energy flux $\boldsymbol{F}$.

The linear balanced flow is the progenitor of nonlinear vortex dynamics because it is closely linked to the vorticity and PV in (2.3). In fact, it is possible to compute the balanced flow from knowledge of the PV. Specifically, one can introduce a mass stream function ψ' such that

$$(-\psi'_y, +\psi'_x) = (Hu'_b, Hv'_b) \quad \text{and} \quad \nabla\cdot\left(\frac{1}{H}\nabla\psi'\right) = Hq'. \tag{2.9}$$

The first equation ensures that (2.8) holds and the second equation follows from (2.3). With $H > 0$ and appropriate boundary conditions for ψ' this elliptic equation can be uniquely inverted to find ψ' from q'; this kind of process is called PV inversion. Thus knowledge of PV plus the assumption of balance allows the computation of the balanced flow fields, and one can say that the PV controls the balanced flow. In linear dynamics the PV is constant in time but in nonlinear dynamics the PV changes due to advection, which is the hallmark of nonlinear vortex dynamics.

Now, the unbalanced part of the flow is unsteady and therefore (2.3) implies that the PV of the unbalanced flow must be zero, i.e., the velocity must be irrotational. This leads to

$$\nabla\times\boldsymbol{u}'_w = 0 \quad \text{and} \quad \frac{\partial^2 h'_w}{\partial t^2} - \nabla\cdot(gH\nabla h'_w) = 0, \tag{2.10}$$

which describes gravity waves in shallow water, or sound waves in the gas-dynamical analogy. For constant H this second-order PDE for h'_w is the standard wave equation with wave speed $c = \sqrt{gH}$. Both the energy E and the energy flux $\boldsymbol{F}$ are non-zero for gravity waves.

Using linear thinking, we conceive of the total flow as the sum of a balanced and an unbalanced part, i.e., $h' = h'_b + h'_w$ and $\boldsymbol{u}' = \boldsymbol{u}'_b + \boldsymbol{u}'_w$. In the present case we have the relations

$$h' = h'_w, \quad \boldsymbol{\nabla} \cdot (H\boldsymbol{u}') = \boldsymbol{\nabla} \cdot (H\boldsymbol{u}'_w), \quad \text{and} \quad \boldsymbol{\nabla} \times \boldsymbol{u}' = \boldsymbol{\nabla} \times \boldsymbol{u}'_b, \tag{2.11}$$

which allow us to decompose any given flow into its balanced and unbalanced components. For the velocity this reduces to the standard Helmholtz decomposition of two-dimensional vector fields if H is constant.

The solution to the general initial-value problem for the linear equations can now be described in terms of the balanced and unbalanced parts of the initial conditions. First, the initial $\boldsymbol{u}'_b$ is computed via (2.9) from the initial PV, which is constant in time and therefore remains the correct PV at all later times. Second, the initial $\partial h'_w/\partial t = -\boldsymbol{\nabla} \cdot (H(\boldsymbol{u}' - \boldsymbol{u}'_b))$ is computed. This provides the needed initial conditions for the second-order PDE in (2.10) and the initial-value problem can now be solved.

In an unbounded domain with constant H and compact initial conditions the procedure just described makes clear that the time-dependent flow converges to the steady balanced flow as $t \to \infty$. This follows from the properties of the wave equation, which predicts the inevitable propagation of all disturbances away to infinity and this implies the non-uniform convergence $h' \to h'_b = 0$ and $\boldsymbol{u}' \to \boldsymbol{u}'_b$ at every location $\boldsymbol{x}$. The convergence is non-uniform because it involves the propagation of wave disturbances. In GFD this convergence scenario is typical for *Rossby adjustment* problems, which are variants of the present problem that involve Coriolis forces and dispersive waves.

2.1.3 Kinematics of plane waves

Here we introduce the generic structure of plane waves, so we do not use any specific properties of the shallow-water system. For constant H, (2.2) becomes a system of linear PDEs with constant coefficients. The general solution to such a system with periodic boundary conditions can be written in terms of complex exponentials, or *plane waves* in the present context. Thus, a plane wave is a solution of the form

$$(h', u', v') = (\hat{h}, \hat{u}, \hat{v}) \exp(i(kx + ly - \omega t - \alpha)) \tag{2.12}$$

where the coefficients $(\hat{h}, \hat{u}, \hat{v})$, the *wavenumber vector* $\boldsymbol{k} = (k, l)$, the *frequency* ω and the phase shift α are all constants and possibly complex. The different sign of the frequency term is conventional and convenient for wave problems.

It is worth pointing out that working with complex-valued plane waves is the natural approach when dealing with linear PDEs, even if the PDEs have real coefficients. First, any complex solution of the type (2.12) contains two real-valued solution as its real and imaginary parts. This is because the complex conjugate of a solution is again a solution if the linear PDE has real coefficients, and therefore taking the real or imaginary part yields two independent solutions. Second, complex plane waves are the building blocks of arbitrary solutions via Fourier series and in this case a real-valued solution to a given initial-value problem, say, emerges naturally once all the complex modes are added up.

In this section we deal with individual plane waves and hence we are interested in the real part of the expressions in (2.12). Thus, a typical plane wave with real $(\boldsymbol{k}, \omega, \alpha)$ has the sinusoidal structure $h' = \hat{h}\cos\theta$ where the *phase*

$$\theta = kx + ly - \omega t - \alpha \tag{2.13}$$

and α is a constant phase shift. Note that flipping *all* the signs of (k, l, ω) does not make a difference to the plane wave, but that the individual signs do have intrinsic meaning. The lines of constant phase are called *wave fronts*, and because (2.13) implies that

$$\boldsymbol{\nabla}\theta = \boldsymbol{k} \quad \text{and} \quad -\theta_t = \omega \tag{2.14}$$

it is clear that at a fixed time the wavenumber vector is normal to the wave fronts. Two wave fronts on which the phase differs by 2π are separated by a *wavelength*, which is $\lambda = 2\pi/\kappa$ where $\kappa = |\boldsymbol{k}|$. Lines of constant phase move in time according to

$$d\theta = k\,dx + l\,dy - \omega\,dt = 0 \quad \Rightarrow \quad k\frac{dx}{dt} + l\frac{dy}{dt} = \boldsymbol{k}\cdot\boldsymbol{c}_p = \omega. \tag{2.15}$$

This is the definition of the *phase velocity* $\boldsymbol{c}_p$, which is clearly non-unique in two or more spatial dimensions. Two definitions are in practical use. First, the phase speed at a fixed y-position, say, arises naturally in wave observations along a line of constant y. This leads to the *scalar phase velocity*

$$c_p = \frac{\omega}{k} \tag{2.16}$$

such that the intersection of a wave front with the x-axis moves with velocity c_p. The second definition assumes that $\boldsymbol{c}_p$ is parallel to $\boldsymbol{k}$, i.e., that $\boldsymbol{c}_p$ is normal to the wave fronts of constant θ. This leads to the *vector phase velocity*

$$c_p = \frac{\boldsymbol{k}}{\kappa}\frac{\omega}{\kappa}. \tag{2.17}$$

Note that this vector has an intrinsic meaning because it is insensitive to overall sign changes of $(\boldsymbol{k}, \omega)$. If $\omega > 0$ then $\boldsymbol{c}_p$ points in the same direction as $\boldsymbol{k}$, which is the reason for the sign convention in (2.13).

Finally, the sinusoidal structure suggests averaging over a wavelength in space or time. For plane waves this is clearly equivalent to the *phase average* over the range $\alpha = [0, 2\pi]$ of the phase shift:

$$\overline{(\ldots)} = \frac{1}{2\pi}\int_0^{2\pi} (\ldots)\, d\alpha. \tag{2.18}$$

When applied to a plane wave field this yields zero, of course, but the result can be non-zero for products of wave fields. For example, the virial theorem applies to phase-averaged energies. A useful formula in this context is that if

$$A = \Re\,\hat{A}\exp(i\theta) \quad \text{and} \quad B = \Re\,\hat{B}\exp(i\theta) \quad \text{then} \quad \overline{AB} = \frac{1}{2}\Re\,(\hat{A}^*\hat{B}) \tag{2.19}$$

where the star denotes complex conjugation and $\Re$ denotes taking the real part. For example, $\overline{h'^2} = \frac{1}{2}|\hat{h}|^2$ and if $\hat{u} = i\hat{h}$, say, then $\overline{h'u'} = 0$.

2.1.4 Shallow-water plane waves

Substituting (2.12) in (2.2) with constant H reduces the linear PDEs to linear algebraic equations, because each differential operator, when applied to a plane wave, is equivalent to multiplication by a corresponding wavenumber or frequency factor according to

$$\frac{\partial}{\partial x} = +ik, \quad \frac{\partial}{\partial y} = +il, \quad \text{and} \quad \frac{\partial}{\partial t} = -i\omega. \tag{2.20}$$

The spatial operators are conveniently summarized by $\boldsymbol{\nabla} = +i\boldsymbol{k}$ such that $\boldsymbol{\nabla}\cdot\boldsymbol{u}$ and $\boldsymbol{\nabla}\times\boldsymbol{u}$ become $i\boldsymbol{k}\cdot\widehat{\boldsymbol{u}}$ and $i\boldsymbol{k}\times\widehat{\boldsymbol{u}}$, respectively. The resulting set of equations is

$$\begin{pmatrix} -\omega & kH & lH \\ gk & -\omega & 0 \\ gl & 0 & -\omega \end{pmatrix}\begin{pmatrix} \hat{h} \\ \hat{u} \\ \hat{v} \end{pmatrix} = 0. \tag{2.21}$$

Non-trivial solutions arise only if the determinant of this matrix is zero, i.e., only if

$$\omega\left(\omega^2 - gH(k^2 + l^2)\right) = 0. \tag{2.22}$$

This is the *dispersion relation*, which is the central object in linear wave theory. Any complex triple (k, l, ω) that satisfies (2.22) corresponds to a solution of the linear equations. For plane waves we usually consider real wavenumbers as given and look for suitable frequencies that satisfy the dispersion relation. Real frequencies correspond to neutral waves with time-constant amplitudes whereas imaginary frequencies correspond to growing or decaying waves.

In the present case there are three real roots

$$\omega_0 = 0 \quad \text{and} \quad \omega_{1,2} = \pm c\kappa \quad \text{where} \quad c = \sqrt{gH} \tag{2.23}$$

is the shallow-water wave speed and $\kappa = |\boldsymbol{k}|$ is the wavenumber magnitude. The corresponding phase velocities are

$$\boldsymbol{c}_{p0} = 0 \quad \text{and} \quad \boldsymbol{c}_{p1,2} = \pm c\frac{\boldsymbol{k}}{\kappa} \quad \text{such that} \quad |\boldsymbol{c}_{p1,2}| = c. \tag{2.24}$$

This illustrates that the wave equation describes waves that all move at the same speed regardless of wavelength. This is the definition of a *non-dispersive* wave system. An equivalent criterion for a non-dispersive system is that the phase velocity agrees with the *group velocity* defined by

$$\boldsymbol{c}_g = \frac{\partial \omega}{\partial \boldsymbol{k}}. \tag{2.25}$$

Basically, linear wave theory shows that the phase velocity describes how individual wave crests move whereas the group velocity describes how the envelope of a wave group moves. We will study dispersive waves in §4.

Now, the zero root in (2.23) corresponds to the balanced mode (2.8) with eigenvector

$$(\hat{h}_b, \hat{u}_b, \hat{v}_b) = \left(0, \frac{-l}{\kappa}c, \frac{+k}{\kappa}c\right) M_0 \tag{2.26}$$

where M_0 is a non-dimensional modal amplitude. Thus $\boldsymbol{u}'_b$ and $\boldsymbol{k}$ are at right angles, which is the structure of a *transversal wave*. On the other hand, the two wave modes have eigenvectors

$$(\hat{h}_w, \hat{u}_w, \hat{v}_w) = \left(\pm H, \frac{k}{\kappa}c, \frac{l}{\kappa}c\right) M_{1,2}. \tag{2.27}$$

Thus $\boldsymbol{u}'_w$ is parallel to $\boldsymbol{k}$ and therefore these are irrotational *longitudinal waves*, as is obvious from the gas-dynamical analogy with sound waves.

Clearly, in a periodic domain we can solve the general initial-value problem explicitly by determining a set of amplitudes (M_0, M_1, M_2) from the initial conditions. This is achieved by inverting

$$\begin{pmatrix} \hat{h}(\boldsymbol{k},0) \\ \hat{u}(\boldsymbol{k},0) \\ \hat{v}(\boldsymbol{k},0) \end{pmatrix} = \begin{pmatrix} 0 & +H & -H \\ -lc/\kappa & kc/\kappa & kc/\kappa \\ kc/\kappa & lc/\kappa & lc/\kappa \end{pmatrix} \begin{pmatrix} M_0(\boldsymbol{k}) \\ M_1(\boldsymbol{k}) \\ M_2(\boldsymbol{k}) \end{pmatrix} \tag{2.28}$$

for each choice of admissible wavenumber vector $\boldsymbol{k}$ in the periodic domain. Here the left-hand side contains the Fourier coefficients of the initial conditions. This is easily done numerically.[1]

Also, superpositions of counter-propagating plane waves can be used to build *normal modes*, which are the natural vibration patterns in bounded domains, although this works only in rectangular domains. For example, if the bounded domain is defined by $(x,y) \in [0,L] \times [0,D]$, say, then wavelike normal modes have the structure $h'_w = \hat{h}_w \exp(-i\omega t) \cos kx \cos ly$ where (k,l,ω) satisfy (2.22) as well as the admissibility conditions, which are that kL and lD are multiples of π. This yields a discrete spectrum of possible wavenumbers and frequencies, and each normal mode can be viewed as a standing wave consisting of counter-propagating plane waves with $\pm k$ and $\pm l$. Also, the virial theorem applies in this case, which shows that the domain-integrated and time-averaged kinetic and potential energies are equal.

Returning to a single plane wave, it also follows from (2.27) that there is *equipartition of energy* in the phase-averaged form

$$\bar{E} = H\frac{\overline{u_w'^2 + v_w'^2}}{2} + g\frac{\overline{h_w'^2}}{2} = H(\overline{u_w'^2 + v_w'^2}) = g\overline{h_w'^2}. \tag{2.29}$$

The mean disturbance energy flux in a plane wave is

$$\overline{\boldsymbol{F}} = c^2\overline{h_w'\boldsymbol{u}_w'} = \bar{E}c\frac{\boldsymbol{k}}{\kappa} = \bar{E}\boldsymbol{c}_p = \bar{E}\boldsymbol{c}_g. \tag{2.30}$$

The final expression is the most important, because it generalizes to dispersive waves. In fact, it is so important that we re-write the energy equation in the phase-averaged plane-wave form

$$\bar{E}_t + \nabla \cdot (\bar{E}\boldsymbol{c}_g) = 0. \tag{2.31}$$

Comparison with the continuity equation (1.4) shows that mean disturbance energy is transported by the group velocity in the same way as mass is transported by the material velocity. This is the precise meaning of the statement 'energy travels with the group velocity'.

Of course, in the present case of a single plane wave all derivatives in (2.31)

[1] Such statements should always be taken with a grain of salt; typically some care is needed to extract the correct limiting behaviour at discrete wavenumbers from an analytical expression such as (2.28). Here the only problematic wavenumber location is $k = l = 0$, which corresponds to uniform steady flow.

are zero, but later we will see that this equation provides the leading-order amplitude evolution for a slowly varying wavetrain, which is a modulated plane wave.

2.1.5 Refraction

The simple world of plane waves and explicit general solutions is lost once variable water depth H is allowed. Physically, this introduces the phenomenon of *wave refraction*, which is familiar from the casual observation of ocean surface waves that are refracted towards a shoreline (this case will be discussed extensively in §13). Mathematically, variable H means that the system of linear PDEs in (2.2) has non-constant coefficients and therefore complex exponentials are not solutions anymore. Significant analytical progress can still be made in the case of one-dimensional topography, where H depends on y only, say. We will look at this textbook case here and then we will consider the more subtle case of two-dimensional topography using geometric wave theory in §3.

For one-dimensional $H(y)$ we look for solutions that retain their plane-wave structure in x and t, i.e., we assume that

$$(h', u', v') = (\hat{h}(y), \hat{u}(y), \hat{v}(y)) \exp(i(kx - \omega t - \alpha)) \tag{2.32}$$

for suitable values of the real constants k and ω. We are interested in wavelike solutions and therefore we assume $Hq' = \boldsymbol{\nabla} \times \boldsymbol{u}' = 0$. Substituting into the linear equations (2.2) then leads to

$$\hat{h} = \frac{\omega}{gk}\hat{u}, \quad \hat{v} = -\frac{i}{k}\hat{u}_y, \quad \text{and} \quad \hat{h}_{yy} + \frac{H_y}{H}\hat{h}_y + \left(\frac{\omega^2}{gH} - k^2\right)\hat{h} = 0. \tag{2.33}$$

If $y \in [0, D]$, say, then the problem is completed by providing appropriate boundary conditions at $y = 0$ and $y = D$. For instance, at a fixed impermeable wall $v = 0$ and this translates to $\hat{h}_y = 0$, which physically corresponds to zero normal pressure gradient at the wall.

For constant H the ordinary differential equation (ODE) in (2.33c) is just the linear harmonic oscillator equation, which leads to simple exponential solutions and therefore to the plane waves considered before. For non-constant H the equation can be written in self-adjoint form as

$$(H\hat{h}_y)_y + \left(\frac{\omega^2}{gH} - k^2\right) H\hat{h} = 0. \tag{2.34}$$

Explicit solutions to (2.34) are restricted to simple functions $H(y)$ such as piecewise constant H, for instance. This special case can also be used as a building block for a numerical solution of (2.34), as follows. Within every interval of constant H the solution consists of two exponentials $A\exp(ily) + B\exp(-ily)$ with local y-wavenumber given by

$$l = \sqrt{\frac{\omega^2}{gH} - k^2}, \tag{2.35}$$

which could be either real or imaginary. A real wavenumber l grows with decreasing water depth, i.e., the wavelength decreases by refraction. On the other hand, l is imaginary if H is too large and in this case the wave field has an exponentially growing or decaying structure in the y-direction. The wave field is said to be *evanescent* in regions of imaginary l.

The numerical task is to match the two coefficients of these piecewise local solutions across all the points of discontinuity of H, which requires the correct jump conditions for $\hat{h}_y$ at these points. These follow from the self-adjoint form of the equation in (2.34), which shows that $\hat{h}$ and $H\hat{h}_y$ are continuous at these points. Together with the boundary conditions this completes the explicit solution, which can easily be implemented on a computer and then can be used to approximate solutions for arbitrary $H(y)$ by letting the intervals of constant H shrink to very small size.

This numerical procedure allows solving for the wave structure for any particular $H(y)$, but it does not allow drawing general conclusions about the behaviour of the solutions. However, there is one important general statement that we can derive here. This follows either from the disturbance energy law (2.4) or from the second-order ODE (2.34).

Using the latter route first, we recall a standard fact about the *Wronskian* of a self-adjoint second-order ODE: if $\hat{h}_1$ and $\hat{h}_2$ are two solutions to (2.34) then $(HW)_y = 0$, where $W = \hat{h}_{1y}\hat{h}_2 - \hat{h}_1\hat{h}_{2y}$ is the Wronskian. This fact is easily checked by substitution and it provides a non-trivial statement if the two solutions are linearly independent. Now, (2.34) has real coefficients and therefore if we have a complex solution $\hat{h}$ then its complex conjugate $\hat{h}^*$ is also a solution. This leads to the statement that

$$H(\hat{h}_y\hat{h}^* - \hat{h}\hat{h}^*_y) = \frac{i\omega}{g}H(\hat{v}\hat{h}^* + \hat{h}\hat{v}^*) = \frac{2i\omega}{g}H(y)\Re(\hat{h}^*\hat{v}) = \frac{4i\omega^2}{g^2k}H\overline{u'v'} \tag{2.36}$$

is constant in y. Here we used (2.33a,b) and the phase-averaging rule (2.19). In the present situation phase averaging is equivalent to averaging over the cyclic x-coordinate. This is called *zonal averaging* and will be of primary concern in the wave–mean interaction theory described in §5.

With this interpretation (2.36) implies that the zonally averaged momentum flux[2] $H(y)\overline{u'v'}$ is constant. In other words, as $H(y)$ changes, refraction

[2] Specifically, this is the wave-induced flux of x-momentum in the y-direction.

changes the structure of both $\hat{u}(y)$ and $\hat{v}(y)$, but in such a way that the momentum-flux combination (2.36) remains constant. This is an important and non-trivial general fact about the linear solution, which also holds for the sum of several waves with different zonal wavenumbers k due to the orthogonality of Fourier modes.

The derivation based on the Wronskian obscures that (2.36) follows in fact directly, and more simply, from the conservation law for disturbance energy in (2.4). Specifically, after phase averaging the mean disturbance energy is time independent, i.e., $\bar{E}_t = 0$, and hence the mean energy flux $\overline{\boldsymbol{F}}$ must be non-divergent. But, after averaging the x-component of $\overline{\boldsymbol{F}}$ cannot depend on x anymore (because the averaging is equivalent to zonal averaging, i.e., to integration over the cyclic coordinate x) and therefore the energy law reduces to the statement that the y-component of $\overline{\boldsymbol{F}}$ must be constant. This yields

$$gH(y)\overline{h'v'} = \frac{\omega}{k}H(y)\overline{u'v'} = \text{const}, \tag{2.37}$$

which is equivalent to (2.36). This straightforward derivation of this important relation illustrates how useful conservation laws such as (2.4) are.

Finally, we have used the implicit assumption that the solution is smooth in order to derive (2.36) or (2.37). This seemingly harmless assumption fails in problems involving critical layers, as will be discussed in detail in §7.2.

2.1.6 WKB theory for slowly varying wavetrains

The governing ODE in (2.34) can be solved asymptotically in the special case where the solution forms a *slowly varying wavetrain*, which is a smooth solution that at every y looks like a plane wave, and whose amplitude and wavenumber are varying slowly in y in order to account for slow changes in H. The physical image is an oscillator (the plane wave) that is subject to a slowly varying environment, where slowly means that significant changes in the environment accrue only over many periods of the oscillator. For instance, if l is a typical y-wavenumber then H_y/H should be small compared to l. In other words, the distance over which $H(y)$ changes significantly should appear large if measured in terms of the intrinsic length scale $1/l$ of the wavetrain.

This scale separation can be exploited to derive asymptotic equations based on a small parameter. This is the field of high-wavenumber asymptotics. We therefore make the ansatz

$$\hat{h}(y) = A(y)\exp(i(\kappa_0 s(y) - \alpha)) \quad \text{and} \quad \kappa_0 \gg 1 \tag{2.38}$$

where the phase shift α is constant, κ_0 is a typical wavenumber and $s(y)$

is the so-called *eikonal* function, which has units of length. Both s and the amplitude A are assumed to be real and they may vary on the same scale as $H(y)$. On the other hand, the phase $\kappa_0 s$ varies much more rapidly because of the assumed large size of κ_0. This scaling set-up with an explicitly large coefficient in the phase is typical for multiscale approximations of this type, which are often called *WKB approximations.*

Substituting (2.38) into (2.34) and collecting real and imaginary parts of the resulting expression yields the two equations

$$(A_y H)_y + AH\left(\frac{\omega^2}{gH} - k^2 - \kappa_0^2 s_y^2\right) = 0 \tag{2.39}$$

and

$$(AHs_y)_y + A_y H s_y = 0, \tag{2.40}$$

respectively. These coupled equations are still exact and therefore equivalent to (2.34). For instance, after multiplication with A, the second equation can be written as

$$(HA^2 s_y)_y = 0, \tag{2.41}$$

which is equivalent to the exact flux statement (2.36).

We want to exploit the fact that κ_0 is a large number and that therefore the terms in (2.39) are highly unequal in importance. Different limits for ω and k for large κ_0 are possible, and the most general propagating wave is obtained by choosing

$$\omega^2 = c_0^2 \kappa_0^2 \quad \text{and} \quad k = \kappa_0 \cos\theta_0, \tag{2.42}$$

where c_0 and θ_0 are convenient constants. For instance, we can choose $y = 0$ as a reference level such that $c(0) = c_0$ and then θ_0 is the angle between the phase gradient and the x-axis there. The link between high wavenumbers and high frequencies in (2.42a) is typical for non-dispersive wave systems. Thus we obtain the still exact

$$(A_y H)_y + AH\left(\kappa_0^2\left[n^2 - \cos^2\theta_0 - s_y^2\right]\right) = 0 \tag{2.43}$$

where

$$n(y) = \frac{c_0}{c(y)} = \sqrt{\frac{H_0}{H(y)}} \tag{2.44}$$

is the so-called *index of refraction.* We are now ready to exploit the smallness of $1/\kappa_0$.

First, we consider the solution over some range of $1/\kappa_0 \ll 1$ near the origin, i.e., we view $A(y, \kappa_0)$ and $s(y, \kappa_0)$ as functions of *both* y and κ_0 over this range. Second, we assume that the dependence on $1/\kappa_0$ can be approximated by a power series in the small parameter $1/\kappa_0$, i.e., we assume that

$$A(y,\kappa_0) = A_0(y) + \kappa_0^{-1} A_1(y) + \kappa_0^{-2} A_2(y) + \cdots \tag{2.45}$$

$$\text{and} \quad s(y,\kappa_0) = s_0(y) + \kappa_0^{-1} s_1(y) + \kappa_0^{-2} s_2(y) + \cdots \tag{2.46}$$

This is an *asymptotic expansion*, i.e., the series is not expected to converge, but it is expected that each successive term becomes negligible compared to the preceding term as the small parameter goes to zero. This means that the leading-order solution (A_0, s_0) is essentially exact as $\kappa_0 \to \infty$.

Substituting (2.45) into (2.41) and (2.43) and collecting terms at various powers of κ_0 results in a hierarchy of equations for the functions in (2.45). At leading order we obtain the *eikonal equation* and the *transport equation*

$$s_{0y}^2 = n^2 - \cos^2\theta_0 \quad \text{and} \quad (n^{-2} A_0^2 s_{0y})_y = 0, \tag{2.47}$$

respectively. Of course, the transport equation is simply a trivial consequence of the exact (2.41), but the eikonal equation is a genuine asymptotic relation. Indeed, the eikonal equation ensures that the plane-wave dispersion relation (2.35) is satisfied if the local y-wavenumber l is identified with $\kappa_0 s_{0y}$. What makes (2.47) easy to solve is the fact that these equations are decoupled, i.e., it is possible to solve for the leading-order eikonal first and then to solve for the leading-order amplitude. We will drop the subscripts now, as we are only going to consider the leading-order wave field.

If $n > |\cos\theta_0|$ then the eikonal equation has two real solutions with opposite signs. The number of solutions corresponds to the order of the underlying ODE and the solutions differ by the sense of propagation in the y-direction, which is indicated by the sign of the energy flux in the transport equation. For example, in a bounded y-domain with $y \in [0, D]$ both solutions are needed whereas in an unbounded domain $y \in [0, +\infty)$ with a radiation condition that specifies radiation away from a source at $y = 0$, say, it is sufficient to consider only the solution that has a positive group velocity in the y-direction, which corresponds to $s_y > 0$ if we use the convention $\omega > 0$. We now consider this case and with the definition of a local y-wavenumber as the y-derivative of the phase via $l(y) = \kappa_0 s_y$ we obtain

$$l(y) = +\kappa_0\sqrt{n(y)^2 - \cos^2\theta_0} \quad \text{such that} \quad \frac{A(y)}{A(0)} = \frac{n(y)}{n(0)}\sqrt{\frac{l(0)}{l(y)}}. \tag{2.48}$$

Putting this together yields the WKB solution

$$\hat{h}(y) = \hat{h}(0)\,\frac{n(y)}{n(0)}\sqrt{\frac{l(0)}{l(y)}}\,\exp\left(i\int_0^y l(\bar{y})\,d\bar{y}\right) + O(\kappa_0^{-1}), \tag{2.49}$$

where $l(y)$ is short-hand for (2.48a). In a nutshell, the structure of the phase in (2.49) ensures that the local wavenumber is consistent with the plane-wave dispersion relation (2.35) whilst the structure of the amplitude ensures that the energy flux is constant in y.

Notably, there is no mechanism for partial reflection in (2.49), i.e., within WKB theory a wave propagating through a region of variable $H(y)$ is always transmitted fully, at least as long as the solution remains bounded. Of course, this is a consequence of the assumed slow variation of $H(y)$, as is made clear by the textbook counterexample of computing reflected and transmitted waves at a step function in $H(y)$.

Most interestingly, (2.49) indicates that the asymptotic wave amplitude becomes infinite if either $n \to \infty$ or $n \to \cos\theta_0$. These mathematical singularities point to interesting physical effects that occur in these two scenarios. Indeed, it is important to realize that singularities in asymptotic solutions don't usually imply a singularity in the true solution, i.e., more often than not the asymptotic solution diverges at some location because the asymptotic series becomes disordered there in response to a breakdown of the fundamental assumptions that went into forming the series in the first place. Thus, by inspecting the breakdown of asymptotic expansions we learn more about the nature of the problem, i.e., its physics.

Now, the first scenario $n \to \infty$ corresponds to $H \to 0$, which is a wavetrain that is propagating towards a line of vanishing layer depth, such as a shoreline. In this case $l \propto H^{-1/2}$ and $\hat{h} \propto H^{-1/4}$, which means the waves are shortening and growing in amplitude. The amplitude increase is even more pronounced in the appropriate non-dimensional amplitude measure $h'/H \propto H^{-5/4}$. This illustrates the familiar effect that waves shoaling on a beach tend to increase in amplitude.

The second scenario occurs if the layer depth H *increases* sufficiently such that l in (2.48) becomes zero at some $y = y_r$, say, where $n(y_r) = \cos\theta_0$. Propagation past this location is not possible, because it would correspond to imaginary l, and therefore the wavetrain is *evanescent* in $y > y_r$. In full linear theory this corresponds to a situation in which the incoming wave suffers *total internal reflection* at $y = y_r$, but the present WKB theory can not describe this reflection accurately near $y = y_r$ and this manifests itself in the divergence of the leading-order asymptotic solution there. The reason for this is a breakdown of the underlying assumptions: near $y = y_r$ the local wavenumber l goes to zero and therefore the distance to the reflection level does not appear long compared to $1/l$. In other words, the wavetrain is not slowly varying at a reflection line.

It is possible to patch-up the WKB solution in the vicinity of a reflection line by matching it with a second, local asymptotic solution based on the smallness of $y - y_r$. This local solution is usually given by a suitable Airy function centred at the reflection line, which is capable of modelling the transition between oscillatory wave motion in $y < y_r$ and evanescent motion in $y > y_r$. This more detailed procedure shows that the wave amplitude increases by only a modest factor at a reflection level. It also shows that the WKB solution in $y \in [0, y_r)$ has a waveguide structure, i.e., its y-structure consists of a standing wave pattern formed by two counter-propagating waves, and hence the net energy and momentum flux in the y-direction is zero. The latter fact is also obvious from the global validity of (2.37), which in the evanescent zone $y > y_r$ requires that the flux must be zero.

In contrast, the diverging solution at a shoreline where $H \to 0$ cannot be easily patched up. In fact, whether or not WKB theory breaks down in this case depends on the shape of $H(y)$ near the shoreline, for reasons that are discussed fully in the section §7.2 on critical layers. But the inexorable growth of wave amplitude as $H \to 0$ implies that even if WKB theory remains valid, linearization itself must break down! Of course, this is familiar from the usual nonlinear wave breaking of shoaling ocean waves, which will be discussed in detail in §13.

2.1.7 Related wave equations and adiabatic invariance

The wave equation with constant wave speed applies to fluid waves, but also to many other waves such as electro-magnetic or elastic waves. Thus, whatever one learns about the wave equation applies in all these physical situations. However, it is important to note that the relevant equation for *variable* wave speed can differ from one physical situation to the next. This also affects the WKB solution, which is why we briefly discuss this here.

For example, in the present shallow-water (or acoustic) context the relevant equation involves the self-adjoint spatial operator in (2.10b), i.e.,

$$h_{tt} - \nabla \cdot (c^2 \nabla h) = 0. \tag{2.50}$$

On the other hand, if ϕ denotes the transverse displacement of an elastic membrane with non-uniform mass density, then the relevant equation involves the non-self-adjoint operator in

$$\phi_{tt} - c^2 \nabla^2 \phi = 0. \tag{2.51}$$

Physically, this is because in this case c^2 is inversely proportional to the mass density per unit area of the membrane, which is a coefficient of the time derivative and does not enter the balance of elastic forces described by the Laplacian.

In the case of one-dimensional refraction with $c(y)$ it can easily be shown that this leads to a governing ODE for ϕ that differs from (2.33c), namely

$$\hat{\phi}_{yy} + l(y)^2\hat{\phi} = 0 \quad \text{where} \quad l(y)^2 = \omega^2/c(y)^2 - k^2. \tag{2.52}$$

Eventually, this leads to a WKB solution of the form

$$\hat{\phi}(y) = \hat{\phi}(0)\sqrt{\frac{l(0)}{l(y)}}\exp\left(i\int_0^y l(\bar{y})\,d\bar{y}\right) + O(\kappa_0^{-1}). \tag{2.53}$$

The phase shows the same behaviour as in (2.49), but the amplitude factor is different. It is easy to check that, as (2.53) suggests, for slowly varying c we can transform (2.50) into (2.51) at leading order by setting $\phi = hc$.

Incidentally, (2.52) describes a linear harmonic oscillator with slowly varying frequency if y is interpreted as time and $l(y)$ as the temporal frequency. This leads to the classical result known as the *adiabatic invariance* of the so-called *action*, i.e., the asymptotic result that the ratio of energy to frequency of a linear harmonic oscillator subject to slowly varying frequency is approximately constant. This is discussed more fully in the context of wave activity conservation laws in §4.5.1. In terms of (2.53) the oscillator action corresponds to

$$\frac{|\hat{\phi}_y|^2}{l} \approx \frac{l^2|\hat{\phi}|^2}{l} \approx l|\hat{\phi}|^2 \approx \text{const} \tag{2.54}$$

whereas in terms of (2.49) it corresponds to the different

$$\frac{|\hat{h}_y|^2}{ln^2} \approx \frac{l^2|\hat{h}|^2}{ln^2} \approx \frac{l|\hat{h}|^2}{n^2} \approx \text{const.} \tag{2.55}$$

For example, if $H \to 0$ then $l \propto n$ is large and $|\hat{h}|^2 \propto l$ is large as well, whereas if $l \to \infty$ in (2.54) then $|\hat{\phi}|^2 \propto 1/l$ is small.

2.2 Notes on the literature

The classic graduate textbooks for linear and nonlinear waves in fluids are Whitham (1974) and Lighthill (1978). WKB theory and many other asymptotic methods are described succinctly in Hinch (1991).

2.3 Exercises

1. *Shallow water waves with rotation.* The shallow-water system with rotation described by a Coriolis parameter f is

$$h_t + \nabla\cdot(h\boldsymbol{u}) = 0 \quad \text{and} \quad \frac{D\boldsymbol{u}}{Dt} + f\hat{\boldsymbol{z}}\times\boldsymbol{u} + g\nabla h = 0. \tag{2.56}$$

Linearize around a state of rest with constant water depth $h = H$ and derive the dispersion relation

$$\omega\left(\omega^2 - f^2 - gH(k^2 + l^2)\right) = 0. \tag{2.57}$$

Show that if $q = (\boldsymbol{\nabla} \times \boldsymbol{u} + f)/h$ then the linear PV is (up to a constant factor)

$$q'(x, y) = v'_x - u'_y - \frac{f}{H}h' = \psi'_{xx} + \psi'_{yy} - \frac{f^2}{gH}\psi' \tag{2.58}$$

in terms of the stream function $\psi'(x, y)$ for the steady vortical mode such that

$$(u'_b, v'_b, h'_b) = (-\psi'_y, +\psi'_x, f\psi'/g). \tag{2.59}$$

The last term shows that $h'_b \neq 0$ in a rotating frame. The wave modes have zero PV, but show that they have nonzero relative vorticity proportional to fh'_w. This is discussed further in § 4.2.1.

2. Variational property of balanced flow. Show that a balanced flow of the form (2.59) is the flow of least disturbance energy

$$\int \frac{1}{2}\left(Hu'^2 + Hv'^2 + gh'^2\right)\, dxdy \tag{2.60}$$

that is compatible with a given PV distribution $q'(x, y)$. Hint: use calculus of variations with a Lagrange multiplier for the PV constraint. This also works for variable $H(x, y)$ if ψ' is replaced by the mass stream function.

3. Adiabatic invariance vs. the pit and the pendulum. Consider a linear pendulum with angle variable $\phi(t)$ whose length l is gradually increased, which implies a gradual decrease of the frequency $\Omega = \sqrt{g/l}$. How does the phase-averaged energy

$$\bar{E} = \frac{1}{2}\overline{\dot{\phi}^2} + \frac{1}{2}\Omega^2\overline{\phi^2} \tag{2.61}$$

scale with l? How about the maximum of $|\phi|$ over an oscillation period and the maximum speed of the pendulum's end point, which is $l\phi_t$? You can answer these questions by relating these quantities back to $\bar{E}$. Compare with E.A. Poe's short story, *The Pit and the Pendulum* (freely available under `www.literature.org/authors/poe-edgar-allan/pit-and-pendulum.html`). Did the poet get it right?

3
Geometric wave theory

Geometric wave theory is the natural extension of WKB theory to situations in which the still layer depth H (and therefore the wave speed) is a slowly varying function of both x and y, and possibly even of t, although we will not consider that case here. In fact, even for constant H geometric wave theory is useful because it allows the computation of the structure of normal modes in bounded domains with irregular shapes, i.e., shapes for which there is no simple explicit expression for the normal modes.[1]

The basic assumption of geometric wave theory is that there is a scale separation between the rapidly varying phase of the wavetrain on the one hand, and the slowly varying layer depth and wavetrain parameters such as amplitude and wavenumber on the other. Of course, in bounded domains the domain size must also be large compared to the wavelength. This basic assumption leads to a flexible and generic asymptotic procedure for solving for the wave field. Eventually, with the inclusion of dispersive effects, geometric wave theory becomes the ray-tracing method, which is the Swiss army knife for computing the asymptotic behaviour of small-scale waves in many fields of physics, including GFD.

A peculiarity of the progression from one-dimensional WKB theory to two-dimensional geometric wave theory and finally to dispersive ray tracing is that the structure of the theory becomes *easier*, not harder, as its generality increases. However, the price to pay for the ease-of-use of the more general asymptotic theory is that it is much harder to assess the error of the solution, especially in situations where the asymptotic solution diverges, such as on caustics.

[1] The only domain shapes for which the normal modes can be easily computed are those that admit separation of variables in the wave equation relative to a coordinate system suitable for the boundary shape. In practice, this means rectangles and ellipses.

3.1 Two-dimensional refraction

We allow for $H(x, y)$, but because the depth is still time independent we can retain the exponential time dependence of the solution. Thus, we are looking to find the spatial structure of $h' = \hat{h}(x, y)\exp(-i\omega t)$. This structure is governed by the so-called *reduced wave equation*[2]

$$\nabla \cdot (n^{-2}\nabla \hat{h}) + \kappa_0^2 \hat{h} = 0, \tag{3.1}$$

which follows from $\omega = c_0\kappa_0$, (2.50) and (2.44). We generalize the ansatz (2.38) to

$$\hat{h}(x, y) = A(x, y)\exp(i(\kappa_0 s(x, y) - \alpha)) \quad \text{and} \quad \kappa_0 \gg 1. \tag{3.2}$$

Substituting in (3.1) and collecting real and imaginary parts yields the exact pair

$$\nabla^2 A - 2(\nabla \ln n) \cdot \nabla A + \kappa_0^2 A(n^2 - |\nabla s|^2) = 0 \tag{3.3}$$

and

$$A\nabla^2 s + 2\nabla s \cdot \nabla A - 2A(\nabla \ln n) \cdot \nabla s = 0, \tag{3.4}$$

respectively. After multiplying with A/n^2, the second equation can be re-written in conservation form as

$$\nabla \cdot \left(\frac{A^2}{n^2}\nabla s\right) = 0 \quad \Leftrightarrow \quad \nabla \cdot \overline{\boldsymbol{F}} = 0 \quad \text{with} \quad \overline{\boldsymbol{F}} = \frac{A^2}{n^2}\nabla s. \tag{3.5}$$

Again, this exact equation is implied by the steady energy law with flux $\overline{\boldsymbol{F}}$. Now, we use the expansion (2.45) as before and obtain the two-dimensional eikonal and transport equations

$$|\nabla s|^2 = n(x, y)^2 \quad \text{and} \quad \nabla \cdot \left(\frac{A^2}{n^2}\nabla s\right) = 0 \tag{3.6}$$

for the leading-order solution, with subscripts omitted.

3.1.1 Characteristics and Fermat's theorem

The eikonal equation is a nonlinear first-order PDE, which can be solved by the *method of characteristics*. This involves finding the characteristic curves in the xy-plane along which (3.6a) can be integrated as an ODE. These characteristic curves are also called *rays* in wave theory, and they are determined from the *characteristic system* belonging to (3.6a). This system of

[2] Sometimes this term is reserved for the corresponding equation based on (2.51).

ODEs describes the joint evolution of (x, y, s_x, s_y) along a ray parametrized by the variable τ, say. After re-writing (3.6a) in the standard form

$$F(x, y, s_x, s_y) = \frac{1}{2}\left(s_x^2 + s_y^2 - n^2\right) = 0 \tag{3.7}$$

it follows from the method of characteristics that

$$\begin{aligned} \frac{dx}{d\tau} &= \frac{\partial F}{\partial s_x} = s_x, & \frac{ds_x}{d\tau} &= -\frac{\partial F}{\partial x} = n n_x, \\ \frac{dy}{d\tau} &= \frac{\partial F}{\partial s_y} = s_y, & \frac{ds_y}{d\tau} &= -\frac{\partial F}{\partial y} = n n_y. \end{aligned} \tag{3.8}$$

This is a *Hamiltonian* set of ODEs, with F being the Hamiltonian function and (x, s_x) and (y, s_y) being two pairs of canonical variables.[3] This makes it obvious that the continuous symmetries of F imply conservation laws for (3.8). For example, the Hamiltonian function F has no explicit dependence on τ and therefore the value of F is conserved along a ray, which is consistent with (3.7). Moreover, if the basic state is homogeneous in x, i.e., n depends on y only, then the value of s_x is conserved as well; the same holds for basic-state homogeneity in y and the value of s_y. In other words, the Cartesian components of the local wavenumber vector $\boldsymbol{k} = \kappa_0 \nabla s$ are conserved if the basic state has a corresponding translational symmetry.

The evolution of s along a ray follows from the total differential $ds = s_x\, dx + s_y\, dy$ as

$$\frac{ds}{d\tau} = s_x \frac{dx}{d\tau} + s_y \frac{dy}{d\tau} = s_x^2 + s_y^2 = n^2. \tag{3.9}$$

Thus, the ray increment (dx, dy) is parallel to ∇s and the increment of s per unit Euclidean distance $\sqrt{dx^2 + dy^2}$ is $ds/\sqrt{dx^2 + dy^2} = n$. In other words, the rays are orthogonal to the *wave fronts* $s =$ const and neighbouring wave fronts $s = 0$ and $s = \Delta s$, say, are separated by a distance $\Delta s/n$. Large values of n imply crowding of wave fronts and vice versa.

The rays are straight lines in the homogeneous case $c = c_0$ and $\nabla n = 0$. The nature of the refraction due to non-zero ∇n can be understood by considering a case where $n_x = 0$ and $n_y < 0$, say. Then s_x is constant along a ray whilst s_y decreases. For instance, if $s_x > 0$ and $s_y > 0$ this implies that the ray curves clockwise towards the x-axis, i.e., it curves *towards* the region of high n. As $nc = c_0$, a low index of refraction means a high wave speed and therefore wave rays curve *away* from regions of high wave speed.

[3] This Hamiltonian system is identical to that of a mechanical system describing the two-dimensional motion in time τ of a discrete point mass exposed to an energy potential $-\frac{1}{2}n(x, y)^2$ per unit mass. This is the content of the *classical wave–particle duality*, which links this mechanical system to the behaviour of high-frequency waves in a refractive medium.

More quantitatively, it can be shown that the curvature of the ray equals $|\nabla \ln n| = |\nabla \ln c|$, thus the relative gradient of the wave speed is the curvature of the ray. For example, this explains once more why surface waves near a beach are always refracted towards the shoreline: the gradient of H and therefore of c points away from the shoreline and consequently the wave turns away from the region of high wave speed in deep water.

The curvature of the ray is an intrinsic property of the ray, i.e., it does not depend on the arbitrary parametrization of the ray in terms of τ. A very useful direct link between the intrinsic properties of the ray and the variability of c is provided by *Fermat's theorem*, which asserts that the rays are the *geodesics* of a Riemannian metric with length element

$$n(x, y)\sqrt{dx^2 + dy^2}. \tag{3.10}$$

This Riemannian metric is conformal to the Euclidean metric with conformal factor given by the index of refraction. Fermat's theorem asserts that a ray from point A to point B makes the Riemannian distance

$$\int_A^B n\sqrt{dx^2 + dy^2} = c_0 \int_A^B \frac{\sqrt{dx^2 + dy^2}}{c} \tag{3.11}$$

stationary. The last integral can be interpreted as a constant times the travel time along the ray, because its integrand is the Euclidean distance divided by the wave speed. Thus Fermat's principle in geometric wave theory is often described by saying that rays traverse the distance between A and B along a line of *least travel time*. Of course, this is not fully correct because the rays are only known to be paths of stationary, but not necessary minimal, travel time. However, for close enough points A and B it can be shown that the travel time is indeed minimal.

3.1.2 Ocean acoustic tomography

Geometric wave theory is currently being used to probe the thermal structure of the global ocean using sound waves. The sound speed in water is an increasing function of both temperature and pressure. Near the surface the water is very warm and near the ocean floor (about 4 km deep on average) the water pressure is very high. This means that c has a vertical profile with a minimum at intermediate depth, where the water is not very warm and the pressure is not very high. By Fermat's principle this implies that sound rays turn away from the regions of high c at the top and the bottom. It can be shown that this leads to the formation of a *waveguide* at intermediate depth, with sound rays meandering up and down whilst travelling horizontally along the waveguide.

In ocean tomography a sound source[4] is placed somewhere in the upper ocean and receivers are placed far away at distant coasts. The sound pulses travels from source to receiver along the waveguide and they can travel across the entire globe. The interesting bit is that there can be multiple rays connecting source and receiver. This does not contradict Fermat's principle, which demands only that rays are *local* geodesics. There can be multiple routes between A and B that are each locally optimal. The key point is that the travel time along these multiple routes will be different. This means a pulse will arrive at the receiver with multiple echoes. The spacing of these echoes can be used to infer something about the globally averaged vertical structure of c. This inference process is an example of an inverse problem in remote sensing. This knowledge of c can be converted into knowledge about the global heat content of the ocean and this could be useful for the monitoring of global warming, for instance.

3.1.3 Ray tubes

The wave amplitude along non-intersecting rays can be found by integrating the conservation law (3.6b) over a *ray tube*, which is defined as the area bounded laterally by two neighbouring rays.[5] The neighbouring rays are infinitesimally close, with local distance d, say, and the beginning and the end of a ray tube is formed by two wave fronts of constant s. These wave fronts do not need to be infinitesimally close. The divergence theorem implies that the net wave energy flux through the boundaries of the ray tube must be zero. However, the flux integral is zero on the lateral ray parts of the boundary because there the flux $\propto \boldsymbol{\nabla} s$ is tangent to the boundary by construction. This means the fluxes across the wave front parts of the boundary must balance each other. Using the infinitesimal tube width d this implies the constancy of the cross-sectionally integrated flux

$$\frac{A^2}{n^2}|\boldsymbol{\nabla} s|d = \frac{A^2}{n}d = \text{const} \quad \text{or} \quad A^2 cd = \text{const}, \tag{3.12}$$

where the eikonal equation was used.[6] This simple relation shows that the amplitude along a ray tube changes via two distinct physical mechanisms.

[4] There is legitimate concern that this is not at all liked by sea mammals that use sound for navigation and communication; this is a hard question to decide.

[5] This is a standard construction for non-divergent vector fields; in fluid dynamics this is familiar from stream tubes based on an incompressible velocity field and from vortex tubes based on the automatically incompressible vorticity field.

[6] For the wave equation in the form (2.51) one would obtain $A^2 d/c = \text{const}$.

First, a change of wave speed c along the ray produces a reciprocal change in A^2. This is because the wave energy density must increase if the wave speed decreases in order for the total flux of energy to remain constant. This occurs even for parallel rays, where $d =$ const. Second, changes of d along the ray also lead to inverse changes in A^2 by contracting or expanding the cross-section of the ray tube. This is a geometric effect that can occur even for constant wave speed $c = c_0$.

We illustrate this geometric effect by considering a case of constant $c = c_0$ in which the wave field emanates from a localized wavemaker with boundary given by some smooth curve $\mathcal{C}$, say, as sketched in Figure 3.1 below. This wavemaker marks out the initial wave front from which to start the rays, i.e., on the wavemaker the amplitude and eikonal can be chosen as $A = 1$ and $s = 0$, respectively, and therefore all rays start orthogonally to $\mathcal{C}$. Moreover, by (3.8) and (3.9) the rays are straight lines and $s = r$ where r is the distance to the wavemaker along the rays. Now, the ray tube width d grows or decays linearly with r, depending on the curvature of $\mathcal{C}$ at the foot point of the ray. Specifically, if $d(r)$ is the width as a function of r and the curvature of $\mathcal{C}$ at the foot point of the ray is R then

$$\frac{d(r)}{d(0)} = \frac{R \pm r}{R} \quad \Leftrightarrow \quad \frac{A(r)}{A(0)} = \sqrt{\frac{R}{R \pm r}}. \tag{3.13}$$

Here the upper sign applies if the wavemaker is convex towards the fluid region at the foot point, and vice versa. This shows that far away from a convex wavemaker the amplitude decays as $r^{-1/2}$. It also shows that the amplitude becomes infinite at a distance $r = R$ away from a concave wavemaker, and that no geometric solution is available for $r > R$ in this case.

Clearly, the focusing of rays due to concave wave fronts leads to the intersection of neighbouring rays and to an inevitable breakdown of the earlier assumptions $|\kappa_0 A| \gg |\nabla A|$ etc. that were made at the outset of geometric wave theory. Thus geometric wave theory breaks down at *caustics*, which is the standard term used for the location of the envelope formed by the intersections of neighbouring rays. In the present example caustic line segments are formed at the centres of curvature associated with concave sections of $\mathcal{C}$. In terms of geometry, such a caustic line coincides with the *evolute* of $\mathcal{C}$, which is the envelope of the line normals relative to the wavemaker.

This example illustrates that caustics are by no means ruled out by the eikonal equation, which as a first-order PDE is not guaranteed to have a smooth global solution. Because caustics appear constantly in all but the simplest wave problems it is important to understand the nature of the exact wave solution at a caustic and also how geometric theory, but not linear wave theory, breaks down there. Most importantly, it needs to be understood how the geometric theory can be continued across the caustic location.

3.2 Caustics

We illustrate caustics using a simple situation without refraction, i.e., the wave speed is constant and $n = 1$. The reduced wave equation for the spatial structure of the time-periodic solution $\hat{h}(x, y)\exp(-i\omega t)$ then takes the simple form

$$(\nabla^2 + \kappa_0^2)\hat{h} = 0, \tag{3.14}$$

which has constant coefficients. The presence of a localized wave maker prohibits the use of plane-wave solutions, but the absence of refraction allows using explicit Green's functions for the solution of (3.14) and we can then study the behaviour of this solution at caustics.

3.2.1 Green's function representation

The Green's function $G(\boldsymbol{x}, \boldsymbol{x}')$ is defined by

$$(\nabla^2 + \kappa_0^2)G(\boldsymbol{x}, \boldsymbol{x}') = \delta(\boldsymbol{x} - \boldsymbol{x}') \tag{3.15}$$

where the differential operator acts on the first argument of G. We first consider a domain $\mathcal{D} \subset R^2$ with a smooth boundary $\partial\mathcal{D}$ and assume that (3.14) and (3.15) hold for all interior points $(\boldsymbol{x}, \boldsymbol{x}')$ in the domain. Then we can write

$$\hat{h}(\boldsymbol{x}) = \int_{\mathcal{D}} \hat{h}(\boldsymbol{x}')\delta(\boldsymbol{x} - \boldsymbol{x}')\, dx'dy' = \int_{\mathcal{D}} \hat{h}(\boldsymbol{x}')(\nabla'^2 + \kappa_0^2)G(\boldsymbol{x}', \boldsymbol{x})\, dx'dy' \tag{3.16}$$

where the differential operator again acts on the first argument of G. Integrating by parts twice yields, in virtue of (3.14),

$$\hat{h}(\boldsymbol{x}) = \int_{\partial\mathcal{D}} \left(\hat{h}(\boldsymbol{x}')\boldsymbol{\nabla}'G(\boldsymbol{x}', \boldsymbol{x}) - G(\boldsymbol{x}', \boldsymbol{x})\boldsymbol{\nabla}'\hat{h}(\boldsymbol{x}')\right) \cdot \boldsymbol{n}'\, d\sigma \tag{3.17}$$

where $\boldsymbol{n}'$ is the outward unit vector normal to $\partial\mathcal{D}$ and $d\sigma$ is the line element along $\partial\mathcal{D}$. This equation gives an explicit representation of $\hat{h}$ in terms of its own boundary data and that of G. As an aside we note that the flexibility in choosing boundary conditions for G makes this exact equation a convenient starting point for many wave problems involving boundaries, including wave diffraction.

3.2.2 High-wavenumber boundary-value problem

We now consider a localized smooth wavemaker with boundary $\mathcal{C}$. For instance, $\mathcal{C}$ could be the boundary of a time-periodic wavemaker in the shape of a circle, say. In terms of the domain $\mathcal{D}$ the wavemaker boundary $\mathcal{C}$ forms the inner part of the boundary $\partial\mathcal{D}$, whilst the outer part is given by a large circle centred at the wavemaker, say. The wavemaker boundary $\mathcal{C}$ marks out an initial wave front on which $\hat{h} = 1$, say. Assuming a high-wavenumber wave, we can then derive a second boundary condition for $\hat{h}$ on $\mathcal{C}$.

Specifically, in the present context the assumption of high wavenumber $\kappa_0 = \omega/c$ means that the curvature of $\mathcal{C}$ must be small compared to κ_0. Geometric wave theory without caustics then applies in a small neighbourhood of $\mathcal{C}$ because the eikonal and transport equations are guaranteed to have smooth local solutions.[7] Therefore, $\boldsymbol{\nabla}\hat{h} = -i\kappa_0\boldsymbol{n}'$ on $\mathcal{C}$, where $-\boldsymbol{n}'$ is the unit vector normal to $\mathcal{C}$ pointing *into* the fluid. This completes the boundary data for $\hat{h}$ that is used in (3.17):

$$\hat{h}(\boldsymbol{x}) = \int_{\mathcal{C}} \left(\boldsymbol{n}' \cdot \boldsymbol{\nabla}' G(\boldsymbol{x}', \boldsymbol{x}) + i\kappa_0 G(\boldsymbol{x}', \boldsymbol{x})\right) d\sigma. \tag{3.18}$$

We do not wish to consider other wave sources than $\mathcal{C}$ and this physical statement has the mathematical consequence that we need to chose a Green's function such that if the domain $\mathcal{D} \subset R^2$ is bounded by $\mathcal{C}$ and also by a very large circle then the contribution of the large circle to the integral in (3.17) should vanish as the radius goes to infinity. This is satisfied by the radiating Green's function

$$G(\boldsymbol{x}, \boldsymbol{x}') = G(\boldsymbol{x}', \boldsymbol{x}) = G(r) = \frac{-i}{4} H_0^{(1)}(\kappa_0 r) \quad \text{where} \quad r = |\boldsymbol{x} - \boldsymbol{x}'| \tag{3.19}$$

and $H_0^{(1)}$ is the zeroth-order Hankel function of the first kind. For finite r and large κ_0 we are interested in the far-field asymptotic expansion of $G(r)$ for $\kappa_0 r \gg 1$, which is

$$G(r) = \frac{-i}{4} H_0^{(1)}(\kappa_0 r) \sim \beta \frac{\exp(i\kappa_0 r)}{\sqrt{\kappa_0 r}} \quad \text{where} \quad \beta = \frac{\exp(-i3\pi/4)}{2\sqrt{2\pi}}. \tag{3.20}$$

Thus, at large distances (3.19) describes concentric wave fronts with outward phase progression and radial decay compatible with constant wave energy flux through any wave front.

[7] This argument ignores the possibility that the local solution near some point on $\mathcal{C}$ might consist of the superposition of several rays emanating from different parts of the wavemaker, which could occur if the wavemaker shape is not convex and which leads to the more complicated theory of wave scattering.

Now, if $\boldsymbol{x}$ is in the interior of $\mathcal{D}$ and $\boldsymbol{x}'$ is on $\mathcal{C}$ then the normal derivative in (3.18) becomes

$$\boldsymbol{n}' \cdot \boldsymbol{\nabla}' G(\boldsymbol{x}', \boldsymbol{x}) = \frac{dG}{dr} \cos\alpha \tag{3.21}$$

where α is the angle between the line $\boldsymbol{x}' - \boldsymbol{x}$ and the outward normal $\boldsymbol{n}'$. For large $\kappa_0 r$ the derivative is $i\kappa_0 G$ and therefore we obtain the final expression

$$\hat{h}(\boldsymbol{x}) = i\kappa_0 \beta \int_{\mathcal{C}} (\cos\alpha + 1) \frac{\exp(i\kappa_0 r)}{\sqrt{\kappa_0 r}} \, d\sigma + O(\kappa_0^{-1}). \tag{3.22}$$

Although we have used high-wavenumber asymptotics to derive (3.22), this equation is still considerably more general than the eikonal and transport equations. This is because here we only assumed that the wavelength is small compared to the curvature of the wavemaker, whereas in geometric wave theory we assumed that the wavelength is small compared to all other scales at every point in the fluid domain. The second assumption breaks down at a caustic, but not the first.

3.2.3 Stationary phase approximation

At first sight, (3.22) suggests that the solution at an interior point $\boldsymbol{x}$ depends on the entire wavemaker boundary $\mathcal{C}$. However, this is misleading because most of the integral amounts to nothing because the integrand contains the rapidly varying phase factor $\exp i\kappa_0 r(\sigma)$. Specifically, as the arc length σ runs along $\mathcal{C}$, the distance $r(\sigma)$ varies in some proportion and for large κ_0 this leads to rapid cancellations in the integral unless there is a point on $\mathcal{C}$ where the derivative $r'(\sigma) = dr/d\sigma$ vanishes and hence the phase is stationary. The dominant contribution to the integral in (3.22) will therefore come from the neighbourhood of such stationary points, which is the usual *stationary phase* argument.

Geometrically, this means that the domain of dependence on the boundary $\mathcal{C}$ of the wave field at any interior point $\boldsymbol{x}$ reduces to the neighbourhoods of the boundary points $\boldsymbol{x}' \in \mathcal{C}$ with stationary distance $r = |\boldsymbol{x} - \boldsymbol{x}'|$ to $\boldsymbol{x}$, as would be the case at the point of minimal distance between $\boldsymbol{x}$ and $\mathcal{C}$, for instance. This is an echo of Fermat's theorem in §3.1.1. Of course, for complicated wavemaker shapes it is possible that more than one stationary point contributes for a given $\boldsymbol{x}$ and in this case the various contributions are added up to get $\hat{h}(\boldsymbol{x})$. We can investigate the contribution stemming from the neighbourhood of a stationary point located at $\sigma = 0$, say, by performing the integral in (3.22) over the interval $\sigma \in [-\epsilon, \epsilon]$ where $\epsilon \ll 1$ is a small constant. First of all, at a stationary point the angle factor $\cos\alpha = \pm 1$

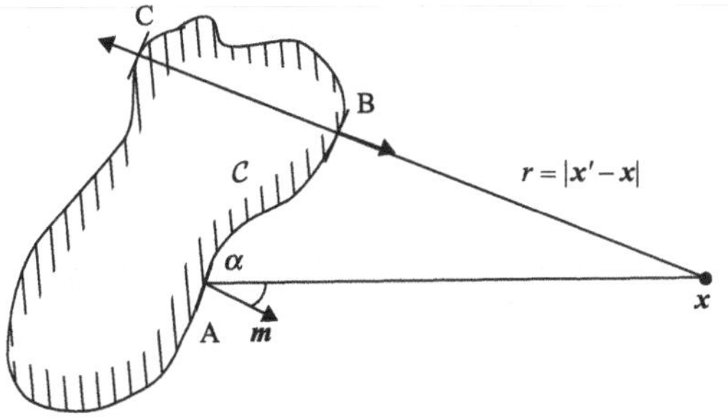

Figure 3.1 Wave field at $\boldsymbol{x}$ due to smooth wavemaker with boundary $\mathcal{C}$. Three boundary points $\boldsymbol{x}'$ are marked, with corresponding normal vectors $\boldsymbol{m} = -\boldsymbol{n}'$ and angles α. Point A is a regular point, which makes no leading-order contribution to the wave field at $\boldsymbol{x}$ for large κ_0. Points B and C are stationary points where $\cos\alpha = \pm 1$, respectively. Clearly, only point B is relevant for the leading-order evaluation of (3.22) for the chosen $\boldsymbol{x}$.

by construction. The positive sign occurs when the inward normal vector $-\boldsymbol{n}'$ points towards $\boldsymbol{x}$ and vice versa. If the negative sign occurs then the integrand is zero, which makes sense because in this case $\boldsymbol{x}$ is in the shadow of the wavemaker relative to the source point $\boldsymbol{x}'$. For instance, if $\mathcal{C}$ is a circle around the origin with radius R and $\boldsymbol{x} = (x, 0)$ with $x > R$ then there will be two stationary points $\boldsymbol{x}'_1 = (R, 0)$ and $\boldsymbol{x}'_2 = (-R, 0)$ but only the first will contribute to the integral. Thus we may assume without loss of generality that $\cos\alpha = 1$ at a relevant stationary point. This generic situation is illustrated in Figure 3.1.

Writing $r_0 = r(0), r'_0 = r'(0)$ etc., the Taylor expansion of r near the critical point $\sigma = 0$ is

$$r(\sigma) = r_0 + \frac{r''_0}{2}\sigma^2 + \frac{r'''_0}{6}\sigma^3 + \cdots \tag{3.23}$$

We can replace r by r_0 everywhere in the integrand except in the rapidly varying phase factor, where we need to include the first non-vanishing power of σ from (3.23). Under the assumption $r''_0 \neq 0$ this leads to

$$2i\beta \frac{\exp(i\kappa_0 r_0)}{\sqrt{r_0}} \int_{-\epsilon}^{\epsilon} \exp\left(i\kappa_0 \left(\frac{r''_0}{2}\sigma^2\right)\right) \sqrt{\kappa_0}\, d\sigma. \tag{3.24}$$

After the substitution $\tilde{\sigma} = \sqrt{\kappa_0}\sigma$ the integral can easily be evaluated for large κ_0 because the new integration limits then tend to infinity. Using the standard integral

$$\int_{-\infty}^{+\infty} \exp(is^2/2)\, ds = \sqrt{2\pi}\exp(i\pi/4) \tag{3.25}$$

the result is

$$\hat{h}(\boldsymbol{x}) = 2\sqrt{2\pi}i\beta\frac{\exp(i\kappa_0 r_0)}{\sqrt{r_0}}\frac{\exp(i\mathrm{sgn}(r_0'')\pi/4)}{\sqrt{|r_0''|}}. \tag{3.26}$$

Using the definition of β we can re-write this as

$$\hat{h}(\boldsymbol{x}) = \frac{\exp(i\kappa_0 r_0)}{\sqrt{r_0}}\frac{\exp(-i\alpha)}{\sqrt{|r_0''|}} \quad \text{with} \quad \alpha = \begin{cases} 0 & r_0'' > 0 \\ \pi/2 & r_0'' < 0 \end{cases}. \tag{3.27}$$

Notably, the phase depends on the sign of r_0''. This will be crucial for the continuation of a solution across caustics.

3.2.4 Curved wave fronts and focusing

To obtain explicit results we consider a wavemaker $\mathcal{C}$ that passes through the coordinate origin $\boldsymbol{x}' = (0,0)$ and we assume that $\mathcal{C}$ is tangent to the y-axis at this point. This point will then be the relevant stationary point for all points $\boldsymbol{x}$ along the ray $(x,0)$ with $x > 0$. By construction, the distance $r_0 = x$ and the asymptotic solution is controlled by r_0'', which depends on the curvature of $\mathcal{C}$ at the origin. For example, for a flat, zero-curvature wavemaker the distance $|\boldsymbol{x}' - \boldsymbol{x}|$ is

$$r(\sigma) = \sqrt{x^2 + \sigma^2} = x + \frac{1}{2}\frac{\sigma^2}{x} + \cdots \tag{3.28}$$

near $\sigma = 0$. Thus $r_0 = x$ and $r_0'' = 1/x$, which is positive, and therefore

$$\hat{h}(x,0) = \frac{\exp(i\kappa_0 x)}{\sqrt{x}}\sqrt{x} = \exp(i\kappa_0 x). \tag{3.29}$$

This recovers a plane wave emanating from a flat wall and propagating at fixed amplitude to the right. On the other hand, if the wavemaker is convex towards the fluid region, with radius of curvature $R > 0$, then the wavemaker boundary at the stationary point is locally given by $x = -\sigma^2/(2R)$. This leads to

$$r(\sigma) = \sqrt{(x + \sigma^2/(2R))^2 + \sigma^2} = x + \frac{\sigma^2}{2}\left(\frac{1}{x} + \frac{1}{R}\right) + \cdots \tag{3.30}$$

and therefore $r_0 = x$, $r_0'' = (R + x)/(xR)$ such that

$$\text{locally convex:} \quad \hat{h}(x,0) = \exp(i\kappa_0 x)\sqrt{\frac{R}{R+x}}. \tag{3.31}$$

As $R \to +\infty$ for fixed x this reduces to the plane wave in (3.29). Also, for any $R > 0$ this is identical to the geometric wave solution with eikonal $s = r_0$ and amplitude decaying like the square root of the distance from the centre of curvature at $(-R, 0)$, which lies outside the fluid. This produces an independent check on the derivation of the geometric theory.

The most interesting situation occurs if the wave front is concave at the stationary point, with local radius of curvature $R > 0$ and a centre of curvature $(R, 0)$ that now lies within the fluid. This centre of curvature is on a *caustic*, being the focus of neighbouring rays emanating from the boundary point. The initial wave front is given by $x = +\sigma^2/(2R)$ and therefore $r_0'' = (R-x)/(xR)$. There are two cases to consider. For points between the wavemaker and the focus we have $x < R$ and therefore $r_0'' > 0$, which leads to

$$\text{locally concave and } x < R\text{:} \quad \hat{h}(x,0) = \exp(i\kappa_0 x)\sqrt{\frac{R}{R-x}}. \tag{3.32}$$

The amplitude *increases* with $r_0 = x$, and diverges as the focus at $(R, 0)$ is approached. This again agrees with geometric wave theory. However, unlike in geometric wave theory, we can also evaluate the solution behind the focus, i.e., for $x > R$. In this case $r_0'' < 0$ and therefore

$$\text{locally concave and } x > R\text{:} \quad \hat{h}(x,0) = -i\exp(i\kappa_0 x)\sqrt{\frac{R}{x-R}}, \tag{3.33}$$

which exhibits a phase-shifting factor $-i = \exp(-i\pi/2)$ as well as decay with increasing x.

3.2.5 The phase shift across caustics

The phase shift between (3.32) and (3.33) appears harmless in the present setting of a monochromatic wave with a single frequency ω, but it is not. Indeed, the phase shift becomes very important once a general time-dependent solution is build up via Fourier series in time and hence superposition of modes with different $\omega = c\kappa_0$. Under the high-wavenumber assumption, all these modes will have similar spatial structures and they will all acquire the same phase shift across a focus.

Now, for a Fourier component $\Re\hat{h}\exp(-i\omega t)$ with $\omega > 0$ the phase shift in (3.33) means that a signal $\cos\omega t$ just before the focus turns into $-\sin\omega t$ just behind the focus, whilst $\sin\omega t$ turns into $\cos\omega t$. This is the signature of (minus) a *Hilbert transform* in time.[8] This is a nonlocal transformation of the signal because a Hilbert transform is equivalent to convolving the time-dependent signal by $1/(\pi t)$.

[8] Formally, the Hilbert transform is a linear operator on functions that turns $\cos\alpha t$ into $\sin\alpha t$ and $\sin\alpha t$ into $-\cos\alpha t$ for any $\alpha > 0$; on the real line this transform is equivalent to convolution with $1/(\pi t)$ (and taking principal values for the integral) whilst in Fourier space it is equivalent to multiplication with the phase-shifting factor $-i\text{sgn}(\omega)$.

Specifically, if at the concave wavemaker a particular high-frequency signal $h'(0,0,t) = f(t)$ is chosen then the signal will propagate along the ray $(x,0)$ along the x-axis in a straightforward fashion until it reaches the focus, i.e.,

$$x < R: \quad h'(x,0,t) = f(t - x/c)\sqrt{\frac{R}{R-x}}. \tag{3.34}$$

Thus, if one observed the wave field at any point $(x,0)$ before the focus then the signal would arrive with some delay due to the propagation and with some overall change in amplitude due to focusing, but otherwise it would be completely recognizable. However, behind the focus the signal will have been scrambled by (minus) the Hilbert transform, i.e.,

$$x > R: \quad h'(x,0,t) = -\frac{1}{\pi}\int \frac{f(s - x/c)}{t-s}\,ds\,\sqrt{\frac{R}{x-R}}, \tag{3.35}$$

where the integral is defined by its principal value. For example, consider the case where $f(t)$ contains a discontinuous wave signal such as a localized hat function. This is a bounded signal before the focus, but the Hilbert transform of a step function leads to a logarithmic singularity together with a long-range tail that decays as $1/t$. So the phase shift across the caustic maps a compact and bounded function into a non-compact and unbounded function. Clearly, this may violate the small-amplitude assumption on which our linear wave theory is based and hints at the generic importance of nonlinear effects near caustics.

It might also be thought (and indeed I did so when this book was first printed) that the non-local integral in (3.35) would affect smooth signals such as beautiful musical pieces in ways that might be perceptible to the human ear. This, however, does not appear to be the case, mostly because the human ear is not very sensitive to phase shifts and, presumably, also because beautiful music does not contain many hat functions.

3.2.6 Solution directly on the caustic

The asymptotic solution considered so far is not valid directly on the caustic, e.g., for a concave wavemaker it is not valid directly on the focus $\boldsymbol{x} = (R,0)$. This is because $r_0'' = 0$ there and hence the stationary phase approximation to the integral (3.22) does not begin with the quadratic term σ^2. Thus, we need to rescale in order to capture the first non-vanishing power of σ in (3.23). Specifically, if this power is p, say, then the stationary phase approximation in (3.24) involves a term $\kappa_0\sigma^p$, which leads to a rescaling of the integral in terms of $\tilde{\sigma} = \kappa_0^{1/p}\sigma$. Factors and phase shifts aside, inspection of the integral leads to

$$\text{locally concave and } x = R\text{:} \quad |\hat{h}| = O(\kappa_0^{1/2-1/p}). \tag{3.36}$$

Away from the caustic, $p = 2$ and we recover the geometric wave solution, whose amplitude indeed did not depend on κ_0. However, on the caustic $p > 2$ and therefore the solution there does increase with κ_0, although it remains finite for finite wavenumber. One could say that geometric wave theory fails at a caustic because the leading-order asymptotic solution there depends on wavenumber, and for this no allowance was made in the asymptotic expansion (2.45). It is possible to recast the asymptotic expansion such that a leading-order term corresponding to (3.36) is included, and in this way a uniformly valid asymptotic expansion can be formulated, albeit at the price of being considerably more complicated.

Now, the generic scaling on a caustic is $p = 3$ and therefore $|\hat{h}| = O(\kappa_0^{1/6})$, which increases quite slowly with wavenumber. This is the typical scaling along a caustic line given by the evolute, i.e., the line formed by the centres of curvature of $\mathcal{C}$. The extreme other case is $p = \infty$, which means that $r(\sigma)$ is constant for a finite σ-interval around the origin. This means the boundary segment around the stationary point is a perfect circle over some finite neighbourhood. This leads to a focal point with the most extreme scaling $|\hat{h}| = O(\kappa_0^{1/2})$.

In summary, the amplitude of high-wavenumber waves increases towards caustics but remains finite on the caustic for finite wavenumber. The amplitude scaling is sub-linear in κ_0 and is at most proportional to $\sqrt{\kappa_0}$. In practice, more important than the amplitude increase might be the phase shift across a caustic, which can lead to substantial changes in the wave signal for non-monochromatic waves and may even lead to unbounded solutions near the caustic. Again, this hints at the importance of nonlinear effects near caustics.

3.2.7 Non-smooth wavemakers and diffraction

Because of its intrinsic mathematical and physical interest we briefly mention the area of wave diffraction, although this will not play a role in the rest of this book.

We have assumed that the wavemaker is smooth and does not have sharp edges and so on, which meant that it was possible to assume that the wavenumber κ_0 was large compared to the curvature of the wavemaker at all points on $\mathcal{C}$, for instance. The presence of a sharp edge, say at a convex corner joining two smooth parts of the wavemaker with different slopes, leads to wave *diffraction*, which broadly speaking is the generation, propagation and scattering of waves due to boundary features that are small compared to the wavelength. This is the point of departure for diffraction theory, a classic example of which is the computation of scattered waves in the shadow zone behind a flat semi-infinite screen.

Actually, diffraction effects are not restricted to interactions with boundaries that include sharp edges. More precisely, diffraction can occur whenever the boundary curvature is large compared to the gradient of the wave phase *normal* to the boundary. This allows for diffraction of a slowly varying wavetrain that is at grazing incidence on a smooth convex boundary, which means that the gradient of the leading-order wave phase is tangential to the boundary at the grazing point. This is relevant to understanding why the shadow zone behind a smooth body is not completely dark, for example.

Now, diffraction invalidates both geometric wave theory and the Green's function solution in its present form. Because of this it is reassuring that the averaged energy flux conservation law $\boldsymbol{\nabla}\cdot\overline{\boldsymbol{F}} = 0$ remains valid for a steady wave field even in the presence of diffraction effects, which once again demonstrates the broad utility of conservation laws.

In light of the complexity of diffraction phenomena, it is remarkable that a geometric theory of diffraction can be formulated, which once again uses high-wavenumber asymptotics. This more general theory, which has found wide-ranging applications in the field of remote sensing and radar engineering, includes the rays of geometric wave theory, but it augments them with a number of new rays called diffractive rays that emanate from the relevant boundary features such as sharp edges and so on. For instance, these diffractive rays fill the shadow regions behind bodies that would otherwise be void of wave activity according to geometric wave theory. In addition, diffractive rays are capable of a special kind of creeping propagation tangent to a smooth boundary. Once the diffractive rays have detached from the boundary they can again be described by standard ray tracing, but their generation and creeping boundary propagation can not.

3.3 Notes on the literature

Geometric wave theory is described in Keller (1978), which also provides an account of the geometric theory of diffraction written by its main architect. The method of characteristics for first-order PDEs is described in the classic text Courant and Hilbert (1989) and in many other textbooks. A brief introduction to this method and to geometric wave theory can also be found in Bühler (2006), which in addition provides further links to the methods of classical mechanics.

3.4 Exercises

1. Wave scattering. The set-up of §3.2.1 can also be used to study wave scattering by immersed bodies. For example, consider an incoming plane acoustic wave moving from right to left with complex field $u_1 = \exp(-i\kappa_0 x)$, say. The total wave field $u = u_1 + u_2$ is then the sum of the incoming wave u_1 plus an outgoing scattered wave u_2 to be determined by the requirement that the boundary condition $\boldsymbol{n}' \cdot \boldsymbol{\nabla} u = 0$ holds at the boundary of the body. Use the Green's function approach of §3.2.1 to formulate an integral expression for the outgoing scattered field u_2.

2. Focusing of scattered waves. In the high-wavenumber regime incoming wave rays are converted into scattered rays via the familiar equal-angle reflection rule: the incoming ray makes the same angle with the scattering boundary as the outgoing ray. Find the scattered rays in the case of horizontal incoming rays $y = b$ moving from right to left, say, which are reflected at a parabolic scattering body with shape $y^2 = ax$ where $a > 0$. Verify that the scattered rays are given by

$$y = b + \left(x - \frac{b^2}{a}\right)\frac{4ab}{4b^2 - a^2} \tag{3.37}$$

and hence that all scattered rays are focused at the single point $(x, y) = (a/4, 0)$. Hint: you might find the formula $\tan 2\theta = 2\tan\theta/(1 - \tan^2\theta)$ useful.

4

Dispersive waves and ray tracing

Non-dispersive plane waves are very special because in this case the roots of the dispersion relation take the linearly homogeneous form $\omega = c\kappa$, which implies that the frequency does not depend on the orientation of $\boldsymbol{k}$ and that all wavelengths travel with the same speed c. In particular, for non-dispersive waves the group velocity

$$\boldsymbol{c_g} = \frac{\partial \omega}{\partial \boldsymbol{k}} \tag{4.1}$$

is equal to the vector phase velocity $\boldsymbol{c_p} = \omega \boldsymbol{k}/\kappa^2$. For dispersive plane waves none of these properties need to apply, and this leads to the study of dispersive waves, which by and large is the study of group-velocity effects.

We start by considering a generic one-dimensional dispersive wave system to become thoroughly familiar with the group-velocity concept and then we look at simple examples of dispersive waves that are of geophysical relevance. Ray tracing is the dispersive generalization of geometric wave theory and it is particularly important in problems involving moving media. Finally, we consider various $O(a^2)$ wave activity measures for dispersive waves, such as wave action and pseudomomentum, and their conservation properties, which again is a central and non-trivial task in the case of moving media.

4.1 Facets of group velocity

We consider a one-dimensional field $u(x,t)$, say, that is described by a PDE with constant coefficients. Our only use of the underlying PDE is that it provides the dispersion relation $D(k,\omega) = 0$, say, if a plane-wave solution is assumed in the form $u = \hat{u}\exp(i[kx - \omega t])$. Depending on the situation at hand, we can use the dispersion relation to yield functions $\omega(k)$ or $k(\omega)$, either of which may involve multiple solution branches. For example, if

$$u_t = u_{xxx} \quad \text{then} \quad D = \omega - k^3 = 0, \quad c_p = k^2, \quad c_g = 3k^2. \tag{4.2}$$

This is clearly a dispersive wave system as $c_g = 3c_p$. Moreover, the dispersion relation implies one frequency branch $\omega(k) = k^3$ but three wavenumber branches $k(\omega) = \omega^{1/3}$.

4.1.1 Beat waves

Beat waves are amplitude-modulated plane waves that arise if two plane waves of nearby wavenumber and frequency are added together. Specifically, if the respective wavenumbers and frequencies are $k \pm \Delta k$ and $\omega \pm \Delta\omega$ then for two plane waves with equal amplitude we obtain the exact superposition

$$\begin{aligned}&\exp(i[(k+\Delta k)x - (\omega+\Delta\omega)t]) + \exp(i[(k-\Delta k)x - (\omega-\Delta\omega)t]) \\ &\quad = 2\cos(\Delta k x - \Delta\omega t)\exp(i[kx - \omega t]). \end{aligned} \tag{4.3}$$

This is a plane wave, with wavenumber k and frequency ω, that is modulated in amplitude by the cosine factor. At any fixed time the amplitude waxes and wanes over a 'beat' distance proportional to $1/\Delta k$, which contains many wavelengths if $\Delta k \ll k$. An analogous beating occurs over time at any fixed x, which is the more familiar phenomenon from noisy passenger cabins in aircraft or cars.

Any point of constant phase $kx - \omega t$ moves with the phase velocity $c_p = \omega/k$, but this not the speed with which the modulation amplitude moves. Instead, the amplitude is constant on points of constant $\Delta k x - \Delta\omega t$, which for small Δk implies a speed

$$\frac{dx}{dt} = \frac{\Delta\omega}{\Delta k} = \frac{d\omega}{dk}(k) + O(\Delta k) = c_g(k) + O(\Delta k). \tag{4.4}$$

This uses the Taylor series $\Delta\omega = \Delta k \omega'(k) + O(\Delta k^2)$. Thus, for small detuning Δk the amplitude moves with the group velocity c_g whilst the phase moves with the phase velocity c_p. This simple example contains the intrinsic core of all group-velocity effects.

4.1.2 Boundary forcing and radiation condition

Next we consider a boundary-value problem involving a localized wavemaker generating waves that then propagate freely into a previously quiescent medium. Specifically, we consider time-periodic waves that are generated at $x = 0$ and radiate along the positive x-axis, say. This example introduces the important concepts of a *causal solution* and of the corresponding *radiation condition*.

The boundary condition is $u(0,t) = \exp(-i\omega t)$ with some frequency ω and the solution for $x \geq 0$ consists of a plane wave with wavenumber $k(\omega)$

consistent with the dispersion relation. If only one branch contributes then the unique solution is

$$x \geq 0: \quad u(x,t) = \exp(i[k(\omega)x - \omega t]). \tag{4.5}$$

However, if there are several wavenumber branches then the solution is a non-unique superposition of several plane waves. We will now seek an argument that restricts the possible branches by making precise the notion that the waves are meant to propagate into a medium that was originally at rest.

To begin with, we can rule out any complex-valued branch $k(\omega)$ that has a negative real part, because this would lead to unbounded growth of $\exp ikx$ as $x \to \infty$. On the other hand, any plane-wave branch has real k and therefore this argument does not work. However, there is an ingenious device we can use here, namely we can introduce a small positive imaginary part to the frequency ω, i.e., we make the causal perturbation

$$\omega \to \omega + i\epsilon \quad \text{with} \quad 0 < \epsilon \ll |\omega|. \tag{4.6}$$

This means that $u(0,t) = \exp(-i\omega t)\exp(\epsilon t)$ and therefore the boundary forcing is growing slowly in time. Here 'slowly' means that significant changes in the amplitude accrue only over many oscillations with frequency ω. By construction, the amplitude was zero in the distant past at $t = -\infty$ and is unity now at $t = 0$. The beauty of linear theory is that the spatial structure of this growing wave can still be found from the dispersion relation simply by computing a complex wavenumber from the function $k(\omega + i\epsilon)$, provided that the function $k(\omega)$ is analytic in a complex neighbourhood of the real ω. For small ϵ this leads to

$$k(\omega + i\epsilon) = k(\omega) + i\epsilon \frac{dk}{d\omega}(\omega) = k(\omega) + \frac{i\epsilon}{c_g(\omega)} \tag{4.7}$$

and therefore the solution is

$$u(x,t) = \exp(\epsilon[t - x/c_g(\omega)]) \exp(i[k(\omega)x - \omega t]). \tag{4.8}$$

Once again the amplitude is non-uniform and points of constant amplitude move with the group velocity c_g, which shows that the result (4.4) was not restricted to beating waves. But more can be said! In particular, for non-zero $\epsilon > 0$ we can at every finite time t demand that $u(+\infty, t) = 0$ in order to obtain the solution that corresponds to waves radiating into a quiescent medium. However, (4.8) then shows that

$$\epsilon > 0: \quad \lim_{x \to +\infty} u(x,t) = 0 \quad \Rightarrow \quad c_g(\omega) > 0. \tag{4.9}$$

This is the sought-after *radiation condition*: the group velocity of causal waves must point away from the wavemaker. In many problems this radiation condition is sufficient to make the boundary-value problem unique. Once we select the correct branch using the radiation condition then we can let $\epsilon \to 0$ and obtain what is called the *causal* solution to the boundary-value problem.

In summary, the radiation condition restricts the possible branches of $k(\omega)$ to those that guarantee decay as $x \to +\infty$ under the causal perturbation (4.6). For plane waves this restricts to branches with group velocity pointing away from the wavemaker and for complex $k(\omega)$ this restricts to branches with exponential decay into the fluid medium. The radiation condition restricts the possible branches of $k(\omega)$, but it does not necessarily make the solution of a boundary-value problem unique. For example, in (4.2) there are three wavenumber branches $k(\omega) = \omega^{1/3}$ such that for $\omega = 1$, say, we have the three roots

$$k_1 = 1, \quad k_2 = \exp(i2\pi/3) \quad \text{and} \quad k_3 = \exp(-i2\pi/3). \tag{4.10}$$

The first root satisfies the radiation condition whilst the third root can be ruled out because it yields unbounded exponential growth as $x \to +\infty$. However, the second root decays exponentially for positive x and is therefore equally acceptable as a causal solution. Thus the boundary-value problem as stated has no unique causal solution and more boundary data is needed. For instance, if both u and u_x are specified at $x = 0$ then this would yield a unique causal solution in $x > 0$.

Interestingly, if the causal solution were sought in the half line $x \leq 0$, then the first and second root in (4.10) are both inadmissible and hence the unique solution is given by the third root, which yields an evanescent wave tail. This means that the solution to the boundary-value problem in $x \leq 0$ is uniquely determined once $u(0, t)$ is specified whereas the solution in $x \geq 0$ requires $u_x(0, t)$ as well.

4.1.3 Asympotic solution to initial-value problem

This is a key problem for dispersive waves because it illustrates the generic emergence of a slowly varying wavetrain out of fairly arbitrary initial conditions. We consider the initial-value problem on the real line with initial condition $u(x, 0) = u_0(x)$ such that for wave systems with a single frequency branch $\omega(k)$ the solution

$$u(x, t) = \frac{1}{2\pi} \int_{-\infty}^{+\infty} \exp(+i[kx - \omega(k)t])\hat{u}_0(k)\, dk. \tag{4.11}$$

This uses the Fourier transform pair

$$\hat{u}_0(k) = \int_{-\infty}^{+\infty} \exp(-ikx) u_0(x)\, dx \tag{4.12}$$

and
$$u_0(x) = \frac{1}{2\pi} \int_{-\infty}^{+\infty} \exp(+ikx) \hat{u}_0(k)\, dk. \tag{4.13}$$

If there is more than one frequency branch $\omega(k)$ then further initial conditions must be specified (such as $u_t(x,0)$ etc.) and further terms must be added in (4.11). For sake of definiteness we restrict to a single branch now, but the frequently occurring case of two equal-and-opposite frequency branches $\pm\omega(k)$ will be considered in §4.1.5.

We are interested in the solution for large values of t in the special case of localized initial conditions, i.e., $u_0(x)$ has compact support such that $u = 0$ outside some finite interval surrounding the origin. Later we will see how compactness can be relaxed to sufficiently fast decay as $|x| \to \infty$. To exploit the asymptotic regime of large t we evaluate (4.11) along a line $x = ct$ based on some observation speed c, say, which leads to

$$u(ct, t) = \frac{1}{2\pi} \int_{-\infty}^{+\infty} \exp(+it[kc - \omega(k)]) \hat{u}_0(k)\, dk. \tag{4.14}$$

This integral is of the standard stationary phase form, i.e., for large t significant contributions accrue only in the neighbourhood of wavenumbers k_0 that are stationary points of the phase $kc - \omega(k)$. These stationary points satisfy the *group-velocity condition*

$$c = \frac{d\omega}{dk}(k_0), \quad \text{i.e.,} \quad \frac{x}{t} = c_g(k_0). \tag{4.15}$$

Thus the wavenumber k_0 that is observed at speed $x/t = c$ is determined by a match with the group velocity, i.e., the wavenumber travels with the group velocity. Thus for given x and t we should expect to see contributions from the spectral content of the initial conditions $\hat{u}_0(k)$ at precisely such wavenumbers k_0 whose associated group velocity was just right to travel the distance from the origin to the location x during the time t.

Even in case of a single frequency branch $\omega(k)$ there will usually be more than one root $k_0(c)$ in (4.15), which means that the various contributions must be added up to yield the correct asymptotic solution. In particular, a real-valued solution $u(x,t)$ implies the reality conditions $\hat{u}(k,t) = \hat{u}^*(-k,t)$ and $\omega(k) = -\omega(-k)$, where the star denotes complex conjugation. This implies $c_g(k) = c_g(-k)$ and hence each stationary point $k_0(c)$ always appears with both signs, which can be combined as follows. If the same stationary phase computation as in §3.2.3 is applied (with $-t\omega_0'' = -t\omega''(k_0)$ replacing r_0'' and so on) then we obtain at k_0 the contribution

$$\sqrt{\frac{2\pi}{t|\omega_0''|}}\frac{\hat{u}_0(k_0)}{2\pi}\exp(it[k_0c-\omega_0]-i\text{sgn}(\omega_0'')\pi/4), \tag{4.16}$$

provided that $\omega_0'' \neq 0$. Now, for a real solution the contribution from $-k_0$ is the complex conjugate of (4.16) and therefore the total contribution from both points is twice the real part of (4.16). Moreover, that real part does not depend on the sign of k_0, so we may assume $k_0 > 0$ only, say.

We have found, then, that for large t the asymptotic solution is

$$u(x,t) = \frac{2}{\sqrt{2\pi t|\omega_0''|}}\Re\left\{\hat{u}_0(k_0)\exp\left(i\left[k_0x-\omega_0t-\text{sgn}(\omega_0'')\frac{\pi}{4}\right]\right)\right\}+O(t^{-1}). \tag{4.17}$$

where k_0 is the positive root of (4.15). Thus, the solution looks locally like a modulated plane wave with wavenumber k_0 and frequency ω_0, which depend on x/t according to the positive root of (4.15). Comparing with the beat waves, here the modulation is in both amplitude and frequency. Along a *group-velocity ray* of constant $x/t = c_g(k_0)$ the wavenumber and frequency are constant, but the amplitude decays as $1/\sqrt{t}$.

We have shown the remarkable fact that in a dispersive wave system any compact initial condition $u_0(x)$ fans out into a slowly varying wavetrain. This highlights the importance of plane waves as the natural asymptotic state of dispersive wave solutions. Note that the detailed structure of the asymptotic solution uses the assumption $\omega_0'' = c_g'(k_0) \neq 0$, because otherwise (4.17) would be infinite and further terms would have to added the stationary phase approximation. The corresponding group-velocity rays $c_g(k_0) = x/t$ such that $c_g'(k_0) = 0$ are examples of *dispersive caustics* and the procedure for finding the correct asymptotic solution there mirrors the procedure for non-dispersive caustics in §3.2.6.

We illustrate (4.15) and (4.17) with the example (4.2). Inspecting (4.15) for $x > 0$ yields $k_0 = \pm\sqrt{x/3t}$ and we choose the upper sign by convention. On the other hand, for $x < 0$ there is no real root k_0. Thus the asymptotic solution for $x > 0$ is

$$u(x,t) \sim \sqrt{\frac{1}{\pi\sqrt{3xt}}}\Re\left\{\hat{u}_0(\sqrt{x/3t})\exp(i[2x\sqrt{x/27t}-\pi/4])\right\} \tag{4.18}$$

and zero if $x < 0$. This is illustrated in Figure 4.1 using step-like initial conditions, which have a broad wavenumber spectrum. Near $x = 0$ the

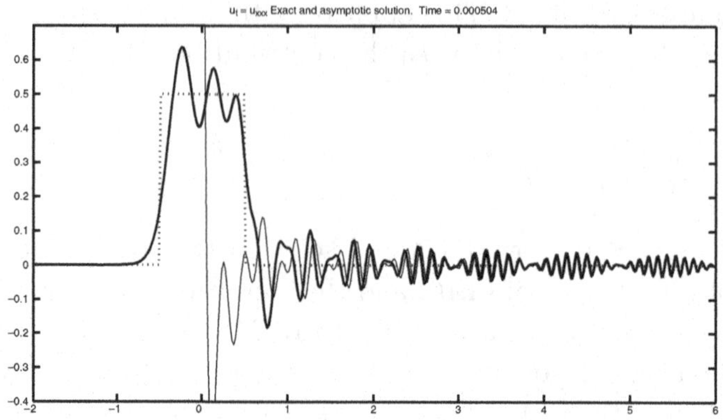

Figure 4.1 Snapshots of exact and asymptotic solution in bold and thin lines, respectively, to $u_t = u_{xxx}$ with compact initial data given by the dotted line. The asymptotic solution is not valid near $x = 0$, but it agrees remarkably well with the exact solution already for moderately large $x > 0$. The 'exact' numerical solution is obtained using Fourier series in a periodic domain that is much larger than the plotted interval; for small enough times such a periodic solution approximates the solution on the real line.

stationary phase solution fails because this is a dispersive caustic, but for $x > 0$ the asymptotic solution (4.18) is remarkably accurate.

It is worth pausing to notice at which step leading up to (4.17) we actually used the assumption of compact initial conditions. In fact, this occurred implicitly when we assumed that $\hat{u}_0(k)$ is a slowly varying function of k compared to the rapidly varying phase $t[kc - \omega(k)]$. For instance, if the initial condition is shifted such that $u_0(x) \to u_0(x - d)$ for some distance d, then $\hat{u}_0 \to \exp(-ikd)\hat{u}_0$ and this exhibits a phase kd. This will vary slowly compared to kct only if $ct \gg d$, i.e., only if the travel distance is large compared to the shift in the initial conditions. Thus all shifted compact initial conditions are eventually described by (4.17), but tighter localization around the origin implies earlier validity of this asymptotic solution. More generally, it follows from the definition of the Fourier transform that

$$\left|\frac{d\hat{u}_0}{dk}\right| \leq \int_{-\infty}^{+\infty} |xu_0(x)|\, dx, \tag{4.19}$$

which is bounded if the initial conditions decay faster than x^{-2} as $|x| \to \infty$. This shows that compactness of the initial conditions can be relaxed to such a decay condition.

We conclude by pointing out two aspects of the asymptotic phase $\theta_0(x,t) = k_0 x - \omega_0 t$ in the present context. In the example $\theta_0 = 2x\sqrt{x/27t}$ and this function is first-degree homogeneous in x and t, i.e., if $(x,t) \to (\lambda x, \lambda t)$

for any constant $\lambda > 0$ then $\theta_0 \to \lambda\theta_0$. By Euler's theorem[1] this implies $\theta_0 = x\theta_{0x} + t\theta_{0t}$ and therefore

$$\frac{\partial\theta_0}{\partial x} = k_0 \quad \text{and} \quad \frac{\partial\theta_0}{\partial t} = -\omega_0 \tag{4.20}$$

hold even though (k_0, ω_0) are *not* constant, but depend on both x and t. That this is a general feature of $\theta_0(x,t)$ can be seen from

$$\begin{aligned} d\theta_0 &= d(k_0 x - \omega_0 t) = k_0\,dx - \omega_0\,dt + x\,dk_0 - t\,d\omega_0 \\ &= k_0\,dx - \omega_0\,dt + (x - tc_g(k_0)\,dk_0 = k_0\,dx - \omega_0\,dt \end{aligned} \tag{4.21}$$

after using the crucial relation (4.15).

The other feature is that figure 4.1 makes clear that the phase θ_0 varies over a large range of values across the space–time domain of interest. For instance, slowly varying wavetrains are often studied in 'slow' coordinates $X = \epsilon t$ and $T = \epsilon x$ where the small parameter $\epsilon \ll 1$ means that $T = O(1)$ if $t = O(1/\epsilon)$ is the large unscaled time. For example, in geometric wave theory x and y were slow coordinates and $1/\kappa_0$ corresponded to ϵ. Thus, in slow coordinates the space–time domain of interest has size $O(1)$ whilst

$$\theta_0 = 2x\sqrt{\frac{x}{27t}} = \frac{2X}{\epsilon}\sqrt{\frac{X}{27T}} \tag{4.22}$$

and therefore the range of $\theta_0 = O(\epsilon^{-1})$ is explicitly large over the domain of interest.

4.1.4 Asymptotic wave energy dynamics

The asymptotic solution (4.17) consists of a plane-wave exponential (whose wavenumber is determined from the intuitive group-velocity condition (4.15)) that is multiplied by an amplitude factor whose structure at first sight has no intuitive explanation. However, we can gain precisely such an intuitive explanation of this factor by considering the energy density of the solution, which for a system with a single frequency branch is simply the square of the solution. The integrated energy density is constant in time, because the magnitude of a plane wave is constant in time and by orthogonality the same is true for the integral of the squared magnitudes in (4.11), provided the integral exists. More briefly, *Parseval's theorem* applied to (4.11) yields

$$\int_{-\infty}^{+\infty} |u(x,t)|^2\,dx = \frac{1}{2\pi}\int_{-\infty}^{+\infty} |\hat{u}_0(k)|^2\,dk = \int_{-\infty}^{+\infty} |u_0(x)|^2\,dx. \tag{4.23}$$

[1] $F(\boldsymbol{x})$ is nth-degree homogeneous in $\boldsymbol{x}$ if $\forall\alpha > 0$: $F(\alpha\boldsymbol{x}) = \alpha^n F(\boldsymbol{x})$; then $\boldsymbol{x}\cdot\nabla F(\boldsymbol{x}) = nF(\boldsymbol{x})$.

We now look at the phase-averaged energy density implied by the asymptotic solution (4.17), which using (2.19) is

$$\overline{u(x,t)^2} = \frac{|\hat{u}_0(k_0)|^2}{\pi t |c_g'(k_0)|}. \tag{4.24}$$

Here the function $k_0(x,t) \geq 0$ is known from the group-velocity condition (4.15). Now, if at fixed t we vary x by dx then this condition implies that k_0 varies by $dk_0 = dx/(tc_g')$, and thus we can re-write (4.24) as

$$\overline{u(x,t)^2} = \frac{1}{\pi} |\hat{u}_0(k_0)|^2 \text{sgn}(c_g'(k_0)) \frac{dk_0}{dx}. \tag{4.25}$$

This shows that the square of the pre-factor in (4.17) is the Jacobian of the map between x and k_0 induced by the group-velocity condition (4.15) at fixed t. This fact is especially useful because it carries over to wave problems in higher dimensions, where the pre-factor becomes a more complicated-looking expression involving the determinant of the matrix of second derivatives of $\omega(\boldsymbol{k})$, but the interpretation of the factor remains the same.

Integrating (4.25) between two points x_1 and x_2 yields

$$\int_{x_1}^{x_2} \overline{u(x,t)^2}\, dx = \frac{1}{\pi} \int_{k_{01}}^{k_{02}} |\hat{u}_0(k_0)|^2 \text{sgn}(c_g'(k_0))\, dk_0 \tag{4.26}$$

where the integration limits for k_0 are determined by the group-velocity condition at x_1 and x_2. It is easy to show that the asymptotic result (4.26) is consistent with global energy conservation, so no energy is lost in the asymptotic procedure. For example, in case of the system (4.2) with $x_1 = -\infty$ and $x_2 = +\infty$ (the asymptotic solution is zero for $x < 0$, so we can include the left half line for free here) we obtain $k_{01} = 0$ and $k_{02} = +\infty$ and therefore

$$\int_{-\infty}^{+\infty} \overline{u(x,t)^2}\, dx = \frac{1}{\pi} \int_0^{+\infty} |\hat{u}_0(k)|^2\, dk = \frac{1}{2\pi} \int_{-\infty}^{+\infty} |\hat{u}_0(k)|^2\, dk, \tag{4.27}$$

which recovers (4.23). Here the last step used the reality condition on $\hat{u}_0$.

Now, if we allow x_1 and x_2 to move in time with the local group velocity then the relevant range of k_0 on the right-hand side of (4.26) is constant, and therefore we have shown that the energy contained between two points moving with the group velocity is constant at large times. This statement is equivalent to the differential conservation law

$$\frac{\partial \bar{E}}{\partial t} + \frac{\partial \bar{E} c_g}{\partial x} = 0 \quad \text{where} \quad \bar{E} = \overline{u^2} \quad \text{and} \quad c_g = x/t. \tag{4.28}$$

Finally, it is instructive to re-write this equation in terms of a directional time derivative that follows a group velocity ray as

$$\left(\frac{\partial}{\partial t} + c_g \frac{\partial}{\partial x}\right) \bar{E} = -\bar{E}\frac{\partial c_g}{\partial x} = -\frac{\bar{E}}{t}, \tag{4.29}$$

which implies the decay $\bar{E} \propto 1/t$ along such rays. In n spatial dimensions analogues of (4.28) and (4.29) hold with $\boldsymbol{c}_g = \boldsymbol{x}/t$ and x-derivatives replaced by divergence operators, which yields $\bar{E} \propto 1/t^n$ along rays. The more rapid decay is related to the more rapid dilution by spreading of the wavefield in higher dimensions.

4.1.5 The case of equal-and-opposite frequencies

The case of two equal-and-opposite frequency branches $\pm\omega(k)$ occurs often in fluid dynamics, which is why we assemble the relevant formulas here. No new theory is involved, but these formulas are useful in practice. For instance, they can be used for *surface gravity waves* on water with constant depth H, which satisfy the dispersion relation

$$\omega^2 = g\kappa \tanh \kappa H \tag{4.30}$$

where $\kappa = |k|$ and $g > 0$ is gravity. This includes the limiting regimes of dispersive deep-water waves, in which $\kappa H \gg 1$ and $\omega^2 = g\kappa$, and of non-dispersive shallow-water waves, in which $\kappa H \ll 1$ and $\omega^2 = gH\kappa^2$.

To solve the initial-value problem we denote by $\omega(k) \geq 0$ the positive root of (4.30) and hence the general solution can be written as

$$u(x,t) = \frac{1}{2\pi}\int_{-\infty}^{+\infty} e^{ikx}\left(e^{-i\omega(k)t}A_1(k) + e^{+i\omega(k)t}A_2(k)\right) dk. \tag{4.31}$$

Fitting the initial conditions for u and u_t determines the spectral coefficients as $A_{1,2} = \frac{1}{2}(\hat{u}_0 \pm i\hat{u}_{t0}/\omega)$ and therefore (4.31) can be summarized by

$$\hat{u}(k,t) = \hat{u}_0(k)\cos\omega(k)t + \frac{\hat{u}_{t0}(k)}{\omega(k)}\sin\omega(k)t. \tag{4.32}$$

This concise expression is particularly convenient for the numerical evaluation of the solution using fast Fourier transforms. Of course, (4.32) can also be derived directly by noting that $\hat{u}(k,t)$ satisfies the linear harmonic oscillator equation

$$\hat{u}_{tt} + \omega(k)^2\hat{u} = 0. \tag{4.33}$$

Now, we can simplify (4.31) further for real-valued $u(x,t)$ by exploiting the redundancy inherent in the map from the two complex functions $\hat{u}_0$ and $\hat{u}_{t0}$

to the two real functions u and u_t. This is particularly simple if the positive root $\omega(k) = \omega(-k)$ is an even function. This corresponds to a reflection symmetry in x of the physical system, which is often satisfied for waves with two equal-and-opposite frequency branches. For instance, (4.30) clearly has this symmetry. We restrict to this case so that

$$\omega(k) \geq 0 \quad \text{and} \quad \omega(k) = \omega(-k) \tag{4.34}$$

both hold and then it turns out that it is sufficient to keep only the A_1 term in (4.31), provided we take twice the real part of the result, i.e.,

$$u(x,t) = \frac{1}{2\pi}\Re\left\{\int_{-\infty}^{+\infty} e^{i[kx-\omega(k)t]}\left[\hat{u}_0(k) + \frac{i}{\omega(k)}\hat{u}_{t0}(k)\right] dk\right\}. \tag{4.35}$$

This expression clearly satisfies the dispersion relation (and therefore the underlying PDE) and it also satisfies the two initial conditions for u and u_t. The latter fact depends on $i\hat{u}_{t0}/\omega$ being the Fourier transform of an imaginary function by (4.34b). This representation of the solution is also economical for a numerical computation, because the single complex function in square brackets captures both parts of the real initial data.

Another advantage is that the asymptotic solution for large t can proceed as in the single-frequency-branch case, with the quantity in square brackets replacing the previous $\hat{u}_0(k)$. Moreover, because ω is an even function the group velocity c_g is now an odd function of k and therefore here we don't have the complication of double roots $\pm k_0$ for the group-velocity condition (4.15). Indeed, a negative k_0 now corresponds to negative $c_g = x/t$ and vice versa. This leads to the asymptotic formula

$$u(x,t) \sim \frac{1}{\sqrt{2\pi t|\omega_0''|}}\Re\left\{\left[\hat{u}_0(k_0) + \frac{i\hat{u}_{t0}(k_0)}{\omega_0}\right] e^{i\theta_0 - i\,\mathrm{sgn}(\omega_0'')\frac{\pi}{4}}\right\} \tag{4.36}$$

where the asymptotic phase θ_0 is the same as in (4.17). Instead of the factor 2 in (4.17) there are now two contributions to the spectral amplitude in square brackets. The computation of mean square values also proceeds as before, e.g.,

$$\overline{u(x,t)^2} = \frac{1}{2}\frac{1}{2\pi t|\omega''(k_0)|}\left|\hat{u}_0(k_0) + \frac{i\hat{u}_{t0}(k_0)}{\omega(k_0)}\right|^2 \tag{4.37}$$

and analogously for $\overline{u_t^2}$. The conserved energy has a kinetic and potential part such that

$$\mathcal{E} = \frac{1}{2\pi}\int_{-\infty}^{+\infty} \left(|\hat{u}_t(k,t)|^2 + \omega^2|\hat{u}(k,t)|^2\right) dk \tag{4.38}$$

is constant, which also follows directly from (4.33). Finally, in the asymptotic solution there is equipartition between these energy forms at every location x. For example, in the special case $u_t(x,0) = 0$ the initial kinetic energy is zero, but in the asymptotic solution it is

$$\int_{-\infty}^{+\infty} \overline{u_t(x,t)^2}\, dx = \frac{1}{2}\frac{1}{2\pi}\int_{-\infty}^{+\infty} \omega^2|\hat{u}_0(k_0)|^2\, dk = \frac{1}{2}\mathcal{E}. \tag{4.39}$$

In general, half of the initial energy turns into kinetic energy in the asymptotic state, which is consistent with the virial theorem.

4.2 Examples of dispersive waves

We consider the rotating shallow-water equations and two-dimensional *Rossby waves* as examples for dispersive wave systems.

4.2.1 Rotating shallow water

We add Coriolis forces to the momentum equation (1.54) in accordance with a vertical Coriolis vector $\boldsymbol{f} = f\hat{\boldsymbol{z}}$ with constant $f > 0$, say. This affects the functional form of the nonlinear PV, which is now $q = (\boldsymbol{\nabla} \times \boldsymbol{u} + f)/H$, and therefore the linear PV is

$$q' = \frac{\boldsymbol{\nabla} \times \boldsymbol{u}'}{H} - f\frac{h'}{H^2}. \tag{4.40}$$

Both relative vorticity and the three-dimensional stretching of background vorticity implied by varying the layer depth contribute to q'. The basic-state PV is $Q = f/H$, say, which is uniform only if H is uniform. For variable H, on the other hand, there is a basic-state PV gradient and therefore the linear PV evolves according to

$$q'_t + \boldsymbol{u}' \cdot \boldsymbol{\nabla} Q = 0. \tag{4.41}$$

Using particle displacements this can be integrated to

$$q' = -\boldsymbol{\xi}' \cdot \boldsymbol{\nabla} Q \tag{4.42}$$

and combining this with (4.40) leads to

$$\boldsymbol{\nabla} \times \boldsymbol{u}' = f\frac{h'}{H} - H\boldsymbol{\xi}' \cdot \boldsymbol{\nabla} Q. \tag{4.43}$$

Notably, in the special case of constant H (and therefore Q) the linear PV is time independent, but the linear vorticity need not be.

The disturbance energy equation (2.4) remains valid, but the virial theorem (2.7) acquires a new term $H(\boldsymbol{f} \times \boldsymbol{\xi}') \cdot \boldsymbol{u}'$ on the right-hand side. This new term indicates that equipartition of energy does not hold once Coriolis forces are introduced, which is a generic fact in rotating systems.

For simplicity we now restrict to constant H and therefore $q'_t = 0$. The equations now contain a new constant length scale, namely the so-called *Rossby deformation length*

$$L_D = \frac{\sqrt{gH}}{f} = \frac{c}{f}. \tag{4.44}$$

We also define the Rossby deformation wavenumber $\kappa_D = 1/L_D$, which is sometimes more convenient. As will become apparent, Coriolis effects modify the linear dynamics at scales comparable to L_D or larger.

For example, the balanced, vortical mode satisfies

$$g\nabla h'_b = -\boldsymbol{f} \times \boldsymbol{u}'_b. \tag{4.45}$$

This equation expresses *geostrophic balance*, which is the term used for a situation in which horizontal Coriolis forces are balanced by horizontal pressure gradients. This implies that the balanced depth disturbance is non-zero, and that it acts as a stream function for the balanced velocity field, i.e., the balanced stream function is $\psi' = gh'_b/f$ such that $\psi'_x = v'_b$ and $\psi'_y = -u'_b$.

As before, we can use the invariance of PV to solve for the vortical mode in terms of the PV, which leads to the PV inversion equation

$$\left(\nabla^2 - \kappa_D^2\right)\psi' = Hq'. \tag{4.46}$$

This is a modified Helmholtz equation, which is clearly scale-selective. For instance, at small scales this equation reduces to the Poisson equation familiar from the non-rotating case, and comparing with (4.40) then means that the PV is dominated by the relative vorticity $\nabla \times \boldsymbol{u}'_b$. Conversely, at large scales most of the PV is due to the stretching term $-fh'_b/H$. The scale-selective nature of (4.46) is also apparent from its Green's function in an unbounded domain, say, which is a modified Bessel function $K_0(\kappa_D r)$ that for small $r = |\boldsymbol{x} - \boldsymbol{x}'|$ is asymptotically equal to $(1/2\pi)\ln r$, but for large r becomes proportional to $\exp(-r/L_D)$. Thus the velocity field belonging to a localized balanced vortex in rotating shallow water consists of a near field that decays with the usual $1/r$ power law, and a far field that decays much more rapidly, namely exponentially with rate $1/L_D$.

The Coriolis force also affects time-dependent gravity waves, which are called *inertia–gravity waves* to indicate the dual nature of the forces responsible for the wave motion. For plane waves with wavenumber vector $\boldsymbol{k}$ and $\kappa = |\boldsymbol{k}|$ the dispersion relation is

$$\omega^2 = f^2 + c^2\kappa^2, \tag{4.47}$$

which shows that there is now a finite lower bound on the wave frequency, namely the Coriolis frequency. The bound is achieved at $\kappa = 0$, which is a so-called *inertial oscillation*, i.e., a horizontally homogeneous gyration of the entire fluid layer at frequency $\omega = f$. In this oscillation the depth disturbance is zero and hence the energy is entirely kinetic.

Now, the dispersion relation and the structure of plane waves are still isotropic because the vertical Coriolis vector does not upset the rotational symmetry of the fluid system in the horizontal plane. Thus it is sufficient to look at the case $\boldsymbol{k} = (k, 0)$, say, in which the polarization relations between the plane wave fields are

$$u'_w = \frac{\omega}{k}\frac{h'_w}{H} \quad \text{and} \quad v'_w = -i\frac{f}{\omega}u'_w. \tag{4.48}$$

Thus there is now also a transversal velocity component perpendicular to $\boldsymbol{k}$, which is ninety degrees out of phase with the depth disturbance and the longitudinal velocity component. A kinematic consequence of this is that linear particle orbits are now ellipses with aspect ratio f/ω, which are traversed in a clockwise sense if $f > 0$. Also, the phase-averaged energy is partitioned according to

$$\boldsymbol{k} = (k, 0): \quad \bar{E} = H\overline{u'^2_w} = H\overline{v'^2_w} + g\overline{h'^2_w}, \tag{4.49}$$

i.e., the transversal velocity acts like a second term in the potential energy. For general $\boldsymbol{k}$ this means

$$H\overline{u'^2_w} + H\overline{v'^2_w} = \left(1 + \frac{f^2}{\omega^2}\right)\bar{E} \quad \text{and} \quad g\overline{h'^2_w} = \left(1 - \frac{f^2}{\omega^2}\right)\bar{E}. \tag{4.50}$$

This is consistent with the rotating virial theorem and quantifies the lack of equipartition in rotating shallow water.

Now, inertia–gravity waves are clearly dispersive, although for high wavenumbers (4.47) converges to its non-dispersive form. Specifically, the group and phase velocities for $\omega > 0$ are

$$\boldsymbol{c}_g = c\frac{\boldsymbol{k}}{\kappa}\sqrt{\frac{\kappa^2}{\kappa^2 + \kappa_D^2}} \quad \text{and} \quad \boldsymbol{c}_p = c\frac{\boldsymbol{k}}{\kappa}\sqrt{\frac{\kappa^2 + \kappa_D^2}{\kappa^2}}. \tag{4.51}$$

This particular form facilitates comparison with the non-dispersive limit $c\boldsymbol{k}/\kappa$. Clearly, the group velocity is reduced below the non-dispersive speed at small wavenumbers, whereas the phase velocity is increased there. Indeed, we have the simple reciprocal relation $|\boldsymbol{c}_g||\boldsymbol{c}_p| = c^2$.

It is straightforward to check that the generic energy evolution law (2.31) holds with (4.51). Thus in rotating shallow water the energy of low-frequency, large waves travels more slowly than that of high-frequency, short waves. The fact that $\boldsymbol{c}_g$ is parallel to $\boldsymbol{k}$ is a direct consequence of the fact that ω depends only on $\kappa = |\boldsymbol{k}|$. This will be different in the second example, which concerns two-dimensional Rossby waves.

4.2.2 Two-dimensional Rossby waves

Rossby waves are ubiquitous in large-scale atmosphere and ocean dynamics. The essential feature of a Rossby wave is that the restoring mechanism is due to a gradient in the basic PV structure. Viewed from this general perspective, Rossby waves include vortex edge waves, and other waves on shear flows, for instance.

In geophysics, Rossby waves are mostly due to the latitude-dependence of the 'vertical' component of the Coriolis vector, where 'vertical' refers to the radial direction on a sphere. This component of the Coriolis vector, which is called the *Coriolis parameter* in GFD, is zero at the equator and maximal at the poles, with equal-and-opposite magnitudes in the northern and southern hemisphere. The reason why the Coriolis parameter is of generic importance on a rotating stably stratified planet is because of the definition of three-dimensional rotating PV in (1.51). This definition involves the dot product of $\boldsymbol{f}$ and the stratification gradient, and in a strongly stratified basic state of spherically symmetric stratification surfaces this gradient points radially outward. This leads to the generic latitude-dependent basic PV distribution in which q increases monotonically from the south pole to the north pole.

Specifically, the Earth rotates with a constant angular frequency

$$\Omega = \frac{2\pi}{24\,\text{hours}} \approx 7.3\ 10^{-5}\text{s}^{-1} \tag{4.52}$$

around its pole-to-pole axis[2] and if $\boldsymbol{\Omega}$ is the rotation vector with magnitude $|\boldsymbol{\Omega}| = \Omega$ then the Coriolis vector is defined as $\boldsymbol{f} = 2\boldsymbol{\Omega}$ such that the background vorticity (also called the planetary vorticity) is equal to $\boldsymbol{f}$. In many geophysical situations the vertical velocities are small compared to the horizontal velocities and this is exploited in the *traditional approximation* to the Coriolis force. In this approximation only the horizontal Coriolis forces due to horizontal velocities are retained in $\boldsymbol{f} \times \boldsymbol{u}$. This is equivalent to retaining only the vertical component of $\boldsymbol{f}$, which means that $\boldsymbol{f}$ is replaced by

$$(\boldsymbol{f} \cdot \widehat{\boldsymbol{z}})\,\widehat{\boldsymbol{z}} = f\widehat{\boldsymbol{z}} \tag{4.53}$$

[2] Amusingly, the Earth radius is about 6300 km, so the circumferential ground speed due to rotation at the equator is about 460 m/s, which is well over the speed of sound.

where the latitude-dependent *Coriolis parameter* f is defined by

$$f(\theta) = 2\Omega \sin \theta \tag{4.54}$$

in terms of latitude θ. Note that $f(\theta) > 0$ in the northern hemisphere and vice versa in the southern hemisphere.

In a local tangent-plane approximation centred at some latitude θ_0 we can use $f \approx f(\theta_0)$. This f-plane approximation is sufficient for discussing gravity waves, but not for Rossby waves, for which at least a linear term in the expansion of $f(\theta)$ must also be retained.

For consistency with the full spherical problem we use Cartesian coordinates $\boldsymbol{x} = (x, y, z)$ where x is the *zonal*, west-to-east coordinate, y is the *meridional*, south-to-north coordinate and z is the vertical coordinate. Zonal motions in the positive or negative x-direction are also called *prograde* or *retrograde* motions, respectively. In a flat shallow water model we can study Rossby waves by the artifice of replacing f by $f + \beta y$ with a constant $\beta > 0$, which mimics the latitude-dependence of the Coriolis parameter on the sphere. This is called the β-plane model in GFD.

Alternatively, we can retain constant f but make the still water depth H a function of y. This leads to so-called *topographic Rossby waves*, which are particularly useful for laboratory studies. For instance, weak topography changes are described by

$$H(y) = H_0(1 - \mu y) \tag{4.55}$$

with constant H_0 and a small spatial rate of change $\mu > 0$. Thus the water depth decreases slowly with y and the basic PV is approximately

$$Q(y) = \frac{f}{H(y)} \approx \frac{f}{H_0}(1 + \mu y) \quad \Rightarrow \quad Q_y = \frac{f\mu}{H_0}. \tag{4.56}$$

This is analogous to a variable Coriolis parameter if β is identified with $f\mu$. Either way, the linear PV equation (4.41) becomes

$$H_0 q'_t + \beta v' = 0 \quad \Leftrightarrow \quad H_0 q' + \beta \eta' = 0. \tag{4.57}$$

With the generic choice $\beta > 0$ this means that displacing a particle 'northwards', i.e., $\eta' > 0$, leads to a negative PV disturbance and vice versa. This kinematic fact is true regardless of the physical cause of the displacement, which makes it obvious that in the presence of a basic PV gradient even inertia–gravity waves cause some PV disturbance.

In linear theory Rossby waves are connected with the vortical mode, which has zero frequency if $\beta = 0$, but can have non-zero frequency otherwise. For small β the frequency of Rossby waves is also small, and if $f > 0$ then β can be chosen small enough such that a clear frequency gap persists between inertia–gravity waves with $\omega^2 > f^2$ and Rossby waves with $\omega^2 \ll f^2$. It is then possible to show that a good approximation to Rossby waves is obtained by approximating v' with $v'_b = \psi'_x$ in (4.57) and using the inversion relation (4.46) to diagnose ψ' at every instant from the time-evolving PV distribution. Physically, this neglects fluctuations in the PV caused by the faster inertia–gravity waves. There are many ways in which to elaborate on this simple heuristic argument, especially on how to extend it to nonlinear flow dynamics. This is the well-studied topic of *balanced dynamics* in GFD.

Here we will just accept this heuristic argument and see what it implies. Replacing v' by ψ'_x in (4.57), using (4.46),[3] and assuming a plane-wave structure in q' and ψ' with $\boldsymbol{k} = (k, l)$ leads to a transversal plane wave subject to the celebrated Rossby-wave dispersion relation

$$\omega = -\beta \frac{k}{k^2 + l^2 + \kappa_D^2}. \tag{4.58}$$

Every aspect of (4.58) is remarkable. First off, $\omega = 0$ if $\beta = 0$, which reminds us that this wave mode reduces to the steady vortical mode if there is no basic PV gradient. This heritage of (4.58) also makes clear why there is only one frequency branch: this is because a Rossby-wave initial-value problem has only a single degree of freedom and requires only the specification of a single flow field such as q' at the initial time. Of course, although there is only one frequency branch the relation (4.58) is still compatible with a real flow field because $\omega(\boldsymbol{k}) = -\omega(-\boldsymbol{k})$.

Second, (4.58) is anisotropic, which is because the basic PV gradient in the y-direction removed the rotational symmetry of the fluid dynamics in the horizontal plane. In particular, the x-component of the phase velocity is negative for all wavenumbers, which exemplifies the generic fact that Rossby waves always exhibit retrograde intrinsic phase propagation on Earth, i.e., the zonal phase propagation relative to the mean flow is in the direction opposite to the Earth's background rotation.

Third, at large scales $\kappa \ll \kappa_D$ and the dispersion relation reduces to $\omega = -\beta L_D^2 k$, which describes a peculiar non-dispersive propagation that operates in the negative x-direction only. On the other hand, at small scales $\kappa \gg \kappa_D$ and we obtain

$$\omega = -\beta \frac{k}{k^2 + l^2}, \tag{4.59}$$

[3] For simplicity, this ignores variable depth in the PV inversion, which can be justified for weak topography. Also, taking variable depth into proper account in the inversion step does not alter the character of Rossby waves in an essential way.

which does not include any dependence on either f or g anymore. This limit is homogeneous of degree minus one in the wavenumbers, which shows that the Rossby-wave frequency decreases as the wavenumber increases, i.e., small scales are slower than large scales.[4] Fourth, the group velocity $\boldsymbol{c}_g = (u_g, v_g)$ is

$$u_g = \beta \frac{k^2 - l^2 - \kappa_D^2}{(k^2 + l^2 + \kappa_D^2)^2} \quad \text{and} \quad v_g = \beta \frac{2kl}{(k^2 + l^2 + \kappa_D^2)^2}. \tag{4.60}$$

This makes it obvious that there is a substantial angle between $\boldsymbol{k}$ and $\boldsymbol{c}_g$ in general, i.e., the group velocity of Rossby waves differs from the phase velocity not only in magnitude but also in direction. In the small-scale limit the group velocity is homogeneous of degree minus two in the wavenumbers, which shows that small-scale Rossby waves are slow not only in frequency, but also in group velocity.

Fifth, there is a definite link between the angle of the constant-phase lines and the sign of the meridional group velocity, which by (4.60) is

$$\operatorname{sgn}(v_g) = \operatorname{sgn}(kl). \tag{4.61}$$

Hence $v_g > 0$ implies $kl > 0$ and vice versa, i.e., a northward-propagating Rossby wave is associated with a backward-sloping phase line pattern when viewed from above. This allows deducing the meridional direction of propagation from a single snapshot of Rossby-wave phase lines. In contrast, this would not be possible for internal waves.

Finally, inspection of (4.58) and (4.60) shows that $|\omega|$ reaches a maximum at $l = 0$ and $k = \pm\kappa_D$, and therefore $|\omega| \leq \beta L_D/2$. Hence for fixed $f > 0$ the frequency gap between Rossby waves and inertia–gravity waves can be ensured at all scales by making β small enough.

4.3 Ray tracing for dispersive wavetrains

Ray tracing is the generalization of geometric wave theory for the leading-order asymptotic description of a slowly varying wavetrain that is subject to both dispersion and refraction. As noted before, despite this impressive generality ray tracing is disarmingly easy to work with both mathematically and numerically. The price to pay for this is that ray tracing fails at caustics, and also that ray tracing does not provide a generic method to compute correction terms to the asymptotic solution. Thus, when ray tracing works it delivers the leading-order solution easily and reliably, but when it fails

[4] This can be traced back to the Laplacian in the inversion equation (4.46), which smoothes small scales such that small-scale structures in q' lead to weak structures in ψ', and some of this smoothing survives the subsequent x-derivative in $v_b' = \psi_x'$.

one has to go back all the way to the original equations to understand the nature of the problem. With this caveat ray tracing is the most important method for computing small-scale waves, especially in the case of moving media.

We make the familiar assumption that the asymptotic wave solution $u(\boldsymbol{x}, t)$, say, has the functional form of a slowly varying wavetrain, i.e.,

$$u(\boldsymbol{x}, t) = A(\boldsymbol{x}, t) \exp(i\theta(\boldsymbol{x}, t) - i\alpha) \tag{4.62}$$

where A is the slowly varying real-valued amplitude, θ is the rapidly varying real-valued phase and α is a constant phase shift. If there are several wave fields then (4.62) holds with vector-valued u and A. The coordinates $(\boldsymbol{x}, t)$ are defined such that the amplitude A as well as the basic state can change by an $O(1)$ amount as $\boldsymbol{x}$ and t change by an $O(1)$ amount. This means that A, $\boldsymbol{\nabla} A$ and A_t are all $O(1)$. In contrast, we assume that the phase θ changes by a much bigger amount $O(1/\epsilon)$ with $\epsilon \ll 1$, say. Correspondingly, θ, $\boldsymbol{\nabla}\theta$ and θ_t are all $O(1/\epsilon)$. This can be made explicit by writing

$$\epsilon \ll 1: \quad \theta(\boldsymbol{x}, t) = \frac{1}{\epsilon}\Theta(\boldsymbol{x}, t) \tag{4.63}$$

for a scaled phase Θ that has the same scaling as A. For example, this recovers the set-up of geometric wave theory if the eikonal s and the large wavenumber κ_0 are identified with the scaled phase Θ and $1/\epsilon$, respectively.

The key fact about a slowly varying wavetrain is that at every $(\boldsymbol{x}, t)$ the solution looks just like a plane wave. This means it makes sense to *define* a slowly varying wavenumber vector $\boldsymbol{k}(\boldsymbol{x}, t)$ and frequency $\omega(\boldsymbol{x}, t)$ via

$$d\theta = \boldsymbol{k} \cdot \boldsymbol{dx} - \omega\, dt \quad \Leftrightarrow \quad \boldsymbol{k}(x, t) = +\boldsymbol{\nabla}\theta \quad \text{and} \quad \omega(x, t) = -\theta_t. \tag{4.64}$$

The basic structure of ray tracing follows immediately from (4.64) together with the constraint that $(\boldsymbol{k}, \omega)$ must satisfy a given dispersion relation at $(\boldsymbol{x}, t)$. We will justify this assumption in a simple one-dimensional problem and then return to the general case.

Before moving on we note that with (4.63) and (4.64) both $|\boldsymbol{k}|$ and ω are $O(1/\epsilon)$ by construction. This works for wave systems in which high wavenumbers are proportional to high frequencies, but fails in other dispersive cases. For instance, in the case of small-scale Rossby waves obeying the dispersion relation (4.59) the frequency goes *down* as $|\boldsymbol{k}|$ goes up. Another geophysical example are internal gravity waves (which will be studied in §6.2.2), for which the frequency is independent of $|\boldsymbol{k}|$. These examples show that additional scaling of some coordinates or of other physical parameters (such as $\beta = O(1/\epsilon^2)$ in the Rossby-wave case) may be necessary in order to allow ray tracing to apply in the generic form presented here. Of course, such scaling operations contain information about the physical situations in which ray tracing can or cannot be applied.

4.3.1 Model example

We again use the simple dispersive example

$$u_t = u_{xxx} \quad \text{such that} \quad \omega = k^3 \tag{4.65}$$

is the global dispersion relation. Refraction effects will be added later. The dispersion relation makes clear that $\omega = O(1/\epsilon^3)$ if $k = O(1/\epsilon)$, so in order to apply ray tracing in the standard form we need to insert a factor ϵ^2 in front of u_{xxx} that balances the growth of ω and k. In the absence of physical parameters that could have this scaling we can achieve the same by rescaling the coordinates such that $(X,T) = (\epsilon^\alpha x, \epsilon^\beta t)$ with constant α and β. This leads to

$$u_T = \epsilon^{3\alpha-\beta} u_{XXX} \tag{4.66}$$

and therefore any combination such that $3\alpha - \beta = 2$ can be used. For example, if $\alpha = \beta = 1$ then the slow coordinates $(X,T) = (\epsilon x, \epsilon t)$ would bring (4.65) into the appropriate form. We consider this done and return to the original notation (x,t), but with the scaled PDE

$$u_t = \epsilon^2 u_{xxx} \quad \text{such that} \quad \omega = \epsilon^2 k^3 \quad \text{and} \quad c_g = 3\epsilon^2 k^2. \tag{4.67}$$

Substituting from (4.62) and (4.63) and collecting imaginary and real parts yields the exact equations

$$\Theta_t + \Theta_x^3 = \epsilon\Theta_{xxx} + \epsilon\frac{3}{A}(\Theta_x A_x)_x \tag{4.68}$$

and

$$A_t + 3\Theta_x^2 A_x + 3\Theta_x\Theta_{xx}A = \epsilon^2 A_{xxx}. \tag{4.69}$$

As in geometric wave theory, (4.69) implies an exact conservation law for A^2, namely

$$(A^2)_t + (3\Theta_x^2 A^2)_x = \epsilon^2(2AA_{xx} - A_x^2)_x. \tag{4.70}$$

Now, the leading-order asymptotic equations are obtained by retaining only the $O(1)$ terms:

$$\Theta_t + \Theta_x^3 = 0 \quad \text{and} \quad (A^2)_t + (3\Theta_x^2 A^2)_x = 0. \tag{4.71}$$

Both these equations are remarkable. The first equation is a nonlinear first-order PDE for the scaled phase $\Theta = \epsilon\theta$, which can be solved using the method of characteristics. Moreover, using (4.64) this equation is equivalent to the constraint that the dispersion relation applies to $\omega = -\theta_t$ and $k = +\theta_x$. The second equation approximates the exact conservation law (4.70) by keeping only the leading-order flux. This hints at the robustness of this conservation law, because even if the leading-order flux is not exact, the integral conservation of A^2 is. Moreover, the leading-order flux is given by the group velocity:

$$3\Theta_x^2 A^2 = 3\epsilon^2\theta_x^2 A^2 = 3\epsilon^2 k^2 A^2 = c_g A^2. \tag{4.72}$$

In terms of the phase-averaged energy density $\bar{E} = \overline{|\Re u|^2} = \frac{1}{2}A^2$ this is

$$\bar{E}_t + (\bar{E}c_g)_x = 0. \tag{4.73}$$

Once more, the wavetrain energy moves with the group velocity.

It is easy to add refraction by changing (4.67) to

$$u_t = \epsilon^2\alpha(x)u_{xxx} \tag{4.74}$$

with some function $\alpha(x) > 0$ that marks a variable basic state. For constant α the global dispersion relation is $\omega = \epsilon^2\alpha k^3$ and for variable α we will show that this relation holds locally. The derivation follows exactly as before and the leading-order asymptotic equations are

$$\Theta_t + \alpha(x)\Theta_x^3 = 0 \quad \text{and} \quad \left(\frac{A^2}{\alpha(x)}\right)_t + (3\Theta_x^2 A^2)_x = 0. \tag{4.75}$$

The first equation establishes the local validity of the dispersion relation, and re-defining the energy density as $\bar{E} = \overline{|\Re u|^2}/\alpha(x) = \frac{1}{2}A^2/\alpha(x)$ makes it obvious that the second equation again results in (4.73), but this time with the local group velocity $c_g = 3\alpha\Theta_x^2 = 3\epsilon^2\alpha k^2$.

In the presence of a variable basic state it is typical that finding a conserved wave activity density such as $\bar{E}$ is a non-trivial part of the problem, for which there is no generic rule. However, for wave system deriving from physics, it is possible to use symmetry methods to derive such conservation laws systematically. This will be discussed in §4.5.

4.3.2 Generic ray-tracing equations

It is customary and useful to dispense with an explicit small parameter ϵ when describing the generic ray-tracing equations, but we keep in mind that an initial scaling procedure similar to that in §4.3.1 is always part of the procedure when applying ray tracing to a new system. We consider first the generic equations for the phase $\theta(\boldsymbol{x}, t)$ and defer the generic equations for the amplitude $A(\boldsymbol{x}, t)$ until §4.4.4, after the impact of a moving medium has been considered

The ray-tracing equations are generated by a local frequency dispersion function $\Omega(\boldsymbol{k}, \boldsymbol{x}, t)$ that follows from the governing PDE if all coefficients to do with the basic state are treated as constants.[5] For instance, the case (4.74) with $\epsilon = 1$ corresponds to $\Omega(k, x) = \alpha(x)k^3$. We use uppercase notation in order to distinguish clearly between the *function* $\Omega(\boldsymbol{k}, \boldsymbol{x}, t)$ and its local *value* when evaluated using the local wavenumber vector $\boldsymbol{k}(\boldsymbol{x}, t)$. By assumption, this value equals the local frequency $\omega(\boldsymbol{x}, t)$, i.e., the crucial equation is

$$\omega(\boldsymbol{x}, t) = \Omega(\boldsymbol{k}(\boldsymbol{x}, t), \boldsymbol{x}, t). \tag{4.76}$$

Alternatively, using

$$\boldsymbol{k}(\boldsymbol{x}, t) = +\boldsymbol{\nabla}\theta \quad \text{and} \quad \omega(\boldsymbol{x}, t) = -\theta_t, \tag{4.77}$$

the equation (4.76) can be viewed as a *Hamilton–Jacobi equation* for $\theta(\boldsymbol{x}, t)$:

$$\theta_t + \Omega(\boldsymbol{\nabla}\theta, \boldsymbol{x}, t) = 0. \tag{4.78}$$

Notably, θ itself does not appear in (4.78), only its space–time gradient. This is the nonlinear first-order PDE and applying the standard method of characteristics to it yields the ray tracing equations, with characteristics in real space that are traced out by the local group velocity. This is the dispersive generalization of the eikonal equation in geometric wave theory, which also makes clear that there is no guarantee of a globally smooth solution for θ because dispersive caustics can form where neighbouring characteristics touch.

Now, it is also possible to derive the ray tracing equations directly from (4.76) and (4.77), which has the practical advantage that one does not have to remember the standard method of characteristics. To do this we use that cross-differentiation of (4.77) shows a kinematic 'conservation law for wave crests', namely

$$\boldsymbol{k}_t + \boldsymbol{\nabla}\omega = 0. \tag{4.79}$$

Substituting for ω from (4.76), noting that

$$\boldsymbol{\nabla} \times \boldsymbol{k} = \boldsymbol{\nabla} \times \boldsymbol{\nabla}\theta = 0 \quad \Leftrightarrow \quad \boldsymbol{\nabla}\boldsymbol{k} = (\boldsymbol{\nabla}\boldsymbol{k})^T \tag{4.80}$$

5 If multiple branches of the frequency dispersion relation are relevant then multiple wavetrains with multiple functions Ω can be used.

holds by construction, and diligent use of the chain rule then results in

$$\boldsymbol{k}_t + \frac{\partial \Omega}{\partial \boldsymbol{k}} \cdot \boldsymbol{\nabla} \boldsymbol{k} + \frac{\partial \Omega}{\partial \boldsymbol{x}} = 0. \tag{4.81}$$

By defining the local group-velocity vector and its directional derivative as

$$\boldsymbol{c}_g(\boldsymbol{x}, t) = \frac{\partial \Omega}{\partial \boldsymbol{k}}(\boldsymbol{k}(\boldsymbol{x}, t), \boldsymbol{x}, t) \quad \text{and} \quad \frac{d}{dt} = \frac{\partial}{\partial t} + \boldsymbol{c}_g \cdot \boldsymbol{\nabla} \tag{4.82}$$

we can re-write (4.81) as the second part of the pair

$$\frac{d\boldsymbol{x}}{dt} = +\frac{\partial \Omega}{\partial \boldsymbol{k}} \quad \text{and} \quad \frac{d\boldsymbol{k}}{dt} = -\frac{\partial \Omega}{\partial \boldsymbol{x}}. \tag{4.83}$$

This is a Hamiltonian set of ODEs for the canonical variables $(\boldsymbol{x}, \boldsymbol{k})$ evolving according to the Hamiltonian function $\Omega(\boldsymbol{k}, \boldsymbol{x}, t)$.[6] The first equation makes clear that the rays $\boldsymbol{x}(t)$ are trajectories of the local group velocity, and the second equation quantifies how $\boldsymbol{k}(t)$ evolves along a ray in response to refraction caused by the spatial inhomogeneity of the basic state.

Once $(\boldsymbol{x}(t), \boldsymbol{k}(t))$ are known along a ray, the frequency $\omega(t)$ can be computed directly from (4.76). Alternatively, ω can also be found along the ray by integrating

$$\frac{d\omega}{dt} = \frac{\partial \Omega}{\partial \boldsymbol{k}} \cdot \frac{d\boldsymbol{k}}{dt} + \frac{\partial \Omega}{\partial \boldsymbol{x}} \cdot \frac{d\boldsymbol{x}}{dt} + \frac{\partial \Omega}{\partial t} = \frac{\partial \Omega}{\partial t}, \tag{4.84}$$

which follows from (4.83). This alternative is convenient in numerical applications, because then (4.76) provides a useful check for numerical accuracy.

Should the value of the phase θ along a ray be desired then it too can be computed according to

$$\frac{d\theta}{dt} = \theta_t + \boldsymbol{c}_g \cdot \boldsymbol{\nabla}\theta = \boldsymbol{k} \cdot \frac{\partial \Omega}{\partial \boldsymbol{k}} - \Omega. \tag{4.85}$$

This involves the value of the Legendre transform of Ω with respect to $\boldsymbol{k}$. However, the leading-order phase might not be accurate enough in order to allow a prediction of crests and troughs. This is discussed in §4.3.4.

As an aside, we can note a useful fact for the frequently occurring case in which Ω is homogeneous in the components of $\boldsymbol{k}$. Specifically, if Ω is homogeneous of degree p in $\boldsymbol{k}$ then (4.85) and Euler's theorem imply that $d\theta/dt = (p-1)\omega$ along a ray. For a time-independent basic state with $\omega =$ const this implies $\theta(t) = \theta(0) + t(p-1)\omega$ along a ray. Physically, this means that an observer travelling with the group velocity will observe oscillations

[6] The emergence of Ω as a Hamiltonian function is also consistent with the Hamiltonian–Jacobi equation (4.78), whose form is familiar from classical mechanics with θ replaced by the least action, $(\boldsymbol{x}, \boldsymbol{k})$ replaced by the canonical coordinates and momenta, and Ω replaced by the energy, i.e., the classical Hamiltonian.

with the modified frequency $(1-p)\omega$, because of the minus sign inherent in the definition of frequency. For example, in the non-rotating shallow-water system $p=1$ and therefore $\theta=$const along a ray.[7] Another peculiar example relevant for *internal gravity waves* is $p=0$, in which the modified frequency is equal to the original frequency.

4.3.3 Symmetries and ray invariants

The Hamiltonian structure of (4.83) makes it obvious that continuous space–time symmetries of the basic state induce invariants along the rays. For example, if the basic state is symmetric to translational shifts in the x-direction, say, then Ω does not depend on x explicitly, i.e., $\partial\Omega/\partial x=0$, and therefore $k(t)$ is constant along a ray, if k is the x-component of the wavenumber vector $\boldsymbol{k}$. The same consideration applies to a symmetry of the basic state with respect to shifts in time, which leads to $\partial\Omega/\partial t=0$ and therefore ω is constant along a ray. Finally, it is easy to show that a symmetry of the basic state with respect to rotations around the z-axis, say, leads to the invariance of $lx-ky$ along a ray. Clearly, these conservation laws are analogous to the conservation laws for momentum, energy and angular momentum in classical mechanics.

These conservation laws are known a priori and can be used to simplify the computation of rays. For example, we consider the two-dimensional case in which $\boldsymbol{x}=(x,y)$ and $\boldsymbol{k}=(k,l)$. A rotational symmetry around the origin then corresponds to a function Ω whose explicit spatial dependence involves only the distance $r=\sqrt{x^2+y^2}$. In the simplest case Ω depends on none of $(\boldsymbol{x},t)$ and therefore all components of $\boldsymbol{k}$ as well as ω are conserved and the rays are straight lines. Slightly less simple is the common textbook case, in which Ω has no t-dependence and depends on *one* spatial variable out of (x,y,r). This means that ω and one number out of $(k,l,lx-ky)$ are conserved. The remaining unknown wavenumber component can then be deduced from the equation $\omega=\Omega(\boldsymbol{k},\boldsymbol{x})$ without having to integrate the ODEs. In general, this common textbook case corresponds to an integrable Hamiltonian system with n degrees of freedom and n independent conserved quantities formed by ω and $n-1$ wavenumber combinations. Outside textbooks, the ray system (4.83) is usually not integrable in closed form and therefore must be solved numerically.

The Hamiltonian structure of (4.83) also suggests a phase-space view of ray tracing. Here *phase space* is the space spanned jointly by the canonical variables $(\boldsymbol{x},\boldsymbol{k})\in R^{2n}$, say. The ray-tracing equations induce a phase-space

[7] The phase θ and the eikonal s are related by $\theta=\kappa_0 s-\omega t$ so this is consistent with geometric wave theory.

flow that is volume-preserving by Liouville's theorem. This avoids the intersection of rays that may occur in the smaller configuration space spanned by $\boldsymbol{x} \in R^n$ alone. In other words, caustics in $\boldsymbol{x}$-space may be healed in the higher-dimensional $\boldsymbol{xk}$-space. The original ray-tracing dynamics in $\boldsymbol{x}$-space can then be viewed as a projection of the phase-space dynamics. Alternatively, one can also project onto $\boldsymbol{k}$-space and obtain ray-tracing equations that might be more convenient in a particular application. Nevertheless, it is worth keeping in mind that phase-space dynamics is mathematically equivalent to the original ray-tracing equations. Ray tracing, in whatever form, is invalid at caustics.

4.3.4 A note on the asymptotic phase in ray tracing

In ray tracing only the first term in the asymptotic expansion

$$\Theta = \Theta_0 + \epsilon\Theta_1 + \epsilon^2\Theta_2 + O(\epsilon^3) \tag{4.86}$$

for the scaled phase Θ is computed. However, the actual phase $\theta = \Theta/\epsilon$ has the expansion

$$\theta = \frac{1}{\epsilon}\Theta_0 + \Theta_1 + +\epsilon\Theta_2 + O(\epsilon^2), \tag{4.87}$$

which means that the unknown first-order correction Θ_1 leads to an $O(1)$ error in the actual phase and therefore to an $O(1)$ error in $\exp i\theta$. In other words, although ray tracing accurately predicts the leading-order wavenumber $\boldsymbol{k}$ and amplitude A at some location $(\boldsymbol{x}, t)$, it cannot predict whether there is a wave crest or wave trough there. Of course, this is because both $\boldsymbol{k}$ and A are slowly varying whilst θ is rapidly varying.

This makes clear that predicting the precise location of wave crests and troughs requires going beyond standard ray tracing in order to compute the next order correction Θ_1. In the related subject area of oscillatory Hamiltonian systems subject to slowly varying parameters, computing this next order correction is referred to as computing the *geometric phase*.

4.4 Ray tracing in moving media

This is a crucial topic for waves in fluids. As mentioned before, the refraction caused by shearing and straining in a moving medium can lead to fundamentally new wave phenomena such as critical layers, which play a major role in wave–mean interaction theory. Also, wave energy is no longer conserved, requiring us to seek other conservable measures of wave activity.

Despite these dramatic changes to the linear wave dynamics, the formal extension of the ray-tracing equations to moving media is disarmingly simple. Indeed, it is not the equations that are much more complex in moving media, but the *solutions* to these equations.

The point of departure for ray tracing is the inclusion of a uniform basic flow. This leads to a straightforward modification of the dispersion relation for plane waves by including the *Doppler shift* of the frequency due to this basic flow. The extension to non-uniform, but slowly varying, basic flows then follows from the generic ray-tracing equations.

4.4.1 Doppler shifting and the intrinsic frequency

We consider a constant basic flow with velocity $\boldsymbol{U}$. Of course, $\boldsymbol{U}$ and other parts of the basic state must be compatible with the governing equations and the prevailing boundary conditions. For example, in a rotating frame the Coriolis force due to a non-zero $\boldsymbol{U}$ must be balanced by suitable pressure forces. Once this is achieved, the only change in the linear equations is that the $O(1)$ part of the material time derivative is augmented by a term that measures advection by $\boldsymbol{U}$:

$$\frac{\partial}{\partial t} \to \frac{\partial}{\partial t} + \boldsymbol{U} \cdot \boldsymbol{\nabla} \equiv \mathrm{D}_t. \tag{4.88}$$

For instance, the definition of the linear particle displacement becomes $\mathrm{D}_t \boldsymbol{\xi}' = \boldsymbol{u}'$. For a plane wave with phase $\boldsymbol{k} \cdot \boldsymbol{x} - \omega t$ we have

$$\mathrm{D}_t \to -i(\omega - \boldsymbol{U} \cdot \boldsymbol{k}). \tag{4.89}$$

The previously derived plane-wave solutions remain valid provided that the *absolute frequency* ω is everywhere replaced by the *intrinsic frequency* $\hat{\omega}$ defined by

$$\hat{\omega} = \omega - \boldsymbol{U} \cdot \boldsymbol{k}. \tag{4.90}$$

For example, the dispersion relation for non-rotating shallow-water waves in the presence of a constant basic flow is

$$\hat{\omega}^2 = gH\kappa^2 \quad \text{such that} \quad \omega = \boldsymbol{U} \cdot \boldsymbol{k} \pm \sqrt{gH}\kappa. \tag{4.91}$$

Physically, the absolute frequency ω is the wave frequency that is observed by a measurement device fixed to the coordinate system. The intrinsic frequency $\hat{\omega}$, on the other hand, is the frequency that is observed by a device travelling with the basic flow $\boldsymbol{U}$. The two frequencies differ by the *Doppler shift* term $\boldsymbol{U} \cdot \boldsymbol{k}$, which quantifies the frequency shift due to advection of a spatial phase pattern past a static observer. The intrinsic frequency is physically more meaningful than the absolute frequency, because it is $\hat{\omega}$ and not ω

that quantifies the dynamics of fluid particles relative to the broad sweeping motion induced by $\boldsymbol{U}$.

The appearance of two frequencies is inherited by the group velocity, i.e.,

$$\omega = \boldsymbol{U} \cdot \boldsymbol{k} + \hat{\omega} \quad \Rightarrow \quad \boldsymbol{c}_g = \boldsymbol{U} + \widehat{\boldsymbol{c}}_g \tag{4.92}$$

where $\boldsymbol{c}_g$ and $\widehat{\boldsymbol{c}}_g$ are the absolute and intrinsic group velocity such that $\boldsymbol{c}_g = \partial\omega/\partial\boldsymbol{k}$ and $\widehat{\boldsymbol{c}}_g = \partial\hat{\omega}/\partial\boldsymbol{k}$. They differ by the explicit advection of the waves by the basic flow $\boldsymbol{U}$.

The ray-tracing equations (4.83) change only by including $\boldsymbol{U}$ in the definition of $d\boldsymbol{x}/dt$. This corresponds to using the absolute frequency function

$$\Omega(\boldsymbol{k}, \boldsymbol{x}, t) = \boldsymbol{U} \cdot \boldsymbol{k} + \hat{\Omega}(\boldsymbol{k}, \boldsymbol{x}, t), \tag{4.93}$$

where the intrinsic frequency function $\hat{\Omega}$ now captures the local plane-wave frequency relative to the basic flow. For example, in the most general non-rotating shallow-water case (4.91) we would have $\hat{\Omega}(\boldsymbol{k}, \boldsymbol{x}, t) = \pm\sqrt{g(t)H(\boldsymbol{x})}\kappa$.

4.4.2 Refraction by the basic flow

The key assumption of ray-tracing with a slowly varying velocity field $\boldsymbol{U}(\boldsymbol{x}, t)$ (which must still be a solution to the governing equations) is that the leading-order dispersion constraint (4.76) holds using (4.93). The essence of this assumption is that further terms in the linear equations proportional to $\boldsymbol{\nabla U}$ and U_t, say, are negligible compared to the retained terms in (4.93).[8]

The corresponding ray-tracing equations with $\boldsymbol{U}(\boldsymbol{x}, t)$ are

$$\frac{d\boldsymbol{x}}{dt} = \boldsymbol{U} + \frac{\partial\hat{\Omega}}{\partial\boldsymbol{k}} \quad \text{and} \quad \frac{d\boldsymbol{k}}{dt} = -(\boldsymbol{\nabla U}) \cdot \boldsymbol{k} - \frac{\partial\hat{\Omega}}{\partial\boldsymbol{x}} \tag{4.94}$$

where in the second-to-last term $\boldsymbol{k}$ contracts with $\boldsymbol{U}$ and not with $\boldsymbol{\nabla}$. This important new term describes the refraction of the wave phase by the basic flow. For example, in the standard textbook case the basic flow is a steady shear flow in two dimensions such that $\boldsymbol{U} = (U(y), 0)$, say. If there is no refraction due to $\hat{\Omega}$ (i.e., if $\hat{\Omega}(\boldsymbol{k})$ only) then in this case the x-wavenumber k is constant whilst the y-wavenumber $l(t)$ changes according to

$$\frac{dl}{dt} = -U_y k. \tag{4.95}$$

For example, in a region of constant shear the y-wavenumber grows or decays

[8] It is tempting to try to improve on this by including such terms in the definition of Ω, but this easily leads to inconsistencies with the underlying assumption of a slowly varying wavetrain. At any rate, there is no general theory to do this, although useful results may be obtained in special cases.

linearly with time, depending on the sign of kU_y. The absolute frequency ω is also constant along the ray but the intrinsic frequency $\hat{\omega}$ changes according to

$$\frac{d\hat{\omega}}{dt} = \frac{d(\omega - Uk)}{dt} = -k\frac{dU}{dt} = -kv_g U_y. \tag{4.96}$$

Thus $\hat{\omega}$ must change in order for $\omega = \hat{\omega} + U(y)k$ to stay constant.

An intuitive understanding of the basic flow refraction term $-(\boldsymbol{\nabla U})\cdot\boldsymbol{k}$ in the more general case can be obtained by comparing it with the analogous term for the evolution of the gradient $\boldsymbol{\nabla}\phi$ of a passive tracer ϕ in (1.19), i.e.,

$$\mathrm{D}_t\phi = 0 \quad \Rightarrow \quad \mathrm{D}_t(\boldsymbol{\nabla}\phi) = -(\boldsymbol{\nabla U})\cdot\boldsymbol{\nabla}\phi. \tag{4.97}$$

In the wave case $\boldsymbol{k} = \boldsymbol{\nabla}\theta$ by definition and then (4.94) shows that the basic flow refracts the phase θ *as if* the phase was a passive tracer relative to the basic flow $\boldsymbol{U}$. For instance, this analogy implies that the antisymmetric part of $\boldsymbol{\nabla U}$ seeks to rotate $\boldsymbol{k}$ without changing its length around an axis parallel to $\boldsymbol{\nabla}\times\boldsymbol{U}$. On the other hand, the symmetric part of $\boldsymbol{\nabla U}$ seeks to strain the θ-surfaces, which changes both the orientation and the length of $\boldsymbol{k}$. For example, if this latter process continues unabated with constant $\boldsymbol{\nabla U}$, then in two dimensions it leads to an alignment of $\boldsymbol{k}$ perpendicular to the axis of maximal strain and to concomitant exponential growth of κ.

These are useful qualitative insights but it needs to be kept in mind that this is a partial analogy, for two reasons. First, the phase pattern also moves relative to the flow by the intrinsic wave propagation, so it is not a passive tracer. Second, (4.94) and (4.97) agree on the right-hand side but not on the left-hand side because (4.97) involves the leading-order material derivative D_t whereas (4.94) involves the derivative along group-velocity rays. These two operators differ by the advection with the intrinsic group velocity, i.e.,

$$\frac{d}{dt} - \mathrm{D}_t = \widehat{\boldsymbol{c}}_g\cdot\boldsymbol{\nabla}. \tag{4.98}$$

Weak intrinsic wave propagation strengthens the analogy between wave phase and a passive tracer. Indeed, (4.93) without $\hat{\Omega}$ is precisely the characteristic system for a passive tracer. This will be relevant for the phenomenon of *wave capture* discussed in §14.3.

4.4.3 Fermat's principle and the curl–curvature formula

The Hamiltonian structure of the dispersive ray tracing equations (4.83) also gives rise to certain variational statements familiar from classical mechanics.

Perhaps the most useful such statement is *Fermat's principle*, which applies to ray tracing for waves propagating on time-independent basic states.[9] This principle uses the functional

$$S[\boldsymbol{k}, \boldsymbol{x}] = \int_{\boldsymbol{x}_A}^{\boldsymbol{x}_B} \boldsymbol{k} \cdot d\boldsymbol{x}. \tag{4.99}$$

based on two functions $\boldsymbol{k}(s)$ and $\boldsymbol{x}(s)$ with parameter $s \in [0, 1]$. The variational statement is that the ray solution is a critical point of S such that the corresponding critical value of the functional S^*, say, satisfies

$$S^* = \min_{[\boldsymbol{x}(s):\boldsymbol{x}(0)=\boldsymbol{x}_A, \boldsymbol{x}(1)=\boldsymbol{x}_B]} \ \max_{[\boldsymbol{k}(s):\Omega(\boldsymbol{k},\boldsymbol{x})=\omega]} S[\boldsymbol{k}, \boldsymbol{x}]. \tag{4.100}$$

Here ω is the constant frequency along the ray and the minimum over $\boldsymbol{x}(s)$ is really just seeking a critical point. To check that (4.100) is consistent with the ray tracing equations one can use a Lagrange multiplier function for the frequency constraint, which leads to

$$\min_{[\boldsymbol{x}(s):\boldsymbol{x}(0)=\boldsymbol{x}_A, \boldsymbol{x}(1)=\boldsymbol{x}_B]} \ \max_{\boldsymbol{k}(s)} \int_0^1 \left(\boldsymbol{k} \cdot \frac{d\boldsymbol{x}}{ds} - \lambda(s)(\Omega(\boldsymbol{k}, \boldsymbol{x}) - \omega) \right) ds \tag{4.101}$$

with associated Euler–Lagrange equations

$$\frac{d\boldsymbol{x}}{ds} = \lambda \frac{\partial \Omega}{\partial \boldsymbol{k}} \quad \text{and} \quad \frac{d\boldsymbol{k}}{ds} = -\lambda \frac{\partial \Omega}{\partial \boldsymbol{x}}. \tag{4.102}$$

With $\lambda(s)\, ds = dt$ we recognize the ray-tracing equations in parametrized form.

The principle is useful in practice if the maximization over $\boldsymbol{k}$ can be performed analytically. In this case the principle reduces to a variational statement for the ray path $\boldsymbol{x}(s)$ alone, which often yields interesting insights into the geometry of the rays. For example, for non-dispersive waves[10] with $\omega = c(\boldsymbol{x})\kappa$ we have $\boldsymbol{k} \cdot d\boldsymbol{x} = \kappa |d\boldsymbol{x}| \cos\alpha = (\omega |d\boldsymbol{x}|/c) \cos\alpha$, where α is the angle between $\boldsymbol{k}$ and $d\boldsymbol{x}$, and $|d\boldsymbol{x}|$ is the Euclidean length of $d\boldsymbol{x}$. This is trivially maximized by $\alpha = 0$ and therefore $\boldsymbol{k} \cdot d\boldsymbol{x} = \omega |d\boldsymbol{x}|/c$. The remaining minimization problem over $\boldsymbol{x}$ is

$$\min_{[\boldsymbol{x}(s):\boldsymbol{x}(0)=\boldsymbol{x}_A, \boldsymbol{x}(1)=\boldsymbol{x}_B]} \int_{\boldsymbol{x}_A}^{\boldsymbol{x}_B} \frac{|d\boldsymbol{x}|}{c(\boldsymbol{x})} \tag{4.103}$$

where the constant ω plays no role and has been omitted. This is the problem of finding a geodesic relative to a Riemannian metric in which the length of a line element is the Euclidean length $|d\boldsymbol{x}|$ divided by the conformal factor

[9] This principle is known as Maupertuis's principle in classical mechanics.
[10] An application of Fermat's theorem to dispersive Rossby waves on shear flows is given in §4.6.

c. Because $|\boldsymbol{dx}|/c$ is the phase travel time along the ray, Fermat's principle for non-dispersive waves is known as the *principle of least time*.

In the presence of a steady basic flow $\boldsymbol{U}(\boldsymbol{x})$, Fermat's principle becomes unwieldy even for non-dispersive waves, but for weak $\boldsymbol{U}$ there is a surprising result that is worth knowing. The same steps as before now yield

$$\boldsymbol{k} \cdot \boldsymbol{dx} = \frac{\omega |\boldsymbol{dx}| \cos \alpha}{c + |\boldsymbol{U}| \cos \beta}, \tag{4.104}$$

where β is the angle between $\boldsymbol{U}$ and $\boldsymbol{k}$, and $\omega = c\kappa + \boldsymbol{U} \cdot \boldsymbol{k}$ has been used. Treating $\boldsymbol{U} \neq 0$ as a perturbation from the previous result and working to first order in $|\boldsymbol{U}|$, we can simplify (4.104) by setting $\cos \alpha = 1$ and $|\boldsymbol{U}| \cos \beta = \boldsymbol{U} \cdot \boldsymbol{dx}/|\boldsymbol{dx}|$. The last step works because $\boldsymbol{k}$ and $\boldsymbol{dx}$ are parallel in the unperturbed problem. A Taylor expansion then yields

$$\boldsymbol{k} \cdot \boldsymbol{dx} = \omega \left(\frac{|\boldsymbol{dx}|}{c} - \frac{\boldsymbol{U} \cdot \boldsymbol{dx}}{c^2} \right) + o(\boldsymbol{U}). \tag{4.105}$$

Thus ray paths are geodesics of this non-Riemannian metric. Moreover, if $\boldsymbol{\nabla} \times (\boldsymbol{U}/c^2) = 0$ then the line integral over the second term is not path-dependent and hence the basic flow drops out of the variational problem for the ray path. This leads to the surprising result that for constant c and irrotational $\boldsymbol{U}$ the ray paths are straight lines to first order in $\boldsymbol{U}$. This is true even though the intrinsic group velocity (and therefore the direction normal to wave crests) is not parallel to the ray path, rather, there is a fortuitous cancellation between $\widehat{\boldsymbol{c}}_g$ and $\boldsymbol{U}$ that makes their sum $\boldsymbol{c}_g$ point in a constant direction. This interesting and useful fact is hard to see directly from the ray tracing equations.

If c is not constant then the ray curvature depends on both $c(\boldsymbol{x})$ and $\boldsymbol{U}(\boldsymbol{x})$ although (4.105) makes clear that at leading order the basic flow enters only via the curl of $\boldsymbol{U}/c^2$, which may be zero even if the curl of $\boldsymbol{U}$ is not. A non-trivial example of this effect for variable $c^2 \propto H$ in the two-dimensional shallow-water system can be formulated for still water depth of the form $H(y)$, say. In this case it is consistent to consider a steady basic shear flow of the form $\boldsymbol{U} \propto (H(y), 0)$, which implies again that $\boldsymbol{U}$ does not affect the shape of the rays at leading order. Of course, in this example rays are not straight lines, because of the variable still water depth $H(y)$, and hence the statement is that adding this particular shear flow will not lead to *additional* ray bending. This could be applied to the refraction of surface waves on beaches with longshore currents (cf. §13.2), for instance, where the classic theory for planar beaches with H a linear function of the distance from the shoreline happens to yield a longshore current of this special form.

Conversely, if c is constant and $\boldsymbol{\nabla} \times \boldsymbol{U}$ is not zero then (4.105) implies that the curvature of the ray is proportional to $\boldsymbol{\nabla} \times \boldsymbol{U}$. Specifically, if the ray is parametrized by $\boldsymbol{x}(s)$ and $\widehat{\boldsymbol{t}} = \dot{\boldsymbol{x}}/|\dot{\boldsymbol{x}}|$ is the unit tangent vector along the ray in terms of the derivative $\dot{\boldsymbol{x}} = d\boldsymbol{x}/ds$, then to first order in $\boldsymbol{U}/c$ the remarkable *curl–curvature formula*

$$\frac{d\widehat{\boldsymbol{t}}}{dr} = \frac{1}{c}(\boldsymbol{\nabla} \times \boldsymbol{U}) \times \widehat{\boldsymbol{t}} \tag{4.106}$$

holds, where d/dr is the derivative with respect to the Euclidean arc length r along the ray, i.e., $dr = |\dot{\boldsymbol{x}}|\, ds$. Therefore, for weak steady basic flows the ray curvature is equal to the curl of the flow field divided by the wave speed, which is a useful and non-obvious result indeed.

The curl–curvature formula (4.106) follows from the variational principle based on (4.105) via the usual Euler–Lagrange equations, which for constant c are

$$\frac{d}{ds}\left(\frac{\dot{\boldsymbol{x}}}{|\dot{\boldsymbol{x}}|} - \frac{\boldsymbol{U}}{c}\right) = -\frac{1}{c}(\boldsymbol{\nabla}\boldsymbol{U}) \cdot \dot{\boldsymbol{x}} \tag{4.107}$$

where $\dot{\boldsymbol{x}}$ on the right contracts with $\boldsymbol{U}$ and not with $\boldsymbol{\nabla}$. Using $d\boldsymbol{U}(\boldsymbol{x})/ds = \dot{\boldsymbol{x}} \cdot \boldsymbol{\nabla}\boldsymbol{U}$ then yields

$$\frac{d}{ds}\left(\frac{\dot{\boldsymbol{x}}}{|\dot{\boldsymbol{x}}|}\right) = \frac{1}{c}\left((\boldsymbol{\nabla}\boldsymbol{U})^T - \boldsymbol{\nabla}\boldsymbol{U}\right) \cdot \dot{\boldsymbol{x}} = \frac{1}{c}(\boldsymbol{\nabla} \times \boldsymbol{U}) \times \dot{\boldsymbol{x}}, \tag{4.108}$$

from which (4.106) follows after using the definitions of $\widehat{\boldsymbol{t}}$ and dr.

Finally, a statement analogous to (4.105) and (4.106) can also be formulated in the general case of an isotropic dispersive system with some intrinsic frequency function $\hat{\Omega}(\kappa)$. Following through the previous steps in this case yields

$$\boldsymbol{k} \cdot d\boldsymbol{x} = \omega\left(\frac{|d\boldsymbol{x}|}{c_{p0}} - \frac{\boldsymbol{U} \cdot d\boldsymbol{x}}{c_{g0}c_{p0}}\right) + o(\boldsymbol{U}). \tag{4.109}$$

Here κ_0 is the root of $\omega = \hat{\Omega}(\kappa_0)$ in the unperturbed problem, $c_{p0} = \hat{\Omega}(\kappa_0)/\kappa_0$ is the unperturbed phase velocity and $c_{g0} = \hat{\Omega}'(\kappa_0)$ is the unperturbed group velocity. This shows that a 'weak basic flow' for dispersive waves means that $|\boldsymbol{U}|$ is small compared to the group velocity. For example, this typically applies to *deep-water surface waves* in the ocean, for which $\hat{\Omega} = \sqrt{g\kappa}$. Thus, (4.109) shows that deep-water surface waves exposed to a weak system of irrotational horizontal currents will propagate in a straight line. *Mutatis mutandis*, for surface waves on horizontal currents with nonzero $\boldsymbol{\nabla} \times \boldsymbol{U}$ it can also be shown that the remarkable curl–curvature formula (4.106) again holds with c replaced by the group velocity c_{g0}.

4.4.4 Wave action conservation and amplitude prediction

The amplitude evolution along non-intersecting rays (i.e., away from caustics) of a slowly varying wavetrain in a moving medium is governed by the conservation law for *wave action*. This follows from the general relationships between conserved wave activities and symmetries of the basic state, which are discussed more fully for linear waves in §4.5 and for nonlinear waves in §10.3.

The spatial density for wave action is $\bar{E}/\hat{\omega}$ where $\bar{E}(\boldsymbol{x},t)$ is the phase-averaged density of disturbance energy and $\hat{\omega}(\boldsymbol{x},t)$ is the intrinsic frequency, *not* the absolute frequency. The conservation law is

$$\frac{\partial}{\partial t}\left(\frac{\bar{E}}{\hat{\omega}}\right) + \boldsymbol{\nabla}\cdot\left(\frac{\bar{E}}{\hat{\omega}}\boldsymbol{c}_g\right) = 0 \quad \Leftrightarrow \quad \frac{d}{dt}\left(\frac{\bar{E}}{\hat{\omega}}\right) + \frac{\bar{E}}{\hat{\omega}}\boldsymbol{\nabla}\cdot\boldsymbol{c}_g = 0. \quad (4.110)$$

If $\boldsymbol{U} = 0$ and the basic state is time independent then $\omega = \hat{\omega} =$ const along a ray, and therefore (4.110b) implies

$$\frac{d\bar{E}}{dt} + \bar{E}\boldsymbol{\nabla}\cdot\boldsymbol{c}_g = \frac{\partial \bar{E}}{\partial t} + \boldsymbol{\nabla}\cdot(\bar{E}\boldsymbol{c}_g) = 0. \quad (4.111)$$

Hence in this special case conservation of wave action implies conservation of disturbance energy. This was the case in all wave examples considered so far. However, this will not be the case anymore once shear flows are allowed. Wave action, not wave energy, is the generic conserved wave activity measure for a slowly varying wavetrain.

A heuristic argument for the conservation of wave action follows from integrating (4.110) over some volume that moves with the local group velocity and that is small enough that variations of $\hat{\omega}$ within the volume are negligible. In this case the total wave action contained in the volume is $\mathcal{E}/\hat{\omega}$, say, where $\mathcal{E}$ is the volume-integrated disturbance energy and $\hat{\omega}$ is the near-constant value of the intrinsic frequency. Action conservation now implies that $\mathcal{E}/\hat{\omega} =$ const, i.e., the total disturbance energy is proportional to the intrinsic frequency. This is reminiscent of the generic result from classical mechanics for a linear harmonic oscillator subject to a slowly varying frequency (cf. the comment below (2.53)), where the action conservation is often referred to as *adiabatic invariance*. Thus wave action generalizes the adiabatic invariance of a linear harmonic oscillator to the continuous case of a slowly varying linear wavetrain.

The second form of (4.110) makes it obvious that it is necessary to solve for a bundle of neighbouring rays in order to compute $\boldsymbol{\nabla}\cdot\boldsymbol{c}_g$. In other words, one cannot compute the wave amplitude along a single ray. However, $\boldsymbol{\nabla}\boldsymbol{c}_g$ (and therefore $\boldsymbol{\nabla}\cdot\boldsymbol{c}_g$) can be expressed in terms of $\boldsymbol{\nabla}\boldsymbol{k}$, and by including the $n(n+1)/2$ independent components of the symmetric $\nabla \boldsymbol{k}$ as additional variables one can write down an extended set of ray-tracing equations that does allow computing the amplitude along a single ray. For example, in one dimension this would lead to one additional equation for k_x, which is

$$\frac{dk_x}{dt} = -\frac{\partial^2 \Omega}{\partial x^2} - 2\frac{\partial^2 \Omega}{\partial x \partial k}k_x - \frac{\partial^2 \Omega}{\partial k^2}k_x^2. \tag{4.112}$$

The price to pay is the appearance of second derivatives of Ω in the equations, the larger set of variables along the ray, and the concomitant problem of specifying consistent initial conditions for them, especially because of the constraint $\nabla \times \boldsymbol{k} = 0$. Notably, this approach to amplitude prediction along a ray is equivalent to (4.110), hence it too must fail at caustics.

Finally, the Hamiltonian structure of (4.83) also gives rise to a phase-space view of wave action conservation. For example, we can define a phase-space density for wave action $b(\boldsymbol{k}, \boldsymbol{x}, t)$, say, such that

$$\frac{\bar{E}}{\hat{\omega}} = \int b(\boldsymbol{k}, \boldsymbol{x}, t)\, dk_1 \ldots dk_n \quad \text{and} \quad \frac{db}{dt} = 0. \tag{4.113}$$

The second equation uses d/dt as the derivative following a ray in phase space and is consistent with (4.110) by Liouville's theorem: in phase space the wave action density is invariant along rays because of the incompressibility of the phase-space volume $dx_1 \ldots dx_n\, dk_1 \ldots dk_n$. However, whilst b may remain finite at a caustic, the spatial density $\bar{E}/\hat{\omega}$ as defined in (4.113a) still diverges there. Ray tracing, in whatever form, fails at caustics.

4.5 Wave activity conservation laws

Physical conservation laws are fundamentally linked to the continuous symmetries of the system under consideration. For example, continuous symmetries of a system under spatial translations or under shifts in time lead to the conservation of total momentum and total energy, respectively, and a continuous symmetry with respect to a spatial rotation leads to the conservation of total angular momentum.

In wave dynamics a similar link between symmetries and conservation laws for wave activities can be established, but in this case it is the symmetries of the basic state that are important, not the symmetries of the entire fluid system. We have already encountered an example of this in the ray tracing equations based on the dispersion function $\Omega(\boldsymbol{k}, \boldsymbol{x}, t)$: a continuous symmetry of the basic state with respect to shifts in x or t, say, is reflected in the lack of an explicit dependence of Ω on these coordinates, and this leads to the invariance of k or ω along a ray. Analogous conservation laws can be found for the wave amplitude, as we shall see here in the case of linear waves and later, in §10.3, in the case of nonlinear waves.

The most general mathematical framework in which to establish the link between continuous symmetries and conservation laws is that of *Hamiltonian mechanics*, i.e., the formulation of the mechanical laws in terms of a variational principle to which *Noether's theorem* can then be applied. For discrete mechanical systems this most general approach is also the most easy to use for practical computations. However, for continuous mechanical systems such as fluid dynamics there are some technical issues because one either has to use Lagrangian variables or one has to use the formalism of so-called *non-canonical Hamiltonian mechanics*. This is because the usual Eulerian variables such as velocity and density do not fit the canonical framework, which is based on explicit particle positions.

Non-canonical Hamiltonian mechanics has become the method of choice for studying wave activity conservation laws, and over the last two decades the use of such methods has matured to the point where it covers most situations of practical interest.

On the other hand, in my view there is no particular utility of Hamiltonian methods for the study of nonlinear wave–mean interactions, whereas the use of Lagrangian variables such as the particle displacement field $\boldsymbol{\xi}'$ turns out to be very useful in this regard. For this reason, we will study wave activity laws using Lagrangian particle displacements rather than the methods of Hamiltonian mechanics. In fact, we will only use a single result from Hamiltonian mechanics at all, namely the formulation of an abstract conservation law based on an ensemble of solutions in phase space. As we shall see, the presence of symmetries of the basic state makes this abstract conservation law visible even for a single flow solution, and this leads directly to the various wave activities in common use.

Before moving on it is important to note that conservation laws are not the only reason to study wave activities. For instance, the central wave activity in wave–mean interaction theory is the *pseudomomentum vector*, whose components are conserved only if the basic state has a corresponding translational symmetry. However, as we shall see, pseudomomentum is intrinsically linked to Kelvin's circulation and it plays an important role in wave–mean interaction theory regardless of whether it is conserved or not.

4.5.1 Ensemble conservation law in discrete mechanics

It is convenient to first recall the derivation of an exact ensemble invariant in finite-dimensional discrete mechanics. Thus, let $(\boldsymbol{q}, \boldsymbol{p}) \in \mathcal{D} \subset R^{2n}$ be canonical variables in a phase space $\mathcal{D}$ that are governed by the canonical equations

$$\dot{\boldsymbol{q}} = +\nabla_{\boldsymbol{p}} H \quad \text{and} \quad \dot{\boldsymbol{p}} = -\nabla_{\boldsymbol{q}} H \tag{4.114}$$

where $H(\boldsymbol{q}, \boldsymbol{p}, t)$ is the Hamiltonian function and dots denote time derivatives. In obvious analogy with Kelvin's circulation theorem we can define a closed loop $\mathcal{C} \in \mathcal{D}$ that moves with the phase-space flow and a corresponding phase-space circulation Γ via

$$\Gamma = \oint_{\mathcal{C}} \boldsymbol{p} \cdot d\boldsymbol{q} = -\oint_{\mathcal{C}} \boldsymbol{q} \cdot d\boldsymbol{p}, \tag{4.115}$$

where the second form follows from integration by parts. For example, we can think of the loop at $t = 0$ as being a set of initial conditions for our dynamical system. Using (4.114) and (4.115) as well as another integration by parts it follows that

$$\dot{\Gamma} = \oint_{\mathcal{C}} -\nabla_{\boldsymbol{q}} H \cdot d\boldsymbol{q} - \nabla_{\boldsymbol{p}} H \cdot d\boldsymbol{p} = -\oint_{\mathcal{C}} dH = 0. \tag{4.116}$$

Thus Γ is conserved for arbitrary H, even those including an explicit time-dependence. This is the exact conservation law that we are after.

If we introduce a real ensemble parameter $\alpha \in [0, 2\pi]$ along $\mathcal{C}$ then we can write

$$\Gamma = \frac{1}{2\pi} \int_0^{2\pi} \boldsymbol{p} \cdot \boldsymbol{q}_\alpha \, d\alpha = \overline{\boldsymbol{q}_\alpha \cdot \boldsymbol{p}} \tag{4.117}$$

where the integral is the definition of *ensemble averaging*. Of course, using $\boldsymbol{q}_\alpha$ we can derive the conservation of Γ directly from the canonical equations (4.114). Indeed, contracting the equation for $\dot{\boldsymbol{p}}$ with $\boldsymbol{q}_\alpha$ and rearranging the terms yields

$$\frac{d}{dt}(\boldsymbol{q}_\alpha \cdot \boldsymbol{p}) = (\dot{\boldsymbol{q}} \cdot \boldsymbol{p})_\alpha - (H)_\alpha \quad \Rightarrow \quad \frac{d}{dt}(\overline{\boldsymbol{q}_\alpha \cdot \boldsymbol{p}}) = 0. \tag{4.118}$$

Crucially, here we used that $H(\boldsymbol{q}, \boldsymbol{p}, t)$ has no explicit dependence on the ensemble parameter α. As it stands, the ensemble conservation law (4.118) has no meaning for a single solution. However, in the case of a rapid oscillation subject to a slowly varying environment one can approximately identify $\mathcal{C}$ with the closed orbit of a fast oscillation and α with its rapidly varying phase. In other words, in this situation a single fast trajectory plays the role of a whole set of trajectories. This leads to the approximate conservation of the usual *adiabatic invariant*, which is Γ evaluated over an orbit C of the fast oscillation. In the simple case of a linear harmonic oscillator it then follows that Γ equals the action, i.e., the ratio of energy to frequency.

We can repeat the derivation that led to (4.118) for linear fluid waves if we identify $\boldsymbol{q}$ with $\boldsymbol{\xi}'$ and $\boldsymbol{p}$ with $\rho_0\boldsymbol{u}'$. Of course, instead of the finite-dimensional vector fields in this section, we will then be dealing with infinite-dimensional vector fields, namely the continuous functions that describe the fluid fields. Still, the basic idea is the same. Actually, the continuous dependence of the fluid fields on $\boldsymbol{x}$ and t allows a more interesting use of the ensemble conservation law that has no counterpart in the discrete case.

4.5.2 Ensemble conservation law for linear waves

For illustration we use the simple setting of the one-dimensional linear shallow water equations with a resting basic state, but we allow for time-dependent gravity $g(t)$ (which could be realized experimentally in an accelerated tank) and for space-dependent still water depth $H(x)$. The linear equations are

$$h'_t + (H(x)u')_x = 0 \quad \text{and} \quad u'_t + g(t)h'_x = 0. \tag{4.119}$$

Using the particle displacement ξ' the first equation can be integrated to

$$h' + (H(x)\xi')_x = 0. \tag{4.120}$$

We now consider a smooth ensemble of solutions parametrized by an ensemble parameter α such that we are dealing with the functions $h'(x,t,\alpha)$ and $u'(x,t,\alpha)$. The crucial assumption at this stage is that the basic state defined by $H(x)$ and $g(t)$ *does not* depend on the ensemble parameter α.

We now multiply the momentum equation in (4.119) by $H(x)\xi'_\alpha$, which mirrors the manipulations that led to (4.118). By making use of $H(x)$ and $g(t)$ we easily arrive at

$$\frac{\partial}{\partial t}\left(H\xi'_\alpha u'\right) + \frac{\partial}{\partial x}\left(gH\xi'_\alpha h'\right) = \frac{\partial}{\partial \alpha}\left(H\frac{u'^2}{2} - g\frac{h'^2}{2}\right). \tag{4.121}$$

The right-hand side contains the α-derivative of the difference between kinetic and potential disturbance energy, which is consistent with (4.118). Using basic-state symmetries, this α-derivative will be seen to contribute either to the flux or the density of a conserved wave activity. Notably, at this stage we have not yet assumed that the ensemble forms a closed loop in phase space.

We can exploit (4.121) in a number of ways. For instance, we can indeed consider a closed-loop ensemble such that $u'(x,t,\alpha+2\pi) = u'(x,t,\alpha)$, say, and then define ensemble-averaging as in §4.5.1 by

$$\overline{(\ldots)} = \frac{1}{2\pi}\int_0^{2\pi} (\ldots)\, d\alpha \quad \text{such that} \quad \frac{\partial}{\partial\alpha}\overline{(\ldots)} = 0. \tag{4.122}$$

Upon averaging (4.121) we obtain the exact ensemble conservation law

$$\frac{\partial}{\partial t}\left(H\overline{\xi'_\alpha u'}\right) + \frac{\partial}{\partial x}\left(gH\overline{\xi'_\alpha h'}\right) = 0. \tag{4.123}$$

However, a much more direct use of (4.121) is possible in situations in which the basic state has additional space–time symmetries.

4.5.3 Pseudomomentum and pseudoenergy

We first consider an x-symmetric basic state, which means H =const although it is still possible to have $g(t)$. With suitable boundary conditions the wave problem is now symmetric under translations of the wave field in x. In other words, if $u'(x,t)$ is a valid solution then we can *construct* an ensemble of solutions by defining

$$u'(x,t,\alpha) \equiv u'(x-\alpha,t) \tag{4.124}$$

and analogously for h' and ξ'. For each value of α this yields a valid solution to the equations. Now, the crucial point is that for a member of this ensemble a derivative with respect to α is equivalent to minus a derivative with respect to x, i.e.,

$$\frac{\partial}{\partial\alpha} = -\frac{\partial}{\partial x}. \tag{4.125}$$

We now use (4.124) in (4.121), apply (4.125) and finally set $\alpha = 0$. The result is a conservation law for the *Lagrangian pseudomomentum*, namely

$$\frac{\partial}{\partial t}\left(-H\xi'_x u'\right) + \frac{\partial}{\partial x}\left(-gH\xi'_x h' + H\frac{u'^2}{2} - g\frac{h'^2}{2}\right) = 0. \tag{4.126}$$

Note that the pseudomomentum *flux* contains terms that in (4.121) were part of the α-derivative. By convention, we define the Lagrangian pseudomomentum density per unit mass and its flux as

$$\mathsf{p} \equiv -\xi'_x u' \quad \text{and} \quad B \equiv -gH\xi'_x h' + H\frac{u'^2}{2} - g\frac{h'^2}{2} \tag{4.127}$$

such that $(H\mathsf{p})_t + B_x = 0$.

We now repeat the argument for a time-symmetric basic state in which $g =$ const but $H(x)$ is allowed. Defining the ensemble by $u'(x,t,\alpha) \equiv u'(x,t+\alpha)$

(the sign change is conventional) we now have $\partial/\partial\alpha = \partial/\partial t$ and therefore the same construction that led to (4.126) now leads to a conservation law for the *Lagrangian pseudoenergy*, which is

$$\frac{\partial}{\partial t}\left(H\frac{u'^2}{2} + g\frac{h'^2}{2}\right) + \frac{\partial}{\partial x}\left(gHu'h'\right) = 0. \tag{4.128}$$

Note that the pseudoenergy *density* contains terms that in (4.121) were part of the α-derivative. We define the density of pseudoenergy as

$$H\mathsf{e} \equiv H\xi'_t u' - H\frac{u'^2}{2} + g\frac{h'^2}{2} = H\frac{u'^2}{2} + g\frac{h'^2}{2}, \tag{4.129}$$

say. In the present case without a basic flow, pseudoenergy equals disturbance energy, i.e., $H\mathsf{e} = E$, but this is not true otherwise.

It is interesting to compare the evolution of pseudomomentum and pseudoenergy in the presence of an impermeable wall. For instance, if there is a solid boundary at $x = 0$ then this implies that $u'(0,t) = \xi'(0,t) = 0$ for all t. Of course, this breaks the x-symmetry of the basic state, because the definition of a symmetric basic state must include the boundary conditions. Not surprisingly, the flux of pseudomomentum is non-zero at $x = 0$, which means that total pseudomomentum is not conserved. Pseudoenergy, on the other hand, is conserved because the flux in (4.128) is zero if $u' = 0$. This is consistent with the undisturbed t-symmetry of the basic state. Finally, in the case of an oscillating boundary (i.e., a vibrating piston) at $x = 0$ both the symmetries in x and t are disturbed and neither pseudomomentum nor pseudoenergy are conserved. This is consistent with the generation of waves by the oscillating boundary, of course.

4.5.4 Wave action for slowly varying wavetrains

We consider a slowly varying wavetrain given by

$$u'(x,t,\alpha) = \hat{u}(x,t)\,\exp\left(\frac{i}{\epsilon}\,\Theta(x,t) - i\alpha\right). \tag{4.130}$$

This is an asymptotic solution for arbitrary values of the phase shift α and therefore the phase-averaged ensemble conservation law (4.123) holds, with real parts understood. Now, if we can replace the α-derivatives by x- or t-derivatives then this would again yield a conservation law for a single flow solution. This is possible asymptotically because of

$$u'_t = \frac{i}{\epsilon}\Theta_t\,u' + O(1), \quad u'_x = \frac{i}{\epsilon}\Theta_x\,u' + O(1) \quad \text{and} \quad u'_\alpha = -i\,u', \tag{4.131}$$

which together with $k = \Theta_x/\epsilon$ and $\omega = -\Theta_t/\epsilon$ imply that at leading order

$$\frac{\partial}{\partial \alpha} = \frac{1}{\omega}\frac{\partial}{\partial t} = -\frac{1}{k}\frac{\partial}{\partial x} \tag{4.132}$$

holds. This can be used to eliminate the α-derivatives in (4.123) and then α can again be set to zero. Using the first equality in (4.132) yields

$$\frac{\partial}{\partial t}\left(H(x)\frac{\overline{u'^2}}{\omega}\right) + \frac{\partial}{\partial x}\left(g(t)H(x)\frac{\overline{u'h'}}{\omega}\right) = 0. \tag{4.133}$$

Similarly, using the second equality in (4.132) and $-H\xi'_x = h'$ at leading order yields the equivalent relation

$$\frac{\partial}{\partial t}\left(\frac{\overline{h'u'}}{k}\right) + \frac{\partial}{\partial x}\left(g(t)\frac{\overline{h'h'}}{k}\right) = 0. \tag{4.134}$$

Either way, by using the structure of plane waves we find that both agree with the wave action law (4.110) restricted to $\omega = \hat{\omega}$.

If we denote the wave action density per unit mass by $\mathsf{A} = \overline{\xi'_\alpha u'}$ then we can express action conservation in the generic form

$$\frac{\partial}{\partial t}(HA) + \frac{\partial}{\partial x}(HAc_g) = \frac{d}{dt}(HA) + (HA)\frac{\partial c_g}{\partial x} = 0. \tag{4.135}$$

Now, we can use (4.132) to deduce generic expressions for the phase-averaged pseudomomentum and pseudoenergy densities for a slowly varying wavetrain as

$$\bar{\mathsf{p}} = k\mathsf{A}, \quad \bar{\mathsf{e}} = \omega\mathsf{A}, \quad \Rightarrow \quad \bar{\mathsf{p}} = \frac{k}{\omega}\bar{\mathsf{e}}. \tag{4.136}$$

The pseudoenergy relation uses (4.129) and energy equipartition. Substitution in (4.135) leads to

$$\frac{\partial}{\partial t}(H\bar{\mathsf{p}}) + \frac{\partial}{\partial x}(H\bar{\mathsf{p}}c_g) = HA\frac{dk}{dt} = -HA\frac{\partial \Omega}{\partial x} \tag{4.137}$$

and

$$\frac{\partial}{\partial t}(H\bar{\mathsf{e}}) + \frac{\partial}{\partial x}(H\bar{\mathsf{e}}c_g) = HA\frac{d\omega}{dt} = +HA\frac{\partial \Omega}{\partial t}. \tag{4.138}$$

These ray-tracing equations make obvious that in ray tracing the conservation of wave action is the fundamental concept, whilst pseudomomentum and pseudoenergy inherit their conservation properties directly from the ray-tracing equations for k and ω.

Of course, this dominance of wave action is confined to the regime of a slowly varying wavetrain. Indeed, as §4.5.3 showed, pseudomomentum or pseudoenergy might be exactly conserved even for wave solutions that do not resemble a slowly varying wavetrain.

4.5.5 Moving media and several dimensions

We add a non-zero basic flow $U(x)$ and care must be taken that $U(x)$ and $H(x)$ are chosen such that the basic flow is a steady solution to the one-dimensional shallow-water equations. This implies $(HU)_x = 0$ as well as a dynamical constraint related to the Bernoulli function, which is however ignorable if $U^2 \ll gH$, as we shall assume. The linear equations are

$$h'_t + (Hu' + h'U)_x = 0 \quad \text{and} \quad \mathrm{D}_t u' + u'U_x + gh'_x = 0 \qquad (4.139)$$

where $\mathrm{D}_t = \partial_t + U\partial_x$ as before. It is still possible to integrate the continuity equation provided the particle displacement is defined by

$$\mathrm{D}_t \xi' = u' + \xi' U_x, \qquad (4.140)$$

in which the right-hand side is the $O(a)$ approximation to the velocity at the displaced position $x + \xi'$. This definition ensures that $h' = -(H\xi')_x$ still holds.[11] Introducing the ensemble as before then leads to a variant of the ensemble conservation law (4.121) in the form

$$\frac{\partial}{\partial t}\left(H\xi'_\alpha u'\right) + \frac{\partial}{\partial x}\left(HU\xi'_\alpha u' + gH\xi'_\alpha h'\right) = \frac{\partial}{\partial \alpha}\left(H\frac{u'^2}{2} - g\frac{h'^2}{2}\right). \qquad (4.141)$$

The new term is an advective flux term due to the basic flow. The pseudomomentum follows as before with the same density, but with a flux that contains a new term $HU\mathsf{p}$. However, the pseudoenergy has a different density now because of the changed definition in (4.140). This yields

$$H\mathsf{e} = E + HU\mathsf{p} + HU_x\xi' u'. \qquad (4.142)$$

This shows that pseudoenergy and disturbance energy are not equal in a moving medium and that e is not invariant under adding a constant velocity U. In other words, unlike p, the pseudoenergy is frame dependent, which is a very important property.

Further general relations are given in the context of nonlinear GLM theory in §10.3.2, so here we consider only the special case of a slowly varying wavetrain, in which terms such as the last terms in (4.140) and (4.142) are negligible. Phase-averaging the latter equation then yields the generic relation

$$H\bar{\mathsf{e}} = \bar{E} + HU\bar{\mathsf{p}} \qquad (4.143)$$

between the various wave activity densities in a wavetrain. Wave action

[11] Specifically, $(\mathrm{D}_t + U_x)(h' + (H\xi')_x) = 0$, which implies the result subject to suitable initialization.

conservation follows after noting that (4.132) can be extended by a new term based on D_t, namely

$$\frac{\partial}{\partial \alpha} = \frac{1}{\hat{\omega}}\mathrm{D}_t = \frac{1}{\omega}\frac{\partial}{\partial t} = -\frac{1}{k}\frac{\partial}{\partial x}. \tag{4.144}$$

Using $\mathrm{D}_t\xi' = u'$ for a slowly varying wavetrain then yields the wave action density $H\mathsf{A} = H\overline{u'^2}/\hat{\omega} = \bar{E}/\hat{\omega}$ with the all-important intrinsic frequency in the denominator. We also have the generic wavetrain relations

$$H\bar{\mathsf{p}} = kH\mathsf{A} = \frac{k}{\hat{\omega}}\bar{E} \quad \text{and} \quad H\bar{\mathsf{e}} = \omega H\mathsf{A} = \frac{\omega}{\hat{\omega}}\bar{E}. \tag{4.145}$$

In several spatial dimensions the necessary manipulations are more cumbersome (e.g., multiplication of the momentum equation by ξ'_α is replaced by contraction with the vector $\boldsymbol{\xi}'_\alpha$), but the basic ideas are the same. For example, the pseudomomentum *vector* is given by

$$\mathsf{p}_i = -\xi'_{j,i}u'_j, \tag{4.146}$$

with summation understood. For a slowly varying wavetrain the phase-averaged pseudomomentum vector is

$$H\bar{\mathbf{p}} = \boldsymbol{k}H\mathsf{A} = \frac{\boldsymbol{k}}{\hat{\omega}}\bar{E} \tag{4.147}$$

and (4.143) becomes

$$H\bar{\mathsf{e}} = \bar{E} + H\boldsymbol{U}\cdot\bar{\mathbf{p}}, \tag{4.148}$$

which is $H\mathsf{A}$ times $\omega = \hat{\omega} + \boldsymbol{U}\cdot\boldsymbol{k}$. The wave action law is

$$\frac{\partial}{\partial t}(H\mathsf{A}) + \boldsymbol{\nabla}\cdot(H\mathsf{A}\boldsymbol{c}_g) = \frac{d}{dt}(H\mathsf{A}) + (H\mathsf{A})\,\boldsymbol{\nabla}\cdot\boldsymbol{c}_g = 0 \tag{4.149}$$

and the corresponding pseudomomentum and pseudoenergy laws are

$$\frac{\partial}{\partial t}(H\bar{\mathbf{p}}) + \boldsymbol{\nabla}\cdot(H\bar{\mathbf{p}}\boldsymbol{c}_g) = H\mathsf{A}\frac{d\boldsymbol{k}}{dt} = -H\mathsf{A}\frac{\partial\Omega}{\partial\boldsymbol{x}}. \tag{4.150}$$

and

$$\frac{\partial}{\partial t}(H\bar{\mathsf{e}}) + \boldsymbol{\nabla}\cdot(H\bar{\mathsf{e}}\boldsymbol{c}_g) = H\mathsf{A}\frac{d\omega}{dt} = +H\mathsf{A}\frac{\partial\Omega}{\partial t}. \tag{4.151}$$

Finally, in multidimensional problems there is also the possibility of a radially symmetric basic state, which leads to a conservation law for *angular pseudomomentum*. In the case of a slowly varying wavetrain, the angular pseudomomentum density per unit mass is $\hat{\boldsymbol{z}}\cdot(\boldsymbol{r}\times\bar{\mathbf{p}})$ where $\hat{\boldsymbol{z}}$ is the symmetry axis and $\boldsymbol{r}$ measures the distance from this axis. For example, in two dimensions with $\bar{\mathsf{p}} = (k, l)\mathsf{A}$ and radial symmetry around the origin the angular pseudomomentum density is $\mathsf{A}(lx - ky)$, which inherits its conservation from (4.149) together with the invariance along rays of $lx - ky$.

4.6 Notes on the literature

The fundamental theory of dispersive waves and of ray tracing is laid out in the classic textbooks Lighthill (1978) and Whitham (1974); see also Hinch (1991) for a short asymptotic derivation of the ray-tracing equations. Fermat's principle for rays in moving media as well as a derivation of the remarkable formula (4.106) for non-dispersive waves are described in Landau and Lifshitz (1959) whilst a derivation for dispersive surface waves can be found in Dysthe (2001). Some links between Fermat's principle and the Hamilton–Jacobi equation of classical mechanics are explored in Bühler (2006).

Early work on the conservation of wave action and other wave activity measures in fluid dynamics is described in Whitham (1974) and Bretherton and Garrett (1968), for instance, who all contributed significantly to the development of the subject. Using particle displacements in the manner described here was pioneered by Hayes (1970). The modern methods of non-canonical Hamiltonian mechanics are reviewed in Morrison (1998), Shepherd (1990) and Salmon (1998). The geometric phase in wave dynamics is considered from this perspective in Vanneste and Shepherd (1999), which also explains how this topic is related to other areas of physics.

4.7 Exercises

1. *Rossby waves created by sidewall undulations.* Consider two-dimensional Rossby waves with intrinsic dispersion relation (4.59) in the presence of a constant zonal flow U such that the absolute frequency

$$\omega = Uk + \hat{\omega} = \left(U - \frac{\beta}{k^2 + l^2}\right) k. \tag{4.152}$$

The fluid is confined to the upper half plane $y \geq y_s(x)$, say, with an impermeable sidewall located at $y = y_s(x)$. For weak sinusoidal undulations $y_s = a\cos(kx)$ with $a \ll 1$ this becomes a linear wave problem in which a plane Rossby wave is generated at the sidewall and propagates away from it in the positive y-direction. General sidewall shapes $y_s(x)$ can then be constructed using Fourier series in x. Crucially, for steady waves the condition $\omega = 0$ holds. Combine the linear boundary condition $v' = U dy_s/dx$ at $y = 0$ with a suitable radiation condition to find the amplitude of the wave as well as its meridional wavenumber $l(k)$. Show that propagating waves imply the famous Charney–Drazin criterion:

$$0 < U < \beta/k^2. \tag{4.153}$$

Also show that all propagating waves have the same wavelength, regardless of k. Compute the absolute group velocity components $u_g = U + \hat{u}_g$ and $v_g = \hat{v}_g$ and verify that

$$u_g = 2U\frac{k^2}{k^2+l^2}, \quad v_g = 2U\frac{kl}{k^2+l^2}, \quad \text{and} \quad \frac{v_g}{u_g} = \frac{dy}{dx} = \sqrt{\frac{\beta}{Uk^2} - 1}. \tag{4.154}$$

The last expression gives the slope of rays associated with waves of wavenumber k. For example, a localized sidewall undulation near the origin $x = y = 0$ will create a slowly varying wavetrain in the upper half plane such that the local wavenumber $k(x, y)$ is constant along straight lines through the origin with slope given by dy/dx. Note that under ray tracing this slope is non-negative, and therefore the waves are confined to the downstream region $x \geq 0$. A similar construction will be used for internal waves in § 6.4.1.

2. *Fermat's principle for zero-frequency Rossby waves.* Derive the equivalent of (4.103) for zero-frequency Rossby waves in the presence of a slowly varying mean flow $U(y) > 0$ and verify that in this case (4.103) holds with $c(\boldsymbol{x})$ replaced by $\sqrt{U(y)/\beta}$. Repeat the computation with finite deformation wavenumber κ_D.

PART TWO

WAVE–MEAN INTERACTION THEORY

5

Zonally symmetric wave–mean interaction theory

This is the classic body of wave–mean interaction theory that has been developed extensively in atmospheric fluid dynamics since the late 1960s. Here 'zonal symmetry' refers to basic flows that are independent of longitude, which is a natural starting point for analysing large-scale atmospheric flows. As we know, such basic flows induce a conservation law for the zonal component of the pseudomomentum vector, and much of the classic interaction theory is focused on the interplay between zonal pseudomomentum and the zonal component of the mean velocity field.

Many interesting and powerful results are available in this theory, such as so-called 'non-acceleration conditions', which provide criteria for whether the presence of waves may lead to an acceleration of the zonal mean flow. Another example is the 'pseudomomentum rule', which makes precise the impact of wave dissipation on mean-flow acceleration.

Against these obvious successes of the classic theory must be weighed its obvious restrictions to zonally symmetric basic flows. For instance, it has been much harder to apply this theory in the ocean, where mean circulations (with few exceptions such as the Antarctic circumpolar current) are hemmed-in by the continents and therefore are manifestly not independent of longitude. This problem is compounded by the fact that many results of the zonally symmetric theory do not apply even approximately in a situation with a slowly varying mean flow. We will go beyond zonal symmetry in part THREE of this book.

Here, we begin with a brief outline of classical interaction theory for small-amplitude waves, which is then followed by an extensive description of internal gravity waves and their interactions with a zonally symmetric mean flow. This includes vertical shear and critical layers. This material is followed by a more general discussion of interactions in three-dimensional rotating flows. Thereafter we switch gears and introduce the so-called generalized

Lagrangian-mean theory, which allows us to extend the previous results to finite-amplitude waves. As we shall see, this theory will be of great utility once the assumption of zonal symmetry has been dropped.

5.1 Basic assumptions

The classic wave–mean interaction theory is based on a combination of three simplifying assumptions:

1. small-amplitude waves,
2. simple geometry (in a sense to be defined) and
3. zonal averaging linked to the simple geometry.

We will look at the consequence of these assumptions in turn.

5.1.1 Small-amplitude wave–mean interactions

To understand the structure of wave–mean interaction theory for small-amplitude waves it is useful to write the governing equations in the abstract form

$$\frac{\partial U}{\partial t} + \mathcal{L}(U) + \mathcal{B}(U, U) = 0, \tag{5.1}$$

where $U(\boldsymbol{x}, t)$ is a vector representing the flow variables, $\mathcal{L}$ is a linear operator, and $\mathcal{B}$ is a bilinear operator that captures, say, the essential quadratic nonlinearity of fluid dynamics due to the advective derivative. Of course, additional nonlinearities due to boundary conditions or nonlinear equations of state can be added to (5.1) as needed. A perturbation expansion of U in terms of a small parameter $a \ll 1$ measuring the wave amplitude is

$$U = U_0 + aU_1 + a^2 U_2 + \cdots \tag{5.2}$$

and substitution of (5.2) in (5.1) then yields a regular perturbation hierarchy of quasi-linear equations to be solved at ascending powers $O(a^n)$. This standard method converts a nonlinear problem into a hierarchy of linear problems. At lowest order this yields the equation for the $O(1)$ *basic flow*

$$O(1): \quad \frac{\partial U_0}{\partial t} + \mathcal{L}(U_0) + \mathcal{B}(U_0, U_0) = 0. \tag{5.3}$$

The next order yields the equation

$$O(a): \quad \frac{\partial U_1}{\partial t} + \mathcal{L}(U_1) + \mathcal{B}(U_0, U_1) + \mathcal{B}(U_1, U_0) = 0 \tag{5.4}$$

for the $O(a)$ *linear waves*. This is the realm of linear theory as studied before in Part I. At the next order the back-reaction onto the basic flow enters:

$$O(a^2): \quad \frac{\partial U_2}{\partial t} + \mathcal{L}(U_2) + \mathcal{B}(U_0, U_2) + \mathcal{B}(U_2, U_0) = -\mathcal{B}(U_1, U_1). \quad (5.5)$$

Crucially, the forcing term on the right-hand side is already known at this stage from the solution of the $O(a)$ problem (5.4).

The basic task of small-amplitude wave–mean interaction theory is to compute the leading-order flow response U_2 and it is now clear that this merely requires solving two linear problems in sequence, namely (5.4) first and then (5.5). Moreover, these linear problems share the same linear operator on the left-hand side, which is

$$\frac{\partial(\cdot)}{\partial t} + \mathcal{L}(\cdot) + \mathcal{B}(U_0, \cdot) + \mathcal{B}(\cdot, U_0). \quad (5.6)$$

Thus the $O(a^2)$ flow response to the waves itself behaves like a forced linear wave. This indicates that the strongest wave–mean interactions will be associated with resonances between the wave-induced forcing terms and the modes of this linear operator. We can makes this more definite by distinguishing between weak and strong interactions, as follows.

A *weak interaction* leads to an $O(a^2)$ mean-flow response that is bounded uniformly in time. Because $a \ll 1$, this suggests that the $O(a^2)$ mean-flow response may be quite small in this case, and perhaps even negligible. A *strong interaction*, on the other hand, does not have a uniform bound and in this case the mean-flow response typically grows secularly in time such that $U_2 = O(a^2 t)$. This means that over long, amplitude-dependent times $t = O(a^{-2})$ the mean-flow response can grow to $O(1)$ and thus change significantly the basic flow on which the waves propagate.

Now, as noted before, the $O(a^2)$ equation (5.5) is a forced linear equation and therefore strong interactions can be related to resonant forcing. For example, consider a simple case with zero basic flow, i.e., $U_0 = 0$. This means we have the abstract equation

$$\left(\frac{\partial}{\partial t} + \mathcal{L}\right) U_2 = R \quad (5.7)$$

for the response U_2 in terms of the forcing terms $R = -\mathcal{B}(U_1, U_1)$, say. We can assume that $\mathcal{L}$ is skew-symmetric, i.e., its spectrum consists of imaginary numbers, which correspond to neutrally stable waves. The solution for U_2 is formally given by the Duhamel formula

$$U_2(t) = \exp(-t\mathcal{L})U_2(0) + \int_0^t \exp(-(t-s)\mathcal{L})R(s)\, ds \quad (5.8)$$

where the operator exponentials are defined by their Taylor expansion or equivalently by their action on eigenvectors. The first term is bounded because $\mathcal{L}$ is skew and therefore any unbounded growth must be due to the forcing integral. Clearly, this forcing integral will be growing secularly if there is a resonance between $R(s)$ and the spectrum of $\mathcal{L}$.

Particular attention is often paid to steady waves, i.e., to waves with time-independent mean properties. For such waves R is constant in time and therefore there is a resonance precisely if R has a non-zero projection onto a null eigenvector of $\mathcal{L}$. In the simplest case $\mathcal{L}R = 0$, and therefore

$$\exp(-(t-s)\mathcal{L})R = R \quad \text{and} \quad U_2(t) = \exp(-t\mathcal{L})U_2(0) + tR. \qquad (5.9)$$

Now, what is the physical meaning of a zero eigenvector of $\mathcal{L}$? Clearly, such an eigenvector corresponds to a zero-frequency mode of the linear system. This means that we are talking about the steady *vortical mode* of the linear system; it is this mode that is secularly growing in strong wave–mean interactions with steady waves.

This generic argument highlights the importance of the vortical mode for strong interactions with steady wave fields. It also indicates a certain puzzle: the vortical mode is governed by the distribution of the PV thus in order to achieve significant changes in the vortical flow the waves must be able to achieve significant changes in the PV field. On the other hand, the PV is materially invariant in perfect fluid flow, which greatly restricts significant changes in its distribution. More precisely, such significant changes must be related either to large displacements of fluid particles in the presence of strong PV gradients, or they must be related to dissipative effects, which break the material invariance of PV. The former process plays a limited role for zonally symmetric mean flows, but wave dissipation will emerge as a key aspect of strong wave–mean interactions.

5.1.2 Simple geometry

The most important assumption is a severe restriction of the flow geometry. Specifically, in Cartesian coordinates (x, y, z) we may define *simple geometry* by the two assumptions

$$U(x+L, y, z, t) = U(x, y, z, t) \quad \text{and} \quad \frac{\partial U_0}{\partial x} = 0. \qquad (5.10)$$

The first assumption is that the full flow is periodic in the Cartesian coordinate x with finite period length L (it is important that L is finite), and the second assumption is that the $O(1)$ basic flow is x-independent. Alternatively, x may also refer to the azimuthal angle in cylindrical or spherical coordinates, in which case $L = 2\pi$ and simple geometry corresponds to flows with azimuthal or zonal symmetry. This makes simple geometry directly relevant to atmospheric applications, where it applies to a basic flow that is independent of longitude.

Technically, zonal symmetry in Cartesian coordinates leads to the conservation of the zonal component of pseudomomentum whereas zonal symmetry in cylindrical or spherical coordinates leads to the conservation of angular pseudomomentum relative to the Earth's rotation axis. These two situations differ in their details, but broadly speaking all relevant phenomena apply to either situation. For simplicity, we will focus on the Cartesian case.

5.1.3 Zonal averaging

Simple geometry suggests the use of *zonal averaging*, in which any flow variable A is decomposed as

$$A = \bar{A} + A' \tag{5.11}$$

where the *mean part*

$$\bar{A} = \frac{1}{L} \int_0^{+L} A \, dx \tag{5.12}$$

and the *disturbance part* A' is defined by (5.11). In other words, $\bar{A}$ is the coefficient of the zero-mode in a Fourier series in x and A' is the x-dependent remainder. This makes it obvious that zonal averaging is a linear projection, i.e.,

$$\overline{aA + bB} = a\bar{A} + b\bar{B} \tag{5.13}$$

for all flow variables (A, B) and constants (a, b). Also, the projection property

$$\bar{\bar{A}} = \bar{A} \quad \Leftrightarrow \quad \overline{A'} = 0 \tag{5.14}$$

holds, which ensures that the decomposition (5.11) is unique.[1] By the assumption of simple geometry the basic flow has no disturbance part, i.e., $\overline{U_0} = U_0$ and $U_0' = 0$.

Any function $f(A)$ can likewise be decomposed into a mean part $\overline{f(A)}$ and a disturbance part $f' = f(A) - \overline{f(A)}$, but there is usually no simple

[1] This projection property is not satisfied exactly if the averaging operation is realized using smoothing techniques in which flow data is convolved with a compact filtering kernel. This is a relevant case in practice, but because the repeated application of the filter will continue to change the data it is clear that (5.14) will not hold exactly.

relation between $\overline{f}$ and $f(\overline{A})$. An exception is a bilinear term AB, for which the simple relations

$$\overline{AB} = \bar{A}\,\bar{B} + \overline{A'B'} \tag{5.15}$$

$$(AB)' = AB - \overline{AB} = A'\,\bar{B} + \bar{A}\,B' + A'B' - \overline{A'B'} \tag{5.16}$$

hold. A useful consequence of (5.15) is that quadratic terms can be viewed as a sum of mean and disturbance quadratic parts. For example, the average of a kinetic energy term u^2 is

$$\overline{u^2} = \overline{u}^2 + \overline{u'^2}, \tag{5.17}$$

which allows speaking of mean and disturbance energies unambiguously. This property is lost if the energy is not quadratic in the flow fields, as in compressible flows where the kinetic energy density is ρu^2, say.

Zonal averaging commutes with $\boldsymbol{\nabla}$ and $\partial/\partial t$ for periodic fields, i.e.,

$$\overline{\boldsymbol{\nabla} A} \equiv \boldsymbol{\nabla}\bar{A} \quad \text{and} \quad \overline{A_t} = \bar{A}_t. \tag{5.18}$$

In particular, we have the important zonal-mean identities

$$\bar{A}_x = 0 \quad \text{and} \quad \overline{A_x B} = -\overline{A B_x}. \tag{5.19}$$

The restriction to periodic fields is important here. For instance, if (5.12) is taken as the definition of averaging for a non-periodic function $A(x)$ in a bounded domain $x \in [0, L]$ then $(\bar{A})_x = 0$, but $\overline{A_x} = (A(L) - A(0))/L \neq 0$. The mean part of a material derivative is

$$\overline{\frac{DA}{Dt}} = \bar{A}_t + (\overline{\boldsymbol{u}} \cdot \boldsymbol{\nabla})\bar{A} + \overline{(\boldsymbol{u}' \cdot \boldsymbol{\nabla})A'}, \tag{5.20}$$

which shows that the material derivative does not commute with zonal averaging because of the nonlinear terms.

Finally, we can apply zonal averaging at every order of the amplitude expansion (5.2) and the corresponding governing equations. In particular, we are interested in the leading-order mean-flow response $\overline{U}_2$, which is governed by

$$O(\overline{a^2}): \quad \frac{\partial \overline{U}_2}{\partial t} + \bar{\mathcal{L}}(\overline{U}_2) + \mathcal{B}(U_0, \overline{U_2}) + \mathcal{B}(\overline{U}_2, U_0) = -\overline{\mathcal{B}(U_1, U_1)}, \tag{5.21}$$

where $\bar{\mathcal{L}}$ is the restriction of $\mathcal{L}$ to x-independent fields. Here we used that simple geometry implies $U_0' = 0$ and that the linear operator $\mathcal{L}$ has no explicit x-dependence, e.g., $\mathcal{L}$ may include x-derivatives, but it may not include x-dependent coefficients. One attraction of zonal averaging is that the restriction $\bar{\mathcal{L}}$ tends to be significantly simpler than the full operator $\mathcal{L}$.

6
Internal gravity waves

We study the *Boussinesq system*, which is the simplest fluid model that admits *internal gravity waves*. These dispersive waves are ubiquitous in the atmosphere and ocean and they owe their restoring mechanism to the stable stratification in these environments, i.e., to the fact that density decreases with altitude. Internal gravity waves are typically far too small in scale (especially vertical scale) to be resolvable within global numerical models and therefore their dynamics and their interactions with the mean flow must be parametrized (i.e., put in by hand) based on a combination of observations and theory. Consequently, the classic wave–mean interaction theory for internal gravity waves has been extensively developed and this provides a convenient starting point for us.

6.1 Boussinesq system and stable stratification

The simplest fluid model that captures the effect of stable internal stratification is the Boussinesq model, which can be derived from the Euler equations and its dissipative counterparts under the assumption of a small density contrast across the fluid, together with $\nabla \cdot \boldsymbol{u} = 0$. Importantly, the latter constraint filters sound waves and thereby reduces the number of degrees of freedom compared to the full Euler system.

Before writing down the governing equations we note that as a realistic model for atmospheric and oceanic flows the Boussinesq system is mostly limited by its global restriction to small density contrasts across the fluid, and that these limitations are much more severe in the atmosphere than in the ocean. Specifically, in the ocean the global density contrast is only about 3%, whilst in the atmosphere density changes by 100% in the vertical. This limits the Boussinesq system to atmospheric flows with a vertical extent of much less than a density scale height, which is 5–10 km throughout most of the atmosphere. For flows with larger vertical extent the compressibility of the air must be taken into account either by using the full Euler equations or by using extensions of the Boussinesq model such as the slightly more complicated *anelastic* models, which allow for $O(1)$ density contrasts whilst still filtering sound waves (cf. §6.7.2).

For an essentially incompressible medium such as sea water the Boussinesq equations can be derived by re-writing the Euler equation as

$$\frac{\rho}{\rho_*}\frac{D\boldsymbol{u}}{Dt} + \boldsymbol{\nabla}\left(\frac{p}{\rho_*}\right) = -\frac{\rho}{\rho_*}g\widehat{\boldsymbol{z}}. \tag{6.1}$$

Here ρ_* is the constant reference density such that $\rho \approx \rho_*$ throughout the fluid, $g > 0$ is a constant gravity acceleration, and $\widehat{\boldsymbol{z}}$ is the upward unit vector. In the Boussinesq approximation ρ/ρ_* is approximated by unity on the left but kept unchanged on the right. Asymptotically, this corresponds to looking for small density contrasts and weak material accelerations compared to gravity g.

Combining this with the approximation $\boldsymbol{\nabla}\cdot\boldsymbol{u} = 0$ and blithely assuming that therefore the material density equation (1.8) can be approximated by $D\rho/Dt = 0$ leads to the simplest version of the Boussinesq equations, namely

$$\frac{D\boldsymbol{u}}{Dt} + \boldsymbol{\nabla}\tilde{P} = S\widehat{\boldsymbol{z}}, \quad \frac{DS}{Dt} = 0, \quad \text{and} \quad \boldsymbol{\nabla}\cdot\boldsymbol{u} = 0. \tag{6.2}$$

Here $\tilde{P} = p/\rho_*$ and $S = -g\rho/\rho_*$. More complicated versions of the Boussinesq equations can be derived by allowing for two kinds of additional physical effects, namely the small density changes along material trajectories that arise from the corrections to the approximation $\boldsymbol{\nabla}\cdot\boldsymbol{u} = 0$, and the presence of additional thermodynamic variables that describe salinity in the ocean and moisture in the atmosphere, for example. Essentially, in these more complicated models $DS/Dt = 0$ is retained as a model equation for one or more materially invariant thermodynamic variables and the density ρ is then related to these thermodynamic variables via a suitable thermodynamic equation of state. For the ocean, this procedure leads to versions of the Boussinesq equations that are in fact the model of choice for practical numerical simulations of the global ocean circulation.

We will only use the simplest Boussinesq equations in the form (6.2) and note that these also describe the flow of an ideal gas with the identification $S = +gs/s_*$ based on the specific entropy s. Thus, stratification surfaces $S =$const. in the ocean are surfaces of constant density (sometimes called *isopycnal surfaces*) whereas in the atmosphere they are surfaces of constant specific entropy.[1]

In problems with a well-defined background stratification $S_0(z)$ it is more convenient to work with a variant of (6.2) in which deviations from this background are the primary variables, i.e.,

$$\frac{D\boldsymbol{u}}{Dt} + \nabla P = b\hat{\boldsymbol{z}}, \quad \frac{Db}{Dt} + N^2 w = 0, \quad \text{and} \quad \nabla \cdot \boldsymbol{u} = 0. \tag{6.3}$$

Here $P = \tilde{P} - \tilde{P}_0(z)$ where the hydrostatic pressure is defined by $\tilde{P}_{0z} = S_0$, b is the so-called *buoyancy*,[2] and $N(z)$ is the *buoyancy frequency*. Obviously,

$$S = b + S_0(z) = b + \int N^2 \, dz. \tag{6.4}$$

We will prefer this version of the Boussinesq equations. However, the omitted hydrostatic pressure $\tilde{P}_0(z) = \tilde{P} - P$ is essentially the weight of the background stratification $S_0(z)$ and this can matter for the correct force computation at flow boundaries. We shall see an example of this in §6.5.2.

If the fluid is at rest with $b = 0$ then the stratification surfaces are flat horizontal planes and the size of $N^2 > 0$ measures the strength of the stable stratification. If the fluid is disturbed from rest then the material invariance of S makes it obvious that positive buoyancy $b > 0$ is associated with a local lowering of the stratification surfaces, which corresponds to a vertical displacement of fluid particles downwards, and vice versa for negative buoyancy $b < 0$. Therefore, a local lowering of a stratification surface induces an upward acceleration of the fluid due to buoyancy and vice versa, which is the principal new mechanism that the Boussinesq equations capture.

6.1.1 Momentum, energy and circulation

Due to buoyancy, the Boussinesq equations conserve horizontal but not vertical momentum. They also conserve energy with a conservation law

$$\frac{\partial E}{\partial t} + \nabla \cdot (E\boldsymbol{u} + P\boldsymbol{u}) = 0 \tag{6.5}$$

where in the simplest case of constant N the energy density per unit volume

$$E = \frac{1}{2}\left(|\boldsymbol{u}|^2 + \frac{b^2}{N^2}\right). \tag{6.6}$$

[1] These simple definitions of stratification surfaces ignore the effects of moisture in the atmosphere and of salinity in the ocean.

[2] Sometimes S is called the buoyancy and b the buoyancy disturbance.

The quadratic form of E in terms of $\boldsymbol{u}$ and b is one of the advantages of the Boussinesq system. The available potential energy density $\frac{1}{2}b^2/N^2$ makes it obvious that the state $b = 0$ with flat stratification surfaces is the configuration of minimal gravitational energy. This generalizes the minimum-energy configuration of a single-layer shallow-water system to the case of continuous stratification. We note in passing that in the case of variable $N(z)$ a more elaborate construction is required to derive a quadratic energy density, which involves the careful identification of the available potential energy of a given buoyancy field that quantifies the surplus of potential energy of this buoyancy field relative to a state in which the materially invariant stratification surfaces are flattened out adiabatically into their horizontal, minimum-energy configuration.

The stratification surfaces are material surfaces and they are also constant-value surfaces of a thermodynamic variable such as density or specific entropy. This suggests that Kelvin's circulation theorem should apply to closed material loops lying within stratification surfaces, at least if the fluid pressure depends only on two thermodynamic variables. This is indeed the case in the Boussinesq system, as can easily be checked from (6.3), (1.37) and (1.38):

$$\frac{d\Gamma}{dt} = \oint_{\mathcal{C}} b\,dz = \oint_{\mathcal{C}} (S - S_0(z))\,dz. \tag{6.7}$$

The first term vanishes because $\mathcal{C}$ lies within a surface of constant S and therefore S can be pulled out of the integral, which then integrates to zero around the closed loop. The second term depends only on z and is hence obviously a perfect differential (equal to $-d\tilde{P}_0$, in fact), which also integrates to zero. As usual, Kelvin's circulation theorem yields a materially invariant potential vorticity in the form

$$q = (\boldsymbol{\nabla} \times \boldsymbol{u}) \cdot \boldsymbol{\nabla} S \quad \Rightarrow \quad \frac{Dq}{Dt} = 0. \tag{6.8}$$

There is no density term because of incompressibility.

6.2 Linear Boussinesq dynamics

For simplicity, we restrict to constant buoyancy frequency N such that the background stratification is given by $S_0 = N^2 z$. This is only a minor restriction, as the linear theory generalizes straightforwardly to variable $N(z)$, including the form of the wave energy density in (6.24) below. The value of N^{-1} defines an intrinsic time scale for the Boussinesq system and this is the only dimensional parameter that enters the dynamics. We linearize around a state of uniform motion such that

$$\boldsymbol{u} = (U, 0, 0) + a\boldsymbol{u}' + O(a^2) \quad \text{and} \quad S = N^2 z + ab' + O(a^2) \tag{6.9}$$

with constant U and then the $O(a)$ Boussinesq equations are

$$\mathrm{D}_t u' + P'_x = 0, \quad \mathrm{D}_t v' + P'_y = 0, \quad \mathrm{D}_t w' + P'_z - b' = 0, \tag{6.10}$$

$$\mathrm{D}_t b' + N^2 w' = 0 \quad \text{and} \quad \nabla \cdot \boldsymbol{u}' = 0. \tag{6.11}$$

The linear particle displacements $\boldsymbol{\xi}' = (\xi', \eta', \zeta')$ are defined as usual via $\mathrm{D}_t \boldsymbol{\xi}' = \boldsymbol{u}'$ and the vertical displacement ζ' is directly related to b' via

$$\zeta' = -\frac{b'}{N^2}. \tag{6.12}$$

This is an example of how a Lagrangian quantity can be directly observable in terms of Eulerian variables because of the material invariance of S.

Now, the pressure P' can be eliminated by taking the curl and up to a constant factor N^2 the vertical component of this is the invariant linear PV

$$q' = v'_x - u'_y \quad \text{such that} \quad \mathrm{D}_t q'_t = 0. \tag{6.13}$$

The horizontal components of the curl are not invariant, e.g.,

$$\mathrm{D}_t(u'_z - w'_x) = -b'_x = +N^2 \zeta'_x, \tag{6.14}$$

which illustrates the baroclinic generation of vorticity due to sloping stratification surfaces.

The uniform basic flow U does not affect the dynamics, so we will now describe the intrinsic linear dynamics in a frame moving with U. To remind ourselves of this shift we explicitly use the intrinsic frequency $\hat{\omega}$. For plane waves with wavenumber vector $\boldsymbol{k} = (k, l, m)$ and intrinsic frequency $\hat{\omega}$ one obtains a dispersion relation with three frequency branches:

$$\hat{\omega}\left(\hat{\omega}^2 - N^2 \frac{k^2 + l^2}{k^2 + l^2 + m^2}\right) = 0. \tag{6.15}$$

Overall, just as in shallow water, the initial-value problem has three degrees of freedom, namely a vortical mode and two gravity-wave modes.

6.2.1 Vortical mode

The structure of the steady vortical mode follows easily from the linear equations, i.e,

$$\hat{\omega} = 0: \quad w' = 0, \quad \nabla P' = b'\hat{\boldsymbol{z}}, \quad \text{and} \quad u'_x + v'_y = 0. \tag{6.16}$$

Thus, this balanced flow has zero vertical velocity, the pressure and buoyancy can depend only on altitude (and correspond to a trivial adjustment of the hydrostatic rest state), and the horizontal flow is area-preserving. Obviously, the balanced flow is entirely determined by the linear PV in (6.13).

6.2.2 Plane internal gravity waves

The structure of internal gravity waves is significantly different from that of shallow-water gravity waves. To begin with, the incompressibility constraint (6.11b) implies that

$$\boldsymbol{k} \cdot \boldsymbol{u}' = 0 \tag{6.17}$$

here, i.e., the velocity is perpendicular to the wavenumber vector and the particle motion takes place in the plane of constant wave phase (we assume that $\boldsymbol{k}$ is real). This shows that internal waves are *transversal* waves, unlike their longitudinal counterparts in shallow water. One consequence of the transversality is that the advective operator $\boldsymbol{u}' \cdot \boldsymbol{\nabla}$ applied to wave fields is identically zero, which implies the peculiar fact that a single plane internal wave is a trivial exact solution of the nonlinear Boussinesq equations. This is obviously very different from the shallow-water case.

The Boussinesq equations do not depend on the horizontal orientation of $\boldsymbol{k}$ and therefore we can set $\boldsymbol{k} = (k, 0, m)$ without loss of generality. This reduces the problem to a two-dimensional flow in the xz-plane. It then follows from (6.10) and (6.11) that[3]

$$u' = -\frac{m}{k} w', \quad v' = 0, \quad b' = -i\frac{N^2}{\hat{\omega}} w', \quad \text{and} \quad P' = -\frac{m\hat{\omega}}{k^2} w'. \tag{6.18}$$

Thus the velocity lies in the vertical plane containing $\boldsymbol{k}$ and is in phase with the pressure, whilst being out of phase with the buoyancy b'. The associated linear particle displacements describe an oscillation along a straight line perpendicular to $\boldsymbol{k}$. During this oscillation vertical motion is correlated with horizontal motion, which implies that phase-averaged fluxes such as $\overline{u'w'}$ are non-zero for plane internal waves. This can be contrasted with the familiar linear particle trajectories for surface waves in deep water: in this case the particle motion is in circles instead along a straight line, there is no net correlation between horizontal and vertical motion, and hence $\overline{u'w'} = 0$.

The dispersion relation is

$$\omega = Uk + \hat{\omega} \quad \text{with} \quad \hat{\omega} = \pm\hat{\Omega}(k, m) = \pm N \frac{k}{\sqrt{k^2 + m^2}} = \pm N \sin\alpha \tag{6.19}$$

[3] It is convenient in (6.18) to express all fields in terms of w', but there is no actual singularity if $k \to 0$ because in this case $w' \to 0$ as well.

where α is the angle between $\boldsymbol{k}$ and the vertical direction. This remarkable dispersion relation shows that $0 \leq |\hat{\omega}| \leq N$ and that $\hat{\omega}$ depends only on the angle α but not on the magnitude $\kappa = |\boldsymbol{k}|$. In other words, $\hat{\Omega}$ is a zeroth-degree homogeneous function of the components of $\boldsymbol{k}$.[4] By Euler's theorem for homogeneous functions we have that

$$\boldsymbol{k} \cdot \frac{\partial \hat{\Omega}}{\partial \boldsymbol{k}} = 0 \quad \text{or} \quad \boldsymbol{k} \cdot \widehat{\boldsymbol{c}}_g = 0. \tag{6.20}$$

Thus the intrinsic group velocity is perpendicular to the intrinsic phase velocity $\hat{\boldsymbol{c}}_p = \boldsymbol{k}\hat{\omega}/\kappa^2$. Specifically, the intrinsic group velocity is

$$\hat{u}_g = \pm N \frac{m^2}{\kappa^3}, \quad \hat{v}_g = 0, \quad \hat{w}_g = \mp N \frac{km}{\kappa^3}. \tag{6.21}$$

Another fact that follows from the homogeneity of $\hat{\Omega}(\boldsymbol{k})$ is that $\widehat{\boldsymbol{c}}_g(\boldsymbol{k})$ is homogeneous in $\boldsymbol{k}$ of degree minus one, which is apparent from (6.21) because $|\widehat{\boldsymbol{c}}_g| = \sqrt{N^2 - \hat{\omega}^2}/\kappa$. This implies that at fixed intrinsic frequency $\hat{\omega}$ the magnitude of the intrinsic group velocity is proportional to the wavelength of the gravity wave, e.g., at fixed frequency longer waves travel faster than shorter waves.

The ratio of the group-velocity components gives the slope of intrinsic wave propagation, i.e.,

$$\frac{dz}{dx} = \frac{\hat{w}_g}{\hat{u}_g} = -\frac{k}{m} = -\text{sgn}(km)\sqrt{\frac{\hat{\omega}^2}{N^2 - \hat{\omega}^2}}. \tag{6.22}$$

This makes it obvious that there is a direct link between intrinsic frequency and intrinsic propagation slope such that waves with lower frequency travel at a shallower angle to the horizontal than waves with higher frequency.

The intrinsic phase velocity is

$$\hat{u}_p = \pm N \frac{k^2}{\kappa^3}, \quad \hat{v}_p = 0, \quad \hat{w}_p = \pm N \frac{km}{\kappa^3}. \tag{6.23}$$

We see that the choice of branch in (6.19) corresponds to the sign of the horizontal phase and group velocities whilst for either branch the vertical phase and group velocities are always equal-and-opposites of each other. This implies that upward wave radiation in the sense of $\hat{w}_g > 0$ goes together with downward phase propagation $\hat{w}_p < 0$. Historically, this peculiar fact has been important for the correct interpretation of observations of high-altitude internal waves in the atmosphere.

[4] It could have been deduced from dimensional analysis that $\hat{\omega}/N$ can only be a function of k/m.

The fact that the intrinsic frequency has a bounded range is very important for internal wave dynamics. The highest frequency $\hat{\omega} = N$ is achieved when $m = 0$; these are *buoyancy oscillations* in which the particle motion is purely vertical. In the atmosphere and ocean these buoyancy oscillations have typical periods of about 7 minutes and 1 hour, respectively. The lowest frequency $\hat{\omega} = 0$ occurs when $k = 0$; here the particle motion is entirely horizontal and the wave has degenerated into a steady shear flow. This zero-frequency mode will be important for two-dimensional wave–mean interactions involving steady wavetrains.

The expression for the phase-averaged wave energy density $\bar{E}$ can be read off (6.6) with real parts understood, and for a plane wave with $k \neq 0$ it turns out that $\bar{E}$ satisfies equipartition, i.e.,

$$\bar{E} = \frac{1}{2}\left(\overline{|\boldsymbol{u}'|^2} + \frac{\overline{b'^2}}{N^2}\right) = \overline{|\boldsymbol{u}'|^2} = \frac{\overline{b'^2}}{N^2} = N^2\overline{\zeta'^2}. \tag{6.24}$$

Again, this equipartition result does not carry over to internal waves in a rotating frame because of Coriolis forces. It is easy to show that the intrinsic energy flux for a plane internal wave satisfies the generic relation $\overline{P'\boldsymbol{u}'} = \bar{E}\hat{\boldsymbol{c}}_g$.

Using (6.24) the phase-averaged pseudomomentum vector for a plane wave follows from the generic expression (4.146) in §4.5.5 as

$$\bar{\mathsf{p}}_i = -\overline{\xi'_{j,i}u'_j} = +\frac{k_i}{\hat{\omega}}\overline{u'_j u'_j} \quad\Leftrightarrow\quad \bar{\mathsf{p}} = \frac{\boldsymbol{k}}{\hat{\omega}}\bar{E}. \tag{6.25}$$

In particular, the zonal pseudomomentum if $\boldsymbol{k} = (k, 0, m)$ is

$$\bar{\mathsf{p}} = \frac{k}{\hat{\omega}}\bar{E} = \pm\frac{\kappa}{N}\bar{E}. \tag{6.26}$$

Finally, the phase-averaged pseudoenergy is the usual

$$\bar{\mathsf{e}} = \frac{\omega}{\hat{\omega}}\bar{E} = \left(1 + \frac{Uk}{\hat{\omega}}\right)\bar{E} = \bar{E} + U\bar{\mathsf{p}}, \tag{6.27}$$

which, unlike $\bar{E}$ or $\bar{\mathsf{p}}$, depends explicitly on the basic flow U.

6.2.3 Spatial structure of time-periodic waves

Internal waves are peculiar in many ways. For instance, the standard group-velocity considerations turn out to be relevant even for wave fields that do *not* form a slowly varying wavetrain. This becomes clear if we consider the spatial structure of time-periodic waves, i.e., in a frame moving with U we replace D_t by $-i\hat{\omega}$ and obtain

$$w'_{zz} = \mu^2 \left(w'_{xx} + w'_{yy}\right) \quad \text{where} \quad \mu^2 = \frac{N^2 - \hat{\omega}^2}{\hat{\omega}^2}. \tag{6.28}$$

This is the two-dimensional wave equation in (x, y) with z playing the role of time t and with a wave 'speed' equal to $\mu(\hat{\omega})$. This is a hyperbolic equation and it shows that the wave propagation emanating from a source at the origin $\boldsymbol{x} = 0$, say, will be sharply concentrated on the double cone $z = \pm\sqrt{x^2 + y^2}/\mu$. The slope of these cones agrees with the group velocities in (6.22); indeed, for a plane wave $\mu = |m/k|$.

The localization of the wave field on this double cone leads to the famous *St Andrews cross* appearance of two-dimensional internal waves generated by an oscillating cylinder. The hyperbolic nature of (6.28) also leads to the ill-posedness of the general normal-mode problem for bounded domains, a situation that was apparently first studied by Sobolev in connection with rotating fuel tanks, in which inertial waves based on the rotation satisfy an equation analogous to (6.28). This is most easily studied in the two-dimensional version of the problem in which there is no y-dependence. In this case (6.28) becomes the one-dimensional wave equation, which can be solved analytically along characteristics with slope $dz/dx = \pm 1/\mu$. These characteristics undergo reflection at the domain boundary and unless the domain is of a simple shape such as a rectangle or ellipse this does not lead to closed orbits of these characteristics. Open orbits lead to very small structures appearing in the wave field, and ultimately this leads to wave dissipation or breaking. The ill-posedness of the problem is apparent, because a small perturbation to a rectangular domain leads to a dramatic change in the mode structure.

6.2.4 Two-dimensional vertical slice model

The two-dimensional Boussinesq equations follow from (6.3) under the assumption $\partial_y = 0$, i.e.,

$$\frac{Du}{Dt} + P_x = 0 \qquad \frac{Dw}{Dt} + P_z = b \tag{6.29}$$

$$\frac{Db}{Dt} + N^2 w = 0 \qquad u_x + w_z = 0. \tag{6.30}$$

They describe the motion in the xz-plane, which is a vertical slice. This is a straightforward simplification, and plane internal waves were already described using such a slice geometry. However, there is a peculiarity here because the initial-value problem for (6.29) can be posed by specifying only an initial stream function and a buoyancy field. This means that the number of degrees of freedom has been reduced from three to two, which is consistent with the PV being identically zero (if $v = 0$). Indeed, the linear dynamics of (6.29) is entirely described by the two remaining gravity-wave branches of the dispersion relation.

Although there is no vortical mode there is still a zero-frequency wave mode: it is given by the $k = 0$ limit of the internal wave dispersion relation for plane waves. This corresponds to a steady horizontal shear flow, which does not involve any buoyancy disturbance, and which will be seen to play the role of the zero-frequency balanced mode in the wave–mean interaction theory in this two-dimensional system.

6.3 Zonal pseudomomentum of internal waves

Zonal pseudomomentum is central for wave–mean interaction theory in simple geometry because it is both a conserved wave activity measure and because it is intrinsically linked to Kelvin's circulation theorem. To prepare the ground we will introduce both a Lagrangian and a Eulerian version of zonal pseudomomentum based on zonal averaging and the linear equations of motion. Arguably, the Lagrangian version is more fundamental in terms of perfect fluid dynamics whilst the Eulerian version is easier to use in flows involving forcing and dissipation. Also, the Eulerian pseudomomentum is in better harmony with Eulerian definitions of the mean flow and vice versa for their Lagrangian counterparts. So it is a good investment to know both.

Before we begin we note a change in notation: previously we used $\bar{\mathsf{p}}$ to denote the mean zonal pseudomomentum. From now on we will drop the bar, i.e., $\bar{\mathsf{p}} \to \mathsf{p}$. The same applies to the pseudoenergy density $\bar{\mathsf{e}} \to \mathsf{e}$.

6.3.1 Lagrangian and Eulerian pseudomomentum

We continue to assume x-periodicity with finite period L. The *Lagrangian zonal pseudomomentum* in the vertical slice model is defined by

$$\mathsf{p} = -\overline{\xi'_x u'} - \overline{\zeta'_x w'}. \tag{6.31}$$

This is the small-amplitude counterpart of an exact pseudomomentum definition that will be discussed in the context of generalized Lagrangian-mean theory in §10.2.7. In some ways the exact theory has a clearer kinematic interpretation than the small-amplitude theory we work with here, but the close link between p and Kelvin's circulation can already be illustrated based on (6.31). To this end we consider a material stratification line $\mathcal{C}$ defined by $S(x,z,t) = S_0(z_0)$ for some constant z_0, say. Thus, $\mathcal{C}$ traverses the periodic domain, in the absence of waves $\mathcal{C}$ is the flat horizontal line $z = z_0$, and in the presence of small-amplitude waves $\mathcal{C}$ is traced out to $O(a)$ by the linear displacements such that

$$\boldsymbol{r} \in \mathcal{C} \quad \Leftrightarrow \quad \boldsymbol{r} = (x + a\xi'(x, z_0, t), z_0 + a\zeta'(x, z_0, t)). \tag{6.32}$$

Here $x \in [0, L]$ is used as a parameter along the curve. Now, let us compute the circulation $\Gamma(\mathcal{C})$ along the disturbed contour $\mathcal{C}$. Assuming that the basic flow is $\boldsymbol{U} = (U, 0)$ with constant U and Taylor-expanding all fields around the rest location (x, z_0) yields

$$\Gamma(\mathcal{C}) = \oint_{\mathcal{C}} \boldsymbol{u}\cdot \boldsymbol{dx} = \int_0^L U\,dx + a\int_0^L (u' + U\xi'_x)\,dx \tag{6.33}$$

$$+ a^2 \int_0^L (\overline{u}_2 + \xi' u'_x + \zeta' u'_z)\,dx + a^2 \int_0^L (\xi'_x u' + \zeta'_x w')\,dx + O(a^3). \tag{6.34}$$

Here we used that the line element along $\mathcal{C}$ is $\boldsymbol{dx} = dx(1 + a\xi'_x, a\zeta'_x)$ to $O(a)$. The x-integrals in (6.33) are equivalent to zonal averaging times L, which makes it obvious that the $O(a)$ terms are zero. The $O(a^2)$ terms are not zero, however, and the first group contains the leading-order *Stokes correction* $\overline{u}_2^S = \overline{\xi' u'_x + \zeta' u'_z}$ to the zonal mean flow. Stokes corrections will be discussed in detail in §10.1.1, but here it is sufficient to note that $\overline{u}_2^L = \overline{u}_2 + \overline{u}_2^S$ is the $O(a^2)$ part of the Lagrangian-mean zonal velocity, which is generally evaluated by averaging u along the material line $\mathcal{C}$. In the final group of $O(a^2)$ terms we recognize the pseudomomentum from (6.31). Thus we find that

$$\Gamma - LU = a^2 L\left(\overline{u}_2 + \overline{u}_2^S - \mathsf{p}_2\right) = a^2 L\left(\overline{u}_2^L - \mathsf{p}_2\right), \tag{6.35}$$

where we have momentarily reinstated the expansion subscript for p. This shows the intimate relationship between pseudomomentum and circulation. It also shows that in a situation in which Γ is conserved we expect $\overline{u}_2^L - \mathsf{p}_2$ to be constant in time as a consequence of Kelvin's circulation theorem. This will indeed by found to be the case in §6.5.3.

The evolution law for p follows from the linear equations (6.10–6.12) restricted to the slice model by contracting the xz-momentum equations with $-(\xi'_x, \zeta'_x)$ and zonal averaging, which yields

$$\mathrm{D}_t\mathsf{p} + \overline{u'u'_x} + \overline{w'w'_x} - \overline{\xi'_x P'_x} - \overline{\zeta'_x P'_z} = -\overline{\zeta'_x b'}. \tag{6.36}$$

The first term simplifies to p_t because $\mathsf{p}_x = 0$. Similarly, the second and third terms are zonal averages of zonal derivatives and hence they too vanish. The last term also vanishes because of (6.12). The linear particle displacement field inherits incompressibility from the linear velocity field such that $\xi'_x + \zeta'_z = 0$ and therefore the remaining terms can be combined into flux form as

$$\mathsf{p}_t - \overline{\xi'_x P'_x} - (\overline{\zeta'_x P'})_z - \overline{\xi'_{xx} P'} = 0 \quad \Rightarrow \quad \mathsf{p}_t - (\overline{\zeta'_x P'})_z = 0 \qquad (6.37)$$

where the last step used integration by parts in x. This shows that $\mathsf{p}(z,t)$ is conserved, with vertical flux equal to the pressure-related vertical flux of horizontal momentum across an undulating material contour whose rest position is horizontal. We call $-\overline{\zeta'_x P'}$ the *Lagrangian-mean momentum flux*.

The Lagrangian pseudomomentum p explicitly involves the particle displacement fields and this can be circumvented by using an alternative *Eulerian zonal pseudomomentum* $\tilde{\mathsf{p}}$, say, that does not use particle displacements. The derivation exploits the Lagrangian information stored in the buoyancy disturbance $b' = -\zeta' N^2$ together with $\nabla \cdot \boldsymbol{\xi}' = 0$ in order to eliminate $\boldsymbol{\xi}'$ as follows:

$$\mathsf{p} = -\overline{\xi'_x u'} - \overline{\zeta'_x w'} = \overline{\zeta'_z u'} + \overline{\zeta' w'_x} = +(\overline{\zeta' u'})_z - \overline{\zeta'(u'_z - w'_x)}$$
$$\Rightarrow \quad \tilde{\mathsf{p}} = \frac{1}{N^2}\overline{b'(u'_z - w'_x)} = \mathsf{p} - (\overline{\zeta' u'})_z. \qquad (6.38)$$

Clearly, $\tilde{\mathsf{p}}$ and p agree for a plane wave because then the divergence term is zero. Also, because the difference term is a divergence, its time derivative can be viewed as a flux divergence, which makes it obvious that both $\tilde{\mathsf{p}}$ and p satisfy conservation laws.

Moreover, $\tilde{\mathsf{p}}$ is also linked directly to the circulation Γ along the undulating material line Γ. Indeed, by Stokes's theorem we have the *exact* identity

$$\Gamma(\mathcal{C}) = L\overline{u}(z_0,t) - \int (w_x - u_z)\mathrm{sgn}(z - z_0)\, dx\, dz \qquad (6.39)$$

where the integral is extended over the area between $z = z_0$ and $\mathcal{C}$. For $O(a)$ undulations the integral is given to $O(a^2)$ by

$$a^2 \int \zeta'_1 (w'_{1x} - u'_{1z})\, dx = a^2 L \overline{\zeta'_1 (w'_{1x} - u'_{1z})}. \qquad (6.40)$$

Here the disturbance fields have again been Taylor-expanded around (x, z_0). Thus we have

$$\Gamma - LU = a^2 L(\overline{u}_2 - \tilde{\mathsf{p}}_2). \qquad (6.41)$$

This is the Eulerian counterpart to (6.35). Whilst p is based directly on the definition of the circulation, $\tilde{\mathsf{p}}$ exploits the vorticity and Stokes's integral theorem to achieve the same goal.

The evolution law for $\tilde{\mathsf{p}}$ can also be computed directly from the linear equations. Specifically, starting with the linear vorticity equation

$$\mathrm{D}_t(u'_z - w'_x) + b'_x = 0 \tag{6.42}$$

we multiply by b'/N^2 and find after averaging that

$$\tilde{\mathsf{p}}_t + \overline{(u'_z - w'_x)w'} = 0. \tag{6.43}$$

Thus $\tilde{\mathsf{p}}$ evolves in response to a vertical flux of disturbance vorticity. This vorticity flux can be re-written as a momentum flux divergence via

$$\overline{(u'_z - w'_x)w'} = \overline{u'_z w'} = (\overline{u'w'})_z - \overline{u'w'_z} = (\overline{u'w'})_z + \overline{u'u'_x} = (\overline{u'w'})_z. \tag{6.44}$$

The final outcome is the Eulerian–Lagrangian juxtaposition

$$\tilde{\mathsf{p}}_t + (\overline{u'w'})_z = 0 \quad \text{and} \quad \mathsf{p}_t - (\overline{\zeta'_x P'})_z = 0. \tag{6.45}$$

Thus, the vertical flux of $\tilde{\mathsf{p}}$ is equal to the advection-related vertical flux of horizontal momentum across a fixed line $z =$ const, which we recognize as the usual *Eulerian-mean momentum flux* $\overline{u'w'}$; it too is constant for a steady wave field. The Eulerian and Lagrangian mean momentum fluxes differ only by a time derivative, i.e., it follows from (6.38) that

$$(\overline{\zeta' u'})_t = \overline{u'w'} + \overline{\zeta'_x P'}. \tag{6.46}$$

For slowly varying wavetrains we have the generic expressions

$$\tilde{\mathsf{p}} = \mathsf{p} = \frac{k}{\hat{\omega}}\bar{E} \quad \text{and} \quad \overline{u'w'} = -\overline{\zeta'_x P'} = \frac{k}{\hat{\omega}}\bar{E} w_g. \tag{6.47}$$

On the other hand, for irrotational surface waves, which are not slowly varying in the vertical direction, $\tilde{\mathsf{p}} = 0$ whilst p does not. This illustrates that the divergence term in (6.38) can be important for vertically evanescent waves and also that the Lagrangian p is the more fundamental definition of pseudomomentum, because it applies equally to a broad range of waves.

6.3.2 Forcing and dissipation of pseudomomentum

We now consider the impact of forcing or dissipation on the zonal pseudomomentum budget. To this end we augment the two-dimensional Boussinesq equations via

$$\frac{D\boldsymbol{u}}{Dt} + \boldsymbol{\nabla} P - b\hat{\boldsymbol{z}} = \boldsymbol{F} \quad \text{and} \quad \frac{DS}{Dt} = R \tag{6.48}$$

where $\boldsymbol{F} = (F, G)$ is a body force per unit mass (of dissipative origin or otherwise) and R is a heat source term that affects the stratification S. For example, in the atmosphere R could represent diabatic heating or cooling due to radiative energy exchange processes.

The linear equations are changed to

$$\mathrm{D}_t\boldsymbol{u}' + \boldsymbol{\nabla}P' - b'\widehat{\boldsymbol{z}} = \boldsymbol{F}' \quad \text{and} \quad \mathrm{D}_t b' + N^2 w' = R'. \tag{6.49}$$

Repeating the steps that led to (6.45) we now obtain

$$\tilde{\mathsf{p}}_t + (\overline{u'w'})_z = \frac{1}{N^2}\overline{b'(F'_z - G'_x)} + \frac{1}{N^2}\overline{R'(u'_z - w'_x)} = \tilde{\mathcal{F}}. \tag{6.50}$$

Here $\tilde{\mathcal{F}}$ is a source term per unit mass that captures the instantaneous effect of the force curl and of the heating term on the zonal pseudomomentum. For example, in the so-called *Newtonian cooling* model the linear heating term is proportional to minus the buoyancy disturbance and we obtain

$$R' = -\alpha b' \quad \text{and} \quad \tilde{\mathcal{F}} = -\alpha\tilde{\mathsf{p}} \tag{6.51}$$

where $\alpha > 0$ is a constant damping rate per unit time. Newtonian cooling is a crude model for the radiative damping of temperature disturbances in the atmosphere and it obviously leads to a very simple decay law for $\tilde{\mathsf{p}}$.

The situation is slightly more awkward for the Lagrangian zonal pseudomomentum p, which evolves according to

$$\mathsf{p}_t - (\overline{\zeta'_x P'})_z = -\overline{\xi'_x F'} - \overline{\zeta'_x G'} - \overline{\zeta'_x b'} = \mathcal{F}. \tag{6.52}$$

In the first part of the source term $\mathcal{F}$ the same operation is applied to $\boldsymbol{F}'$ as was applied to $\boldsymbol{u}'$ in the definition of p, which is a natural and explicit representation of the Lagrangian pseudomomentum source due to $\boldsymbol{F}'$. On the other hand, the final term was zero previously because of $\zeta' = -b'/N^2$. This link is broken in the presence of heating because

$$\mathrm{D}_t(b' + N^2\zeta') = R'. \tag{6.53}$$

Therefore $-\overline{\zeta'_x b'} = -\overline{\zeta'_x(b' + N^2\zeta')} \neq 0$ does capture pseudomomentum changes due to heating, albeit only in an implicit way. This result is typical for Lagrangian definitions of pseudomomentum in the presence of diabatic processes: strictly speaking, a simple result such as (6.51) is now available only for slowly varying waves, in which case (6.53) can be approximately time-integrated. We note in passing that another option is a re-definition of $\boldsymbol{\xi}'$ such that both $b' = -\zeta'/N^2$ and $\boldsymbol{\nabla}\cdot\boldsymbol{\xi}' = 0$ are maintained in the presence of diabatic heating.[5]

[5] One such choice is $\mathrm{D}_t\boldsymbol{\xi}' = \boldsymbol{u}' + (\psi_z, -\psi_x)$ with $N^2\psi = \int_0^x R'(s, z, t)\,ds$.

6.4 Mountain lee waves and drag force

The classic example of internal gravity waves in the atmosphere is the formation of *lee waves* by air flow over mountains. Such lee waves can propagate vertically to large altitudes[6] and their associated momentum flux means that they are capable of contributing significantly to the angular momentum budget of the atmosphere. How to account for the high-altitude 'wave drag' associated with these lee waves in numerical models was one the first applications of wave–mean interaction theory in atmospheric science.

In the simplest model the air flow is a constant wind U and the mountain is a compact topography disturbance with a simple shape and maximal height H, say. In three dimensions, the flow can go either over the top of the mountain or it can avoid the mountain by flowing around it sideways. The first regime emits much stronger internal waves than the second, so it is important to understand how the regime is chosen. Broadly speaking, the presence of stratification as measured by the buoyancy frequency N tends to inhibit vertical motion because of the associated increase of available potential energy, and therefore the criterion for the flow regime is based on the non-dimensional *Froude number*

$$F = \frac{U}{HN}. \tag{6.54}$$

For small values of F the stratification is strong compared to the kinetic energy of the flow and the flow splits and avoids the mountain by horizontal motion. For large values of F the energy penalty to go over the top of the mountain is relatively small and the flow avoids the mountain by vertical motion. Typical values in the atmosphere are

$$N = 0.01\,\mathrm{s}^{-1} \quad \text{and} \quad U = 10\,\mathrm{ms}^{-1}, \tag{6.55}$$

so $U/N = 1\,\mathrm{km}$ and therefore for a mountain of height $H = 1\,\mathrm{km}$ the Froude number is $F = 1$.

Now, if the topography is modelled with linear theory then $H = O(a)$ whilst $(U, N) = O(1)$, which implies that $F \gg 1$ in linear theory. In other words, the regime of horizontal avoidance cannot be captured by linear theory. Also, in two-dimensional vertical slice models there is no possibility to avoid the mountain by sideways motion, regardless of the Froude number, so even nonlinear models in vertical slice geometry cannot access this regime. This can lead to a significant over-estimation of wave radiation based on such models.

[6] This is known to glider pilots, who exploit lee waves to soar to otherwise unreachable altitudes.

6.4.1 Linear lee waves in two dimensions

The linear lee wave problem with constant U and N is straightforward in principle, although in practice the correct choice of vertical wavenumber (real or imaginary) and of the dispersion relation branch can be confusing at first. We therefore give a bit more detail on this topic.

The slice domain is the xz-plane with $x \in [0, L]$ and $z \in [0, \infty)$. The topography and all flow fields are assumed to be periodic in x with finite period length L. The restriction to finite L is not essential for the $O(a)$ wave problem, but it is essential for the $O(a^2)$ mean-flow response problem that follows. For concreteness, we can think of x as the zonal, east–west coordinate in the atmosphere. The $O(1)$ basic flow is $\boldsymbol{u} = (U, 0)$ with constant $U > 0$ and undisturbed stratification $S = N^2 z$ with constant N. The $O(a)$ linear equations can be combined to yield

$$\mathrm{D}_t \mathrm{D}_t \nabla^2 w' + N^2 w'_{xx} = 0 \tag{6.56}$$

for w' or any other wave field. For a stationary wave field $\partial_t = 0$ and hence $\mathrm{D}_t = U\partial_x$, which leads to

$$\nabla^2 w' + \frac{N^2}{U^2} w' = 0. \tag{6.57}$$

The topography is described by a bottom elevation $h'(x)$ such that the impermeable solid ground lies at $z = h'(x)$, with $h' = O(a)$ small, leading to the linearized boundary condition

$$w'(x, 0) = \mathrm{D}_t h' = U h'_x(x) \quad \Leftrightarrow \quad \zeta'(x, 0) = h'(x). \tag{6.58}$$

The useful second form uses $w' = \mathrm{D}_t \zeta'$; it implies that the topography shape coincides with the lowermost stratification surface $S = 0$.

We write the topography as a Fourier series in x via

$$h'(x) = \frac{1}{L} \sum_{n=-\infty}^{+\infty} \hat{h}(k) \exp(ikx) \quad \text{and} \quad \hat{h}(k) = \int_0^L e^{-ikx} h'(x)\, dx \tag{6.59}$$

where $k = 2\pi n/L$. For real $h'(x)$ we have $\hat{h}^*(-k) = \hat{h}(k)$, but for now we simply compute the solution for a generic component $\hat{h}(k)\exp(ikx)$. We also note Parseval's theorem, which states that for any two complex functions $A(x)$ and $B(x)$ the equality

$$\int_0^L A^* B \, dx = \frac{1}{L} \sum_{n=-\infty}^{+\infty} \hat{A}^* \hat{B} \tag{6.60}$$

holds. Now, assuming no wave reflection in the vertical, the solution for each Fourier component of ζ' consists of a single vertical mode of the form

$$\hat{\zeta}(k,z) = e^{imz}\hat{\zeta}(k,0) = e^{imz}\hat{h}(k). \tag{6.61}$$

It remains to find m from the dispersion relation (6.19):

$$\omega = 0 \quad \Leftrightarrow \quad \hat{\omega} = -Uk = \pm N\frac{k}{\sqrt{k^2+m^2}}. \tag{6.62}$$

Notably, Doppler-shifting maps the single absolute frequency $\omega = 0$ into many intrinsic frequencies $\hat{\omega} = -Uk$. If $U > 0$ then the lower branch in (6.62) *must* be chosen, and we obtain

$$\kappa = \frac{N}{U} \quad \text{or} \quad m^2 = \frac{N^2}{U^2} - k^2. \tag{6.63}$$

If $k^2 < N^2/U^2$ then this yields a real m and vertically propagating waves and otherwise it yields an imaginary m and vertically trapped, or *evanescent waves*. In the unstratified case $N = 0$ and all waves are evanescent, which recovers the familiar result of irrotational flow over topography. For $N > 0$ and fixed U there is a finite range of zonal wavenumbers k that lead to propagating waves. Using (6.55), this criterion limits wave propagation to zonal wavelengths larger than 6 km or so. The same restriction applies to the vertical wavelength $2\pi/m$ as well.

So far, (6.63) fixes the magnitude of m but not its sign. In the propagating case the sign of m is chosen by the radiation condition $w_g > 0$. For the lower branch in (6.21) this implies that k and m have the same sign, i.e., the lines of constant wave phase have a negative slope in the xz-plane. Graphically, the wave crests are tilting against the basic flow $U > 0$. Thus we find that for propagating waves

$$k^2 < \frac{N^2}{U^2} \quad \Rightarrow \quad m = \operatorname{sgn}(k)\sqrt{\frac{N^2}{U^2} - k^2}. \tag{6.64}$$

In the evanescent case the wave field must decay with altitude z and therefore the imaginary part of m must be positive regardless of the sign of k. Thus for trapped waves

$$k^2 > \frac{N^2}{U^2} \quad \Rightarrow \quad m = i\sqrt{k^2 - \frac{N^2}{U^2}}, \tag{6.65}$$

which completes the solution in terms of Fourier modes. Some lee waves patterns are computed numerically and illustrated in figure 6.1.

Several interesting aspects of the lee wave solution can be pointed out.

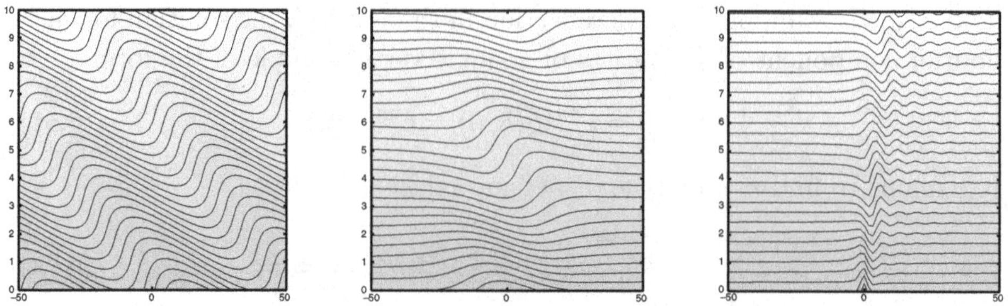

Figure 6.1 Stratification lines S = const for lee waves with topography height $H = 0.75\,\text{km}$, $N = 0.01\,\text{s}^{-1}$ and $U = 10\,\text{ms}^{-1}$. The size of the domain is 100 km in the horizontal and 10 km in the vertical (the numerical domain is 800 km in the horizontal to reduce periodicity effects). Left: sinusoidal topography with 50 km wavelength. This shows the characteristic sloping against the wind of the wave phase pattern. Middle: centred Gaussian topography with width 10 km. The wave pattern is essentially hydrostatic as described in §6.4.2. Right: centred Gaussian topography with width 1 km. The narrower topography excites non-hydrostatic waves, which travel downstream as described by (6.67).

First, (6.63) implies that $\kappa = |\boldsymbol{k}|$ is the same for all waves, i.e., all propagating plane waves have the same two-dimensional wavelength. Second, the absolute group velocity is

$$u_g = U + \hat{u}_g = \frac{U^3}{N^2}k^2 \quad \text{and} \quad w_g = \hat{w}_g = \frac{U^3}{N^2}|k|\sqrt{\frac{N^2}{U^2} - k^2}. \tag{6.66}$$

The absolute zonal group velocity is always positive, which explains the name 'lee wave', because for a localized mountain the waves are found downstream. Specifically, the maximal $u_g = U$ is reached at the propagation threshold $|k| = N/U$; in this limit $u_g - U$ and the wave is simply advected by the wind. On the other hand, the maximal vertical group velocity is $w_g = U/2$ and it is reached at the intermediate wavenumber $|k| = N/(\sqrt{2}U)$, whilst $w_g = 0$ at both end points of $|k| \in [0, N/U]$. The fact that the magnitude of $\boldsymbol{c}_g$ is $O(U)$ is relevant to numerical simulations of lee waves as initial-value problems, because it makes clear that weaker winds U require longer integration times until the wave pattern has established itself.

These group-velocity expressions can be used for asymptotic ray tracing computations as in §4.1.3 (with large time replaced by large z). For instance, if a localized mountain is centred at $x = 0$ then ray tracing predicts that the zonal wavenumber travels along lines of constant slope z/x such that for $x > 0$ and large $z > 0$

$$\frac{z}{x} = \frac{w_g}{u_g} = \sqrt{\frac{N^2}{k^2U^2} - 1} \quad \Rightarrow \quad k(x,z) = \pm\frac{N}{U}\sqrt{\frac{x^2}{x^2+z^2}}. \tag{6.67}$$

This differs from (6.22) because here the absolute, as distinct from intrinsic, group velocities are computed. It shows that over the mountain the horizontal wavelength is larger than further downstream, see figure 6.1. Ray tracing fails in a caustic region directly over the mountain where $x \approx 0$ and the local solution in this caustic region is well captured by the explicit hydrostatic solution in §6.4.2 below. Also, for any fixed z/x the zonal wavenumber increases if U is lowered, i.e., at a fixed observing point the zonal wavelength decreases as U decreases. This illustrates a non-uniform convergence property of this linear wave problem: the steady solution as $U \to 0$ differs substantially from the zero solution for $U = 0$. This is explained in physical terms by the longer and longer waiting times that are needed before the transient wave systems settles to a steady state as $U \to 0$.

Finally, we compute the usual wave properties from the linear solution. This is particularly simple because phase-averaging is equivalent to zonal averaging for Fourier modes on a periodic domain, and the contributions of different Fourier components to the quadratic wave properties then simply add because of orthogonality and Parseval's theorem (6.60). For propagating waves this leads to

$$\bar{E} = N^2 \overline{h_p'^2}, \quad \mathsf{p} = \tilde{\mathsf{p}} = -\frac{N^2}{U}\overline{h_p'^2}, \quad \text{and} \quad \mathsf{e} = 0. \tag{6.68}$$

Of course, near the topography there are additional vertically decaying contributions to $\bar{E}$ and p from the evanescent waves.

Notably, the energy density is the same for *any* $U > 0$, even though $\bar{E} = 0$ if $U = 0$, as then there are no waves. The pseudoenergy $\mathsf{e} = \omega\bar{E}/\hat{\omega} = \bar{E} + U\mathsf{p}$, on the other had, is always zero because $\omega = 0$. This is related to the fact that in the current reference frame the mountain is fixed and therefore it does no work on the fluid. Conversely, in a frame moving with the wind U the basic fluid is at rest but the mountain would be moving backwards with velocity $-U$, and in this frame the mountain would indeed do work on the fluid and the pseudoenergy would then be $\mathsf{e} = \bar{E}$. This illustrates the frame-dependence of e and hints at the potential confusion caused by explaining wave–mean interactions based on energy budgets.

6.4.2 Hydrostatic solution using Hilbert transforms

It is advantageous to note a simple real space solution that is valid in the *hydrostatic wave regime*. In this regime $\mathrm{D}_t w'$ is negligible compared to P_z' or b' and therefore $\mathrm{D}_t w' + P_z' = b'$ is replaced by the *hydrostatic relation*

$$P_z' = b'. \tag{6.69}$$

Even though this removes a time derivative from the governing equations it is easy to check that the basic structure of the initial-value problem with two gravity-wave modes has not changed.[7] An equivalent procedure to (6.69) is to assume that $k^2 \ll m^2$ and then keeping only the first-order terms in k/m. Physically, this corresponds to low-frequency gravity waves with nearly horizontal particle motions, which indeed satisfy (6.69). The hydrostatic dispersion relation is

$$\hat{\omega} = \pm N \frac{k}{|m|}. \tag{6.70}$$

Clearly, (6.70) is accurate for small $|k/m|$, but misses the finite frequency bound $\hat{\omega} \leq N$ for large $|k/m|$. For hydrostatic lee waves this implies that there are no evanescent waves.

The hydrostatic approximation is accurate if the spectral support of h' is confined to the hydrostatic range of horizontal wavenumbers. For example, with N and U given by (6.55) this limits $|k|$ to be less than $1\,\text{km}^{-1}$, which for a Gaussian mountain with half-width R means that $R > U/N = 1\,\text{km}$. On Earth, many mountains satisfy this scale constraint, so lee waves are often modelled using the hydrostatic approximation.

Returning to (6.70), the vertical wavenumber is now

$$m = \text{sgn}(k)\frac{N}{U}, \tag{6.71}$$

which does not depend on $|k|$. This leads to

$$\zeta'(x,z) = \zeta'(x,0)\cos\left(\frac{Nz}{U}\right) + \frac{U}{N}\zeta'_z(x,0)\sin\left(\frac{Nz}{U}\right). \tag{6.72}$$

We know $\zeta'(x,0) = h'(x)$ but still need to find $\zeta'_z(x,0)$. Using Fourier components and (6.61) we have

$$\hat{\zeta}_z(k,z) = im\,\hat{\zeta}(k,z) = i\text{sgn}(k)\frac{N}{U}\,\hat{\zeta}(k,z). \tag{6.73}$$

We recall from §3.2.5 that $-i\text{sgn}(k)$ is the Fourier symbol of a Hilbert transform and hence (6.73) is equivalent to

$$\zeta'_z(x,z) = -\frac{N}{U}\,\tilde{\zeta}(x,z) \tag{6.74}$$

[7] This example serves as an antidote to the simplistic rule that the number of modes in an initial-value problem should always equal the number of time derivatives in the equations, or that number minus the number of constraints.

where $\tilde{\zeta}$ is the Hilbert transform in x of ζ'. This leads to the explicit real-space solution

$$\zeta'(x,z) = h'(x)\cos\left(\frac{Nz}{U}\right) - \tilde{h}(x)\sin\left(\frac{Nz}{U}\right). \tag{6.75}$$

Thus, as a function of z the solution oscillates between two fixed patterns given by $h'(x)$ and its Hilbert transform $\tilde{h}(x)$, cf. figure 6.1. The first occurrence of $\zeta' = -\tilde{h}$ takes place at $z = \pi U/(2N)$, which is about 1.5 km based on (6.55). There is no net drift in the downstream direction, because the horizontal group velocity is zero in the hydrostatic approximation. Indeed, this solution complements the ray tracing solution (6.67) in the narrow caustic zone $x \approx 0$ directly above the topography.

Despite the absence of a net drift, the hydrostatic wave field can be considerably wider in x than the original topography, and might also encroach into the upstream region. For example, for a mountain with compact $h'(x)$ the Hilbert transform $\tilde{h}(x)$ decays only as $1/x$ with distance from the origin. Thus the horizontal extent of the wave field waxes and wanes with altitude.

6.4.3 Drag force and momentum flux

The drag force D is the zonally averaged *horizontal* force exerted on the mountain by the fluid. It is clear that without viscous forces such a horizontal force can only be due to correlations between the surface pressure field and the slope of the topography, and we seek to compute this force correct to $O(a^2)$. As the topography h' and its slope are $O(a)$ this means we need to know the surface pressure $p_s(x)$, say, correct to $O(a)$. Thus at leading order the horizontal drag D is the product of two $O(a)$ fields, which shows that D is a wave property. This is not a trivial fact because wave-induced forces on bodies need not be wave properties in general. For example, the *vertical* force on the mountain is not a wave property, because it requires knowing the pressure field at $O(a^2)$.[8]

Now, p_s at $O(a)$ is P' (scaled by the reference density) evaluated at $z=0$ plus a second term due to the $O(1)$ hydrostatic pressure $\tilde{P}_0(z)$ evaluated at the undulating topography $z = ah'$. A first-order Taylor expansion around $z=0$ yields a correlation with the topography slope of the form $\tilde{P}_{0z}(0)\,\overline{h'h'_x}$ for this term, which is zero by zonal averaging. Thus the hydrostatic pressure field does not contribute to the horizontal drag.

Using $\zeta' = h'$ at $z=0$ we define the mean drag (also called the *form*

[8] This will be computed in §6.5.2 below, so you can test your intuition now by guessing whether the $O(a^2)$ vertical force on the mountain will be positive or negative.

stress) by[9]

$$D = \overline{\zeta'_x P'} \quad \text{at } z = 0 \tag{6.76}$$

where the bar denotes zonal averaging and the total drag force in a periodic domain is DL. This is equal to minus the vertical flux of Lagrangian pseudomomentum p, which has the advantage that D can also be evaluated inside the fluid body, where it quite generally corresponds to the downward flux of horizontal momentum across an undulating material contour. The drag at the fluid bottom is then merely a special case of this Lagrangian-mean momentum flux. Moreover, (6.45) makes clear that $-D$ is also the net flux of Lagrangian pseudomomentum p from the topography into the fluid.

For a single Fourier mode of the steady lee wave problem we find that

$$P' = imU^2\zeta' \quad \text{and} \quad \zeta'_x = ik\zeta' \tag{6.77}$$

and therefore P' and ζ'_x are in phase if m is real, which is the propagating wave case. On the other hand, for a trapped wave the drag is zero. Specifically, for a propagating plane wave with $\zeta'(x,0) = h'(x) = h_0 \cos kx$ we obtain

$$D(k,U) = kmU^2\overline{h'^2} = UN|k|\sqrt{1 - \frac{k^2U^2}{N^2}}\frac{h_0^2}{2}. \tag{6.78}$$

There is a neat independent check on this formula: in a frame moving with the basic flow the mountain moves with speed $-U$ to the left, and according to elementary mechanics it then does work on the fluid at a rate DU. In this frame $\bar{E}$ is the leading-order energy density and therefore we have the energy flux balance $DU = \bar{E}w_g$, which after using (6.66) agrees with (6.78).

Thus $D(k,U)$ has the same sign as U and as a function of U for fixed k it grows linearly from zero, reaches its maximum

$$\max_{U>0} D(k,U) = N^2\frac{h_0^2}{4} \tag{6.79}$$

at $U^2 = \frac{1}{2}N^2/k^2$, and then drops to zero sharply at the propagation threshold $U^2 = N^2/k^2$. By orthogonality, the drag due to an arbitrary $h'(x)$ is the sum over the individual Fourier contributions. Indeed, Parseval's theorem (6.60) implies

$$\overline{\zeta'_x P'} = \frac{1}{L}\int_0^L \zeta'_x P'\,dx = UN\frac{2}{L^2}\sum_{0<k<N/U} |\hat{h}|^2 k\sqrt{1 - \frac{k^2U^2}{N^2}}. \tag{6.80}$$

[9] To obtain a drag with the dimensions of a momentum flux this definition needs to be multiplied by the reference density of the Boussinesq system.

Here the summation is over positive wavenumbers only; the negative wavenumbers make an equal contribution because of the reality condition and this leads to the factor two. Consistent with (6.78), for a plane wave with $h'(x) = h_0 \cos kx$ the only non-zero entry is $\hat{h}(k) = h_0 L/2$. Notably, combining (6.79) and (6.80) leads to a bound on the wave drag, namely

$$D = \overline{\zeta'_x P'} \leq \frac{N^2}{2}\overline{h'^2}. \tag{6.81}$$

In the hydrostatic approximation the drag has the simpler form

$$D = UN\frac{2}{L^2}\sum_{0<k}|\hat{h}|^2 k \tag{6.82}$$

Moreover, there is also an amusing peculiarity in the case of compact topography. If L is much larger than the support of the compact topography then the total drag is

$$DL = UN\frac{1}{\pi}\int_0^\infty |\hat{h}|^2 k\,dk \tag{6.83}$$

where $\hat{h}(x)$ is the usual Fourier *transform* of $h'(x)$, which is defined by (6.59b) with $L \to \infty$; in the inverse formula summation is then replaced by integration and $1/L$ is replaced by $dk/2\pi$. Now, (6.83) has a scaling symmetry such that if $h'(x)$ is replaced by $h'(x/\alpha)$ for any real $\alpha > 0$ then the integral remains unchanged.[10] In other words, the hydrostatic drag on an isolated topography feature is the same for any stretched version of the same topography.

Finally, for a steady wave field there is a simple relation between the shape-hugging form stress D and the Eulerian-mean momentum flux $\overline{u'w'}$ across a fixed level $z = $ const. The steady linear equations imply $Uu' = -P'$ and $U\zeta'_x = w'$ and therefore

$$D = \overline{\zeta'_x P'} = -\overline{u'w'}. \tag{6.84}$$

Thus a positive wave drag on the mountain goes hand-in-hand with a negative wave-induced vertical flux of horizontal momentum and $-D$ is also the the net flux of Eulerian pseudomomentum $\tilde{\mathsf{p}}$ from the topography into the fluid.

6.5 Mean-flow response

We now turn to computing the $O(a^2)$ mean-flow response to the $O(a)$ lee waves. The key question is where the horizontal force that is exchanged (via

[10] This follows from $\hat{h}(k) \to \alpha\hat{h}(\alpha k)$ if $h'(x) \to h'(x/\alpha)$.

the form stress D) between the fluid and the topography manifests itself in terms of zonal accelerations within the fluid body. A reasonable first guess would be that $D > 0$ should lead to an equal-and-opposite negative drag force exerted on the fluid layers near the topography, but this turns out to be far from the truth.

6.5.1 Eulerian-mean equations

The exact mean-flow equations follow by applying the zonal averaging operator to the Boussinesq equations, which involves manipulations based largely on (5.15) that are familiar from the standard Reynolds-averaging for turbulent shear flows. In the present two-dimensional case the resulting equations are disarmingly simple. This is mostly due to the linearity of the incompressibility condition, which upon averaging yields

$$\overline{u}_x + \overline{w}_z = 0 \quad \Rightarrow \quad \overline{w}_z = 0. \tag{6.85}$$

The lower boundary condition prohibits the fluid to move vertically as a whole and therefore (6.85) implies $\overline{w} = 0$, which is a great simplification. Based on this, the mean equations for $(\overline{u}, \overline{b}, \overline{P})$ are

$$\overline{u}_t + \left(\overline{uu}\right)_x + \left(\overline{uw}\right)_z + \overline{P}_x = 0 \Rightarrow \overline{u}_t = -\left(\overline{u'w'}\right)_z \tag{6.86}$$

$$\overline{w}_t + \left(\overline{wu}\right)_x + (\overline{w^2})_z + \overline{P}_z - \overline{b} = 0 \Rightarrow \overline{P}_z - \overline{b} = -(\overline{w'^2})_z \tag{6.87}$$

$$\text{and} \quad \overline{b}_t + \left(\overline{bu}\right)_x + \left(\overline{bw}\right)_z + N^2\overline{w} = 0 \Rightarrow \overline{b}_t = -\left(\overline{b'w'}\right)_z. \tag{6.88}$$

The dynamics of both $\overline{u}$ and $\overline{b}$ is determined by the divergence of their respective disturbance or *eddy fluxes* whilst $\overline{P}$ is given diagnostically. In turbulence theory one needs to devise an ad hoc closure procedure at this stage in order to relate the eddy fluxes to the mean fields.

In small-amplitude wave theory, on the other hand, the leading-order mean-flow equations are systematically obtained by evaluating the eddy fluxes using the linear wave solutions, i.e.,

$$\overline{u_{2t}} = -\left(\overline{u_1'w_1'}\right)_z \quad \text{and} \quad \overline{b_{2t}} = -\left(\overline{b_1'w_1'}\right)_z \tag{6.89}$$

using expansion subscripts for clarity. Not only does this lead to a concrete answer in any particular lee wave problem, but we can also use the linear equations to seek further simplifications, as follows.

6.5.2 Mean buoyancy and pressure response

Using the linear equations the buoyancy flux can be written as a time derivative, namely

$$\overline{b_1' w_1'} = -\frac{1}{N^2}\overline{b_1' \mathrm{D}_t b_1'} = -\frac{1}{2N^2}(\overline{b_1'^2})_t = -\frac{N^2}{2}(\overline{\zeta_1'^2})_t \tag{6.90}$$

and therefore (6.89b) can be time-integrated to yield

$$\overline{b_2} = \frac{N^2}{2}(\overline{\zeta_1'^2})_z. \tag{6.91}$$

For example, in the case of an evanescent lee wave this equation shows that $\overline{b_2} < 0$ whilst in a steady propagating wave $\overline{b_2} = 0$.

There is a simple kinematic explanation for the sign of the buoyancy flux in (6.90) that follows from the material invariance of S together with the vertical stratification gradient of the basic state, which is $\boldsymbol{\nabla} S = N^2\hat{\boldsymbol{z}}$. We consider a material contour $\mathcal{C}$ that in the basic state coincides with the flat horizontal line $z = z_0$, say. Given $S(\boldsymbol{x}, t)$, the material invariance of S then implies that $\mathcal{C}$ is defined by

$$\boldsymbol{x} \in \mathcal{C} \quad \Leftrightarrow \quad S(\boldsymbol{x}, t) = N^2 z_0. \tag{6.92}$$

We now imagine that this material contour is perturbed from rest, i.e., it becomes increasingly undulated such that undulation measures such as $\overline{\zeta'^2}(z_0, t)$ grow in time. To fix ideas, we may imagine that this undulation is due to a vertically propagating transient internal wavepacket that is incident from below at the level $z = z_0$.

The key observation is that an increasing undulation of the material contour implies a *negative* (i.e., downward) buoyancy flux across $z = z_0$ as quantified by (6.90). The kinematic reason for this flux becomes clear once we combine the material advection of S with the fact that all fluid particles below $\mathcal{C}$ satisfy $S < N^2 z_0$ and vice versa for the fluid particles above the same contour. Thus, the fluid particles that travelled upwards and are now lying above $z = z_0$ but below $\mathcal{C}$ carry lower values of S than the fluid particles that went the opposite, downward way. The robust correlation between upward motion and negative buoyancy disturbance means that a growing undulation corresponds to a net *down-gradient* transport of buoyancy across $z = z_0$. This downward flux of buoyancy corresponds to an upward flux of mass.

This kinematic effect can also be applied to other material tracers in the presence of a basic-state gradient, such as the PV in the presence of a β-effect, for instance. This is relevant to growing Rossby waves, which are therefore associated with a down-gradient flux of PV. This robust kinematic

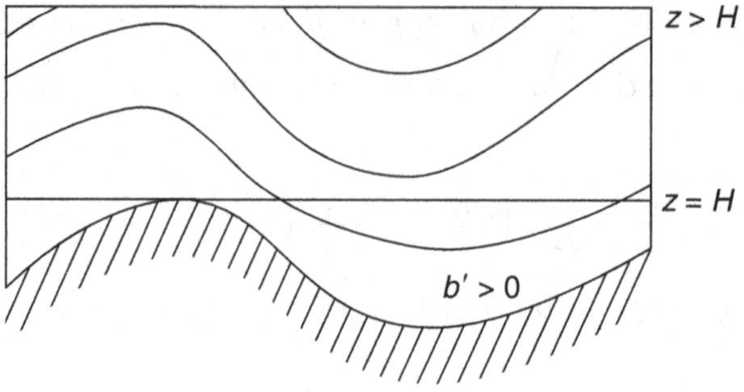

Figure 6.2 Topography with height H and lowermost stratification lines in the lee wave problem. For steady waves the Eulerian-mean flux of vertical momentum across any flat line $z \geq H$ equals the Lagrangian-mean flux of vertical momentum across the undulating topography plus the integral of buoyancy acceleration b' over the enclosed area between the topography and the flat line. As indicated, this integral is positive as $z \to H$ from above, because $b' > 0$ in the trough of the topography. At $O(a^2)$ the Eulerian flux is (6.94) and the Lagrangian flux is (6.95). The sketch also makes clear that the centre of volume of the fluid colum is lowered due to undulating topography.

effect also carries over to finite-amplitude waves, although it may fail when material contours overturn, as would occur in nonlinear wave breaking.

We can now compute the mean pressure response from (6.87) as

$$\overline{P_2} = \frac{N^2}{2}\overline{\zeta_1'^2} - \overline{w_1'^2} = \left(\frac{N^2}{2} - \hat{\omega}^2\right)\overline{\zeta_1'^2}, \tag{6.93}$$

which shows that $\overline{P_2}$ can have either sign depending on the size of $\hat{\omega}$. The $O(a^2)$ Eulerian-mean vertical flux of vertical momentum across a fixed line of constant z is the sum of this pressure term plus an advective term:

$$\overline{P_2} + \overline{w_1'^2} = +\frac{N^2}{2}\overline{\zeta_1'^2}. \tag{6.94}$$

This is positive because the stratification undulations have led to a net upward displacement of mass across any fixed line $z =$ const, which means that a larger momentum flux is needed to support the increased weight of the fluid column.

It is tempting to think that (6.94) is also equal to the $O(a^2)$ downward *vertical* force that the fluid exerts on the topography. After all, it turned out that the *horizontal* form stress D for a steady wave field was equal to the Eulerian momentum flux $-\overline{u'w'}$ at leading order. However, there is a subtlety here because vertical momentum is not conserved, i.e., there is a source term due to buoyancy. This implies that for steady waves the Lagrangian-mean, shape-hugging vertical-momentum flux due to pressure at the topography and the Eulerian-mean vertical-momentum flux due to pressure and advection at fixed z differ by the buoyancy of the enclosed fluid mass, as illustrated in figure 6.2. Moreover, the $O(1)$ hydrostatic pressure $\tilde{P}_0(z)$, which does not enter explicitly the Boussinesq equations in the form used here, also matters at $O(a^2)$ for the vertical-momentum flux when integrated over the undulating topography.

A concrete intuitive formula for the $O(a^2)$ vertical-momentum flux across the topography is $\overline{P_2} + \overline{P'_{1z}\zeta'_1} + \frac{1}{2}\overline{\zeta_1'^2}\tilde{P}_{0zz}$ evaluated at $z = 0$, which can be derived from the exact relations in Lagrangian-mean theory in §10.2.6. The last term arises from the second-order Taylor expansion of $\tilde{P}_0(a\zeta'_1 + a^2\zeta'_2)$; there is also an unkown term $\tilde{P}_{0z}\zeta'_2$, but it averages to zero.

The $O(1)$ and $O(a)$ equations together with $\tilde{P}_{0z} = S_0 = N^2 z$ then yield

$$\overline{P_2} + \overline{P'_{1z}\zeta'_1} + \frac{1}{2}\overline{\zeta_1'^2}\tilde{P}_{0zz} = \overline{P_2} + \overline{w_1'^2} - \frac{N^2}{2}\overline{\zeta_1'^2} - \frac{1}{2}(\overline{\zeta_1'^2})_{tt}, \tag{6.95}$$

which quantifies the difference between the Lagrangian and Eulerian fluxes. For steady lee waves we therefore obtain

$$\overline{P_2} + \overline{P'_{1z}\zeta'_1} + \frac{1}{2}\overline{\zeta_1'^2}\tilde{P}_{0zz} = 0 \tag{6.96}$$

for the $O(a^2)$ downward vertical force on the topography. In hindsight this is obvious, as the topography simply supports the total weight of the fluid column, which has not changed because of the undulations.

As an aside, we consider the intriguing question of what would happen if the hydrostatic pressure $\tilde{P}_0$ were omitted in this computation. Clearly, this would lead to a 'suction' force on the mountain equal to $-\frac{1}{2}N^2\overline{\zeta_1'^2}$. The reason for this is that for steady waves the integral of P over the topography equals minus the integral of b over the fluid column (this assumes no wave motion as $z \to \infty$ for simplicity). Now, with $b = S - N^2 z$ and materially conserved S, changes in minus the total buoyancy integral are proportional to changes in altitude of the centre of volume of the fluid, which is *lowered* in the presence of undulating zero-mean topography (cf. figure 6.2 and also figure 10.3). In other words, the fluid column gains total buoyancy if the topography is slowly grown from zero, and this buoyancy leads to the spurious reduction in vertical force that is computed if $\tilde{P}_0$ is omitted.

This specific example illustrates a couple of general principles. First, the mean vertical flux of horizontal momentum is a wave property and easy to compute, whereas the mean vertical flux of vertical momentum is not a wave property and its correct computation requires great attention to detail such as the inclusion of the hydrostatic pressure field. Second, the differences between Eulerian and Lagrangian fluxes are acutely important if there are source terms such as the buoyancy in the present case of vertical momentum. This also applies in the case of horizontal momentum in a rotating frame, where Coriolis forces then play the role of the source terms. As we shall see later, this leads to the celebrated formula for the *Eliassen–Palm flux*, whose vertical component is the Lagrangian-mean vertical flux of zonal momentum in a rotating system.

6.5.3 Zonal mean-flow response

Somewhat surprisingly, as long as we neglect dissipation the leading-order zonal mean-flow response in (6.89a) can be written as a time derivative based on the Eulerian pseudomomentum in (6.45), which yields

$$\overline{u}_{2t} = -(\overline{u_1'w_1'})_z = \tilde{\mathsf{p}}_{2t} \quad \Rightarrow \quad \overline{u}_2 = \tilde{\mathsf{p}}_2 = \overline{\zeta_1'(w_{1x}' - u_{1z}')}. \tag{6.97}$$

Thus the zonal mean-flow response simply equals the zonal pseudomomentum. For example, for lee waves this leads to

$$\overline{u}_2 = \tilde{\mathsf{p}}_2 = -\frac{N^2}{U}\,\overline{h_p'^2} \tag{6.98}$$

where h_p' is the projection of the topography onto propagating waves as before (this ignores the vertically decaying pseudomomentum of evanescent waves). This shows that the sign of the mean-flow response $\overline{u}_2$ is always opposite to the sign of U, i.e., the mean-flow response retards the basic wind. This is consistent with the general notion of a stable flow situation because the sense of the mean-flow response acts to reduce the flow speed.

It is easy interpret (6.97) based on the discussion of Kelvin's circulation theorem in §6.3.1. For example, if $\Gamma(\mathcal{C})$ is constant in time and equal to its basic-state value LU then (6.41) immediately implies (6.97). This makes it obvious that (6.97) is a very robust result, as long as we continue to neglect dissipation.

6.5.4 Mass, momentum and energy budgets

It proves instructive to examine these global fluid budgets in the lee wave problem. This problem is simple enough that no ambiguities arise as to how to interpret these budgets, yet it is complex enough that it can illustrate the features that are essential to understand more complicated problems, especially those involving shear flows.

To this end we consider a causal initial-value problem, i.e., we assume that the flow started with the undisturbed basic state and that at $t = 0$ the infinitesimal topography is smoothly grown from zero to its terminal height, which it reaches at some $t = \Delta t$, say. For simplicity, we consider a monochromatic topography $h' = h_0 \sin kx$ where k falls in the range of propagating wavenumbers and h_0 varies smoothly from zero to its final value at $t = \Delta t$.

The linear wave field at times $t \gg \Delta t$ then consists of three distinct regions that are demarcated by the wave front[11] position

$$z_f = w_g t \tag{6.99}$$

where w_g is the group velocity belonging to k. In the near-field region $z \ll z_f$ and the wave field is steady and given by the steady state relations in §6.4.1. In the far-field region $z \gg z_f$ the wave has not yet arrived and hence the flow is just the basic flow. Finally, in the region $z \approx z_f$ (which has width proportional to $w_g \Delta t$) there is a smooth transition between the other two regions.

The structure of the mean-flow response is

$$z \gg z_f: \quad \overline{b_2} = 0 \quad \text{with} \quad \overline{u}_2 = 0 \tag{6.100}$$

and

$$z \ll z_f: \quad \overline{b_2} = 0 \quad \text{with} \quad \overline{u}_2 = \tilde{\mathsf{p}}_2 = \frac{k\bar{E}_2}{\hat{\omega}} = -\frac{\bar{E}_2}{U}. \tag{6.101}$$

In the transition region surrounding the wave front, $\overline{u}_2$ varies monotonically between these two values, whereas $\overline{b_2}$ makes an excursion to negative values described by (6.91). This negative excursion, which corresponds to a positive density anomaly, accounts for the increased weight of the fluid column above a fixed control level $z =$ const that was described by (6.94). Moreover, this positive density anomaly at the wave front occurs at an altitude z_f that increases secularly with time, which makes it obvious that the centre of gravity of the fluid is continually rising, which is consistent with the secular increase in total available potential energy that arises because the wavetrain covers more and more of the fluid.

We now turn to the zonal momentum budget. First, in this fluid system there is no ambiguity about the statement that the mean fluid momentum resides in the mean flow.[12] This is simply because the Boussinesq momentum density u is linear in the velocity. Of course, this is not the case in other fluid system such as the shallow-water equations, where the momentum density is hu and therefore $\overline{hu} = \overline{h}\overline{u} + \overline{h'u'}$ has both a mean and a disturbance part. We note in passing that maintaining the rule that mean fluid momentum should

[11] This use of the term 'wave front' differs from the use in terms of wave phase in §2.1.3.

[12] An exception is the detailed distribution of fluid momentum in the troughs of the topography (cf. figure 6.2), where u' can correlate with h' and can hence contribute to the mean momentum in a manner that is familiar from surface waves. However, for propagating internal waves u' is out of phase with h' and this contribution to the mean momentum is zero.

reside unambiguously in the mean flow is one of the goals of Lagrangian-mean theory.

Now, per unit time, the fluid loses zonal momentum as well as zonal pseudomomentum at a rate equal to the drag $D = -\tilde{\mathsf{p}}_2 w_g$ at the lower boundary and this loss is precisely accounted for in the vertical integral of $\overline{u}_2$, which grows in time proportional to w_g. In this way the momentum budget for the fluid is closed, and Newton's law of action equals reaction holds between the drag force on the mountain and the equal-and-opposite mean-flow deceleration of the fluid at the wave front. It is noteworthy that the wave front is found further and further away from the mountain, which illustrates the generic non-local nature of the momentum budget when wave propagation over large distances is concerned. This is worth repeating: it is *not* true that the wave drag is felt by the mean flow in the vicinity of the mountain, rather it is felt at the remote wave front (4.1).

Finally, we consider the energy budget. Unlike the momentum density, the energy density in (6.6) is quadratic in both $|\boldsymbol{u}|$ and b and therefore the mean energy has *both* mean and disturbance parts according to

$$\frac{1}{2}\left(\overline{u}^2 + \frac{\overline{b}^2}{N^2}\right) + \frac{1}{2}\left(\overline{|\boldsymbol{u}'|^2} + \frac{\overline{b'^2}}{N^2}\right). \tag{6.102}$$

At leading order, the disturbance part is the wave energy $a^2\bar{E}_2$. At first sight, it then seems as if the energy density in the near field should be higher than in the far field, because the wave energy density is non-zero only in the former. However, this neglects the impact of the $O(a^2)$ mean-flow changes on the first part of the energy density in (6.102). This matters in the kinetic energy part because the basic zonal velocity is $O(1)$ and therefore the complete $O(a^2)$ change in the mean energy density follows from $\overline{u} = U + a^2\overline{u}_2$ as

$$a^2(U\overline{u}_2 + \bar{E}_2) = a^2(U\tilde{\mathsf{p}}_2 + \bar{E}_2) = a^2\tilde{\mathsf{e}}_2 = 0. \tag{6.103}$$

This shows the surprising fact that the total mean energy density of the fluid does not change in the lee wave problem. Rather, there is a local conversion of mean-flow energy into disturbance energy that occurs in the transition region surrounding the wave front.

Put another way, the expanding wavetrain extracts the necessary energy from the mean flow at the wave front, and not from the wave source at the ground. Indeed, there is no energy flux across the fluid boundary because the topography is fixed and cannot do work on the fluid. As noted before, this statement is sensitive to the frame of reference, and a strikingly different version of the energy budget is obtained in the fluid rest frame, in which the fluid is at rest in the basic state, the topography is moving with speed U to the left, and the $O(a^2)$ energy density is entirely due to the wave energy. In this frame of reference the wave energy is then indeed provided by the wave source and $\tilde{\mathsf{e}}_2 = \bar{E}_2$ holds.

It would be tempting to argue that one should always use this latter frame of reference to get the 'correct' version of the energy budget. However, this would be wrong on two counts: first, it is a basic principle of physics that all inertial frames are equally 'correct' for energy budgets. Second, in situations with non-uniform U there does not even exist a single frame in which the basic flow is at rest. This will occupy us a lot in the case of continuous shear flows, but we can already note a simple example that illustrates this point using essentially a piecewise constant U as in figure 6.3.

Here the topography has been removed and the fluid region has been extended into $z < 0$ such that the basic flow in the lower half plane moves with speed U in the *opposite* direction of the flow in the upper half plane. The two regions can be connected by a smooth shear zone in which the basic zonal velocity changes from $-U$ to $+U$ and if the width of this shear zone is wide enough then this can be achieved with a large basic state *Richardson number* $N^2/|U_z|^2$ that guarantees linear stability of this basic flow. It is then possible to consider a zero-frequency long wave in this setup, where 'long' means that the horizontal wavelength is large compared to the width of the shear layer. The vertical structure of such a long wave away from the shear layer is then given by the lee wave solution we have studied before, except that the lee wave in the lower half plane has the signs of its group-velocity components reversed.

In the causal version of this problem it appears natural to assume that the energy for the growing wavetrain must somehow be provided by the wave source in the shear zone, say by slowly eroding the mean shear there. However, it can be shown that there are no on-going mean-flow changes in the shear zone: just as in the standard lee wave problem the energy for the wavetrain is provided by the mean-flow changes at the wave front.

In practice, it is convenient to use whatever conservation law is at hand to compute the wave field whilst discussing the local physics in terms of intrinsic quantities such as $\bar{E}$, and regardless of whether they are conserved or not. This foreshadows the importance of the pseudomomentum vector $\mathbf{p}$ for wave–mean interaction theory outside simple geometry, where the intrinsic $\mathbf{p}$ is crucial to the problem regardless of whether it is conserved or not.

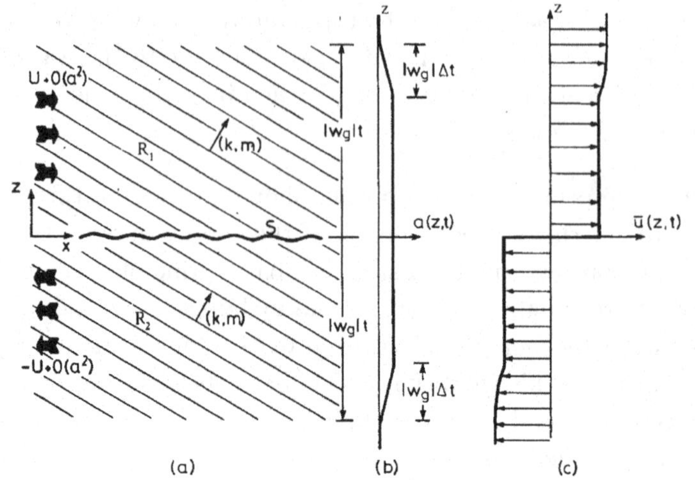

Figure 6.3 Radiation away from a shear layer. Left: basic flow and approximate phase lines of the waves. Middle: time-dependent wave amplitude structure showing wave front propagation with w_g. Right: zonal mean flow correct to $O(a^2)$, showing retardation of the basic flow at the wave front, which provides the energy for the wavetrain. From McIntyre and Weissman (1978)

6.6 Wave dissipation

So far, all interactions have been weak in the sense that they would only lead to a uniformly bounded $O(a^2)$ mean-flow response related to wave transience. This changes once wave dissipation is introduced.

Wave dissipation leads to the attenuation of eddy fluxes and therefore to non-zero flux divergences even for a steady wavetrain. This is the key to strong wave–mean interactions in simple geometry. As expected, the dissipation of pseudomomentum plays a prominent role and it regulates whether or not the mean flow is subject to persistent, cumulative acceleration.

6.6.1 Radiative damping and secular mean-flow growth

The simplest set-up uses radiative damping in the form of Newtonian cooling $R = -\alpha b$ with a constant damping rate per unit time $\alpha > 0$ (cf. §6.3.2). We recall the Eulerian pseudomomentum law as

$$\tilde{\mathsf{p}}_{2t} + (\overline{u_1' w_1'})_z = -\alpha \tilde{\mathsf{p}}_2 = \tilde{\mathcal{F}}. \tag{6.104}$$

For example, for a steady plane wave with weak damping (i.e., $|\alpha\hat{\omega}| \ll 1$) we have $\overline{u_1' w_1'} = \tilde{\mathsf{p}}_2 w_g$, in which case the steady (6.104) can be integrated to yield

$$w_g = \text{const} \quad \Rightarrow \quad \tilde{\mathsf{p}}_2(z) = \tilde{\mathsf{p}}_2(0)\exp\left(-\frac{\alpha z}{w_g}\right). \tag{6.105}$$

Clearly, the damping rate per unit vertical distance is α/w_g. The $O(a^2)$ mean-flow equations are

$$\overline{u}_{2t} = -(\overline{u'_1 w'_1})_z = \tilde{\mathsf{p}}_{2t} + \alpha\tilde{\mathsf{p}}_2 \tag{6.106}$$

and

$$\overline{b}_{2t} + \alpha\overline{b}_2 = -\left(\overline{b'_1 w'_1}\right)_z. \tag{6.107}$$

Using (6.49) the buoyancy flux can be written as

$$\overline{b'_1 w'_1} = -\frac{1}{2N^2}(\overline{b'^2_1})_t - \frac{\alpha}{N^2}\overline{b'^2_1}, \tag{6.108}$$

which shows that damping produces a down-gradient flux of buoyancy. For steady waves the buoyancy response settles down to

$$\text{steady waves} \quad \Rightarrow \quad \overline{b}_2 = \frac{1}{N^2}(\overline{b'^2_1})_z = -\frac{\alpha}{N^2 w_g}\overline{b'^2_1}. \tag{6.109}$$

Overall, we see that damping leads to a bounded response in the Eulerian-mean buoyancy, so there is no strong interaction in this variable.[13]

The situation is very different for the zonal mean flow, which for a steady wave field is subject to

$$\text{steady waves} \quad \Rightarrow \quad \overline{u}_{2t} = -(\overline{u'_1 w'_1})_z = +\alpha\tilde{\mathsf{p}}_2 \tag{6.110}$$

and therefore exhibits secular growth in time. For example, in the causal version of the lee wave problem the initial mean-flow response to the arriving wavetrain is $\overline{u}_2 = \tilde{\mathsf{p}}_2$ and thereafter the mean flow grows according to (6.110). Explicitly,

$$\overline{u}_2(z,t) = \tilde{\mathsf{p}}_2(1 + \alpha t) \tag{6.111}$$

if the wavetrain has arrived at z at time $t = 0$, say. This agrees with the generic resonance structure discussed in §5.1.1, with the zero-frequency shear flow represented by $\overline{u}_2$ replacing the resonantly forced vortical mode.

Experience shows that the tendency indicated by (6.110) is robust: dissipating wavetrains tend to accelerate the zonal mean flow in the direction of the zonal pseudomomentum, and in the absence of other restoring mechanism the concomitant mean-flow response grows without bound.

[13] Here the Newtonian cooling approximation is applied equally to b'_1 and $\overline{b}_2$. More realistic representations of radiative damping might be more scale-selective, in which case large-scale features such as $\overline{b}_2$ would be damped more weakly than small-scale features such as b'_1. This would lead to a larger mean buoyancy response.

6.6.2 Non-acceleration and the pseudomomentum rule

The results in the previous section are the cornerstones of wave–mean interaction theory in simple geometry. They are often summarized in terms of so-called *non-acceleration conditions* and the *pseudomomentum rule*. The statement about non-acceleration follows directly from (6.106), which makes it obvious that the zonal mean flow is not accelerated if the wave field is both steady and non-dissipative. Corresponding statements about the conditions under which $\overline{u}_{2t} = 0$ also hold in more general situations with three-dimensional rotating flows and with variable $N(z)$ and $U(z)$, for example. It is also possible to extend these statement to finite-amplitude waves and we will encounter a very general version of the non-acceleration statement in the context of Lagrangian-mean theory later.

The physical underpinning of non-acceleration statements is invariably Kelvin's circulation theorem: in the absence of dissipation the circulation along stratification lines is conserved and the kinematic construction in §6.3.1 then shows that $\overline{u}_2 - \tilde{\mathsf{p}}_2$ must remain constant. Thus the only possible change in $\overline{u}_2$ is due to wave transience and as such it is bounded in terms of $\tilde{\mathsf{p}}_2$. On the other hand, dissipation allows the circulation to change in time and this change is reflected in (6.106).

The pseudomomentum rule complements the non-acceleration condition by quantifying the effective mean force felt by the zonal mean flow in the case of a steady wavetrain: this effective force is equal to minus the dissipation density of zonal pseudomomentum. This is obvious from the present example, where the pseudomomentum dissipation density in (6.104) is $\tilde{\mathcal{F}} = -\alpha\tilde{\mathsf{p}}$ and the effective force in (6.110) is $+\alpha\tilde{\mathsf{p}} = -\tilde{\mathcal{F}}$. Basically, the pseudomomentum rule elevates the specifics of this situation to a general principle governing strong wave–mean interactions. According to this rule, an effective mean force exerted on the zonal mean flow occurs only where and when the waves break or otherwise dissipate, and then the mean forces arising from dissipation can be equated to minus the rate of dissipation of zonal pseudomomentum.

Because of the underlying exact conservation law for zonal momentum, the pseudomomentum rule also applies to nonlinear wave breaking, even though in this case there is no detailed computation that would produce an equation such as (6.110). This is relevant for the mean-flow response in a *critical layer* to be discussed in §7.2. To see this we consider the zonal mean flow in an layer $\mathcal{D} = [z_1, z_2]$, which changes according to the exact flux balance

$$\frac{d}{dt}\int_{z_1}^{z_2} \overline{u}\, dz = -\left(\overline{u'w'}(z_2,t) - \overline{u'w'}(z_1,t)\right). \tag{6.112}$$

If the wave is incident from below and breaking within $\mathcal{D}$ then we might know that the momentum flux at z_2 is zero. In addition, if z_1 lies below the zone of wave breaking then linear theory might be accurate there and hence the momentum flux can be approximated by the linear solution, which equates it with the pseudomomentum flux. In this way the pseudomomentum rule applies to the control volume, because the vertically integrated effective force equals the incoming pseudomomentum flux.[14]

For example, in a steady lee wave problem the wave field below $\mathcal{D}$ is not subject to dissipation and therefore $\overline{u_1'w_1'} = -D$ in terms of the drag force D on the mountain. Thus, the net mean zonal force exerted on the mean flow in the dissipation region high above is equal-and-opposite to the drag force exerted on the mountain far below. Newton's law of action equals reaction is satisfied, albeit in a non-local manner.

This integral view of mean-flow acceleration also makes clear that the specific dissipation mechanism is not important, e.g., it is irrelevant whether the wave is dissipating due to radiative or frictional effects, and whether or not wave breaking occurs. The equation (6.112) is also useful to understand the interplay between wave transience, dissipation and mean-flow acceleration. For instance, in an unsteady situation involving wavepackets the mean flow in $\mathcal{D}$ may change if a wavepacket arrives at the lower boundary z_1, but has not yet reached the upper boundary z_2. With no dissipation, the mean-flow acceleration that ensues because of this effect is reversible in the sense that it is undone once the wavepacket leaves again through the upper boundary of $\mathcal{D}$. Again, the only lasting change in the mean flow arises from dissipation, because dissipation reduces the momentum flux within a wavepacket and therefore the mean-flow change during the arrival of the wavepacket in $\mathcal{D}$ is *larger* in amplitude than the oppositely-signed mean-flow change when the wavepacket leaves $\mathcal{D}$. In a nutshell, wave dissipation makes permanent a transient mean-flow change that has already occurred during the arrival of the wavepacket.

This makes clear that dissipation *per se* does not lead to mean-flow acceleration, i.e., there is no sudden jolt to the mean flow if a wavepacket suddenly dissipates. The following somewhat extreme example makes this explicit. We consider a wavepacket that is impulsively dissipated over a very short time interval, which could be achieved mathematically by letting α tend to a delta function in time centred at $t = 0$, say. During this rapid dissipation the momentum flux is negligible compared to $\tilde{\mathcal{F}}$ in the pseudo-

[14] Notably, the nonlinear definition of pseudomomentum in Lagrangian-mean theory maintains the equality between these fluxes even at finite wave amplitude.

momentum law (6.104) and thus

$$\tilde{\mathsf{p}}_{2t} = \tilde{\mathcal{F}} = -\alpha\tilde{\mathsf{p}}_2 \quad \text{and} \quad \overline{u}_{2t} = \tilde{\mathsf{p}}_{2t} - \tilde{\mathcal{F}} = 0 \tag{6.113}$$

follows from (6.106) for the duration of the dissipation event. In other words, the mean-flow acceleration directly due to dissipation is *zero*. Of course, the final mean-flow response after the annihilation of the wavepacket is still given by

$$\overline{u}_2(z, t = 0+) = \tilde{\mathsf{p}}_2(z, t = 0-), \tag{6.114}$$

where the pseudomomentum is evaluated just before the dissipation event. Once more, this is the mean-flow response that arose during to the transient arrival of the wavepacket, and which has simply been made permanent by the dissipation event.

This distinction between mean-flow acceleration and wave dissipation is sometimes lost sight of when the pseudomomentum rule is applied in unsteady flow situations. Also, whilst the pseudomomentum rule is the central result in wave–mean interaction theory for the zonally symmetric basic flows of simple geometry, it does not extend to interactions beyond this restrictive geometry. Outside simple geometry, strong interactions are possible that have no essential dependence on dissipation.

6.7 Extension to variable stratification and density

The lee wave problem discussed so far was restricted to constant N and also to the assumption of a nearly constant mass density ρ, which is part of the Boussinesq model. Here we discuss the reasonably straightforward extension of the results to variable $N(z)$ and $\rho(z)$. This introduces two new physical effects: wave reflection in the vertical and amplitude growth.

6.7.1 Variable stratification and wave reflection

The buoyancy frequency N is defined by $N^2 = dS_0/dz$ in the basic rest state, so a variable $N(z)$ corresponds to a $S_0(z)$ that is not simply linear in z. This is relevant for both the atmosphere and the ocean, because N varies by a factor of two in the bulk of the atmosphere (up to about 100 km altitude, say) and by a factor of ten in the ocean. In both cases higher values of N occur in the higher regions. The linear equations in this case do not have constant coefficients in the vertical and therefore plane waves are not global solutions anymore. Still, one can seek modal solutions in which the wave field is $w' = \hat{w}(z)\exp(i[kx + ly - \omega t])$, say, and the vertical structure $\hat{w}(z)$

is determined from an appropriate version of (6.56) together with boundary conditions in the vertical direction. This is straightforward in principle (and with suitable boundary conditions may lead to a complete set of modal solutions), although in practice this computation must be done numerically unless the form of $N(z)$ is very simple.

In the case of an upward-propagating plane wave incident on a single jump in N at $z = z_r$ this leads to the generation of a reflected wave in $z < z_r$ and a transmitted wave in $z > z_r$. Incoming, reflected, and transmitted waves all share the same horizontal wave numbers and frequency, but their vertical wavenumbers differ in order to satisfy the respective radiation conditions ($w_g > 0$ for incoming and transmitted and $w_g < 0$ for the reflected wave) and the dispersion relations in the two regions. The amplitude of the reflected wave is related to the strength of the jump in N; for instance, for hydrostatic waves it is proportional to the jump in N. Of course, such partial internal reflections are absent in the ray tracing approximation.

The most important physical feature here is the possibility of *total internal reflection*, which occurs when N jumps to a lower value in $z > z_r$ such that the local dispersion relation solved for m^2 then yields a negative value, e.g.,

$$m^2(z) = k^2 \frac{N^2(z) - \hat{\omega}^2}{\hat{\omega}^2} < 0 \tag{6.115}$$

in the slice model. Here k and $\hat{\omega}$ are fixed and therefore $m^2 < 0$ occurs if N is lowered below $|\hat{\omega}|$. This means the wave is evanescent in the region $z > z_r$.

How are the evolution laws for $\bar{E}_2$ and $\tilde{\mathsf{p}}_2$ affected by variable $N(z)$? It turns out that the evolution equations (including the mean-flow response equations) we have derived previously remain valid, because we used only that N was constant in the horizontal directions and in time. For example, this implies that for a steady wave field $\overline{u_1' w_1'}$ is still a constant. Now, in the generic case of a single plane lee wave undergoing total reflection at some level $z = z_r$, this leads to a steady wave field with $\overline{u_1' w_1'} = 0$ everywhere. This follows from a detailed study of the linear problem, but we can also deduce it in general from the constancy of the momentum flux. Suppose the wave field is steady and therefore $\overline{u_1' w_1'}$ is constant. Total reflection implies that the waves are evanescent in the region $z > z_r$, and therefore the value of any non-zero wave property is changing there. This obviously implies that $\overline{u_1' w_1'} = 0$ holds throughout.

Now, zero momentum flux also implies a zero mountain drag D. Physically, a non-zero drag D in our finite-L model is possible only if there is a continual flux of pseudomomentum into the fluid region, which is not possible if the wave is steady and trapped in the vertical. The non-generic exception to this scenario is the case where this trapped mode is resonant in the domain, which occurs if the phase progression between the ground and $z = z_r$ is an appropriate multiple of π; for such resonant modes there is no steady state, the wave amplitude grows linearly in time, and the drag D is non-zero.

For atmospheric lee waves total reflection due to reductions in N is seldom relevant, because N usually increases rather than decreases with altitude, e.g., N roughly doubles at the tropopause at 10–12 km altitude. However, reflection due to mean shear is often relevant for these waves, and analogous considerations about $\overline{u_1' w_1'} = 0$ apply in this case as well, as discussed in §7.

6.7.2 Density decay and amplitude growth

In the atmosphere the density decays rapidly with altitude and therefore the Boussinesq equations are not uniformly valid in the vertical. Specifically, their validity is restricted to motions whose vertical scales are small compared to the *density scale height* H_s, say, over which $\rho(z)$ varies significantly.[15] For most of the atmosphere H_s ranges between 5–10 km, and for linear waves the restriction $|mH_s| \gg 1$ is then compatible with typical lee waves, for instance. However, the Boussinesq model can still be misleading for the long-range vertical propagation of lee waves over several scale heights.

To capture this long-range propagation a more accurate fluid model such as the so-called *anelastic equations* must be used. These equations are intermediate between the Boussinesq equations and the full compressible Euler equations, because they retain the reduced number of degrees of freedom of the Boussinesq equations (and thereby filter sound waves) whilst including some of the compressible effects due to the density decay. From an asymptotic point of view these equations are based on a small Mach number as well as on a low-frequency assumption in order to filter the sound waves. There is more than one version of the anelastic equations in common use, depending on precisely how many terms are modified to include compressible effects, but all versions agree on the amplitude increase of internal waves that goes hand-in-hand with the basic density decrease. This amplitude increase must be such that for steady waves the full momentum flux

$$\rho(z)\,\overline{u_1' w_1'} = \text{const.} \tag{6.116}$$

This implies that the wave amplitude is proportional to $\rho^{-1/2}$, which is a

[15] For an isothermal ideal gas under gravity $\rho(z) = \rho(0)\exp(-z/H_s)$ with $H_s = RT/g$, where R is the gas constant and T is the constant temperature. The constant buoyancy frequency N is then related to H_s by $N^2 = (\gamma - 1)g/(H_s\gamma)$, where γ is the ratio of specific heats.

huge increase over the course of several scale heights. Mutatis mutandis, our previous results for the mean-flow response remain valid if $\overline{u_1'w_1'}$ is replaced by $\rho(z)\,\overline{u_1'w_1'}$, e.g.,

$$\overline{u}_{2t} = -\frac{1}{\rho(z)}\,(\rho(z)\overline{u_1'w_1'})_z. \tag{6.117}$$

Now, the simplest anelastic model differs from the Boussinesq equations only in one term, namely $\boldsymbol{\nabla}\cdot\boldsymbol{u} = 0$ is replaced by

$$\boldsymbol{\nabla}\cdot(\rho(z)\boldsymbol{u}) = 0 \quad \Leftrightarrow \quad \boldsymbol{\nabla}\cdot\boldsymbol{u} + w\frac{d\ln\rho(z)}{dz} = 0. \tag{6.118}$$

For an exponential density profile $\rho \propto \exp(-z/H_s)$ this is equivalent to

$$\boldsymbol{\nabla}\cdot\boldsymbol{u} = \frac{w}{H_s}, \tag{6.119}$$

which models the volume dilation of fluid parcels that move upwards.[16] The resulting linear equations again have constant coefficients and therefore admit plane wave solutions. Exponential growth or decay in altitude manifests itself in an imaginary part of the vertical wavenumber m, which can be absorbed in a suitable external factor. Specifically, for a plane wave in the form

$$(u_1', w_1', b_1') = (\hat{u}, \hat{w}, \hat{b})\exp\left(\frac{z}{2H_s}\right)\exp(i[kx + mz - \omega t]) \tag{6.120}$$

the wavenumbers are real and the intrinsic dispersion relation is

$$\hat{\omega}^2 = N^2\frac{k^2}{k^2 + m^2 - \frac{1}{4H_s^2}}, \tag{6.121}$$

which shows a compressible correction that becomes negligible if $\kappa H_s \gg 1$, as expected. Much more important is the explicit amplitude growth with altitude in (6.120); for large z/H_s this must lead to very large wave amplitudes and therefore to wave instability and breaking. This is relevant for high-altitude breaking of waves, say at $z = 60$ km or higher.

For example, the mean retrograde force exerted on the mean flow because of dissipating lee waves (both of gravity-wave and of Rossby-wave type) is crucial for the closure of the *mesospheric jet* at high altitudes of 80 km or so. The realization of the necessity of such a wave drag to in order to achieve

[16] Despite the natural appearance of (6.118) and (6.119), it should be noted that in more complex anelastic models the continuity equation is different. For example, starting from the compressible Euler equations and looking at low-frequency linear waves relative to an isothermal basic state leads to (6.119) with H_s replaced by γH_s. Of course, other parts of the equations are then changed as well such that the overall structure of the waves is again consistent with the crucial condition (6.116).

the correct global-scale behaviour of the jet flow was a major milestone in wave–mean interaction theory applied to atmospheric flows.

If we assume $|mH_s| \gg 1$ then the only changes in the evolution equations for wave properties such as $\tilde{\mathsf{p}}_2$ stem from replacing the pseudomomentum per unit mass $\tilde{\mathsf{p}}_2$ with the pseudomomentum per unit volume $\rho(z)\tilde{\mathsf{p}}_2$, e.g.,

$$\rho(z)\tilde{\mathsf{p}}_{2t} + (\rho(z)\overline{u_1' w_1'})_z = \rho(z)\tilde{\mathcal{F}} \tag{6.122}$$

and so on. In this way the important wave–mean interaction relation

$$\overline{u}_{2t} = \tilde{\mathsf{p}}_{2t} - \tilde{\mathcal{F}} \tag{6.123}$$

remains valid. This is as it should be, because we know that (6.123) derives from Kelvin's circulation theorem, which holds for the circulation along stratification lines even in the fully compressible Euler equations.

6.8 Notes on the literature

Boussinesq models are described in GFD textbooks such as Salmon (1998) and Vallis (2006). A careful asymptotic justification for these equations in the ideal gas case is given in Spiegel and Veronis (1960). The subtle case of sea water with salinity effects, which involves unusual nonlinearities in the equation of state that are of practical importance in ocean dynamics, is discussed in Young (2010), for example.

The peculiar spatial structure of time-periodic internal waves has been investigated recently in a sequence of works starting with Maas and Lam (1995); the original work by Sobolev on this problem is described in Arnold and Khesin (1998). Lee waves are discussed extensively in Baines (1995) and the hydrostatic solution using Hilbert transforms is found in Drazin and Su (1975). The importance of wave drag for the global-scale atmospheric circulation is discussed in Andrews *et al.* (1987).

6.9 Exercises

1. *Fermat's principle for mountain lee waves.* Derive the equivalent of (4.103) for non-hydrostatic mountain lee waves in the presence of a slowly varying mean flow $U(z) > 0$ and buoyancy frequency $N(z)$ and verify that in this case (4.103) holds with $c(\boldsymbol{x})$ replaced by $U(z)/N(z)$. If U is constant and N increases with altitude, does the ray slope of a mountain lee wave become shallower or steeper? Are you surprised by the answer?

2. *Mountain lee waves with a tropopause.* Derive the solution for hydrostatic mountain lee waves in a configuration where N jumps from a lower

value N_1 to a higher value N_2 at a tropopause with altitude $z = H$, say. (For example, in the atmosphere, $N_2 \approx 2N_1$ and $H \approx 10\,\mathrm{km}$ in mid-latitudes.) To do this assume continuity of w' and of w'_z at $z = H$ and follow §6.7.1 by allowing for a superposition of upward and downward waves in the lower region $z < H$. Work out the mean drag force on a sinusoidal mountain range and show that it is now an oscillatory function of the wind speed U. Note that the mean momentum flux is still constant with altitude in this case.

7
Shear flows

A basic shear flow $U(z)$ leads to fundamentally new wave effects, the most important being the possibility of *critical layers* and the concomitant *singular absorption* of waves there. This goes hand-in-hand with strong wave–mean interactions in critical layers, which are especially important for atmospheric gravity waves.

We begin by describing how basic shear affects the linear dynamics in the Boussinesq slice model and then we discuss critical layers in detail. This is followed by a consideration of the long-term evolution of the joint system of waves and mean flows in a simple model for the famous quasi-biennial oscillation in the equatorial atmosphere.

It is worth noting at the outset that, one way or another, almost all of the shear-related effects can be understood by the simple example of a *sheared-over disturbance*. This is a passive tracer with density $\phi(x, z, t)$ that is advected by a steady shear flow $\boldsymbol{u} = (U(z), 0)$, i.e.,

$$\frac{D\phi}{Dt} = \phi_t + U(z)\phi_x = 0. \tag{7.1}$$

The initial-value problem has the simple explicit solution

$$\phi(x, z, t) = \phi_0(x - U(z)t, z) \quad \text{where} \quad \phi(x, z, 0) = \phi_0(x, z). \tag{7.2}$$

For example, if

$$\phi_0 = \exp(i[kx + mz]) \quad \text{then} \quad \phi = \exp(i[kx + mz - kU(z)t]) \tag{7.3}$$

and therefore

$$\phi_x = ik\phi \quad \text{and} \quad \phi_z = i(m - kU_z t)\phi. \tag{7.4}$$

This solution could also be computed from the general refraction equation

$$\frac{D\boldsymbol{\nabla}\phi}{Dt} = -(\boldsymbol{\nabla}\boldsymbol{U}) \cdot \boldsymbol{\nabla}\phi = (0, -U_z\phi_x). \tag{7.5}$$

Either way, this solution makes explicit the secular growth of the vertical wavenumber for large t. Specifically, if $U = \lambda z$ with constant λ then the vertical wavenumber is $m(t) = m - k\lambda t$. The magnitude of $m(t)$ must eventually grow without bound, which is the key effect caused by vertical shear.

As discussed in §4.4.2, similar effects as for ϕ also arise for the wave phase θ, whose gradient $\boldsymbol{k} = \boldsymbol{\nabla}\theta$ is affected by $\boldsymbol{\nabla U}$ in a manner that is analogous to $\boldsymbol{\nabla}\phi$. This makes the example of a sheared-over disturbance relevant to wave dynamics and ray tracing. Finally, for more general $\boldsymbol{U}(\boldsymbol{x}, t)$ it is clear from (7.5) that both components of $\boldsymbol{\nabla}\phi$ are subject to change, which can lead to exponential growth of $|\boldsymbol{\nabla}\phi|$ rather than the algebraic growth apparent in (7.4). This will be important for the wave dynamics outside simple geometry.

7.1 Linear Boussinesq dynamics with shear

We allow for arbitrary shear $U(z)$ and stratification $N(z)$ subject to the stability constraint of large enough *Richardson number*

$$R = \frac{N^2}{U_z^2} \tag{7.6}$$

such that the basic flow is stable. We set $\boldsymbol{u} = (U, 0) + a\boldsymbol{u}_1'$ with small wave amplitude $a \ll 1$ and then the linear Boussinesq equations for $(\boldsymbol{u}', b', P')$ in the vertical slice model are (omitting expansion subscripts)

$$\mathrm{D}_t u' + w' U_z + P'_x = 0, \quad \mathrm{D}_t w' + P'_z - b' = 0, \tag{7.7}$$

$$\mathrm{D}_t b' + N^2 w' = 0, \quad \text{and} \quad \boldsymbol{\nabla} \cdot \boldsymbol{u}' = 0. \tag{7.8}$$

There is only one explicit new term, but of course the operator $\mathrm{D}_t = \partial_t + U(z)\partial_x$ has a variable coefficient now. For general $\boldsymbol{U}(\boldsymbol{x}, t)$ the definition of the linear particle displacements is given by (cf. (4.140))

$$\mathrm{D}_t \boldsymbol{\xi}' = \boldsymbol{u}' + (\boldsymbol{\xi}' \cdot \boldsymbol{\nabla})\boldsymbol{U} \tag{7.9}$$

where all fields are evaluated at $\boldsymbol{x}$. This ensures that $\boldsymbol{x}(t) + a\boldsymbol{\xi}'(\boldsymbol{x}(t), t)$ traces out a material trajectory correct to $O(a)$ if $\boldsymbol{x}(t)$ moves with $\boldsymbol{U}$, i.e., if $\mathrm{D}_t \boldsymbol{x} = \boldsymbol{U}(\boldsymbol{x}, t)$. Note that if $\boldsymbol{\nabla} \cdot \boldsymbol{U} = 0$ then

$$\mathrm{D}_t \boldsymbol{\nabla} \cdot \boldsymbol{\xi}' = \boldsymbol{\nabla} \cdot \boldsymbol{u}' + (\boldsymbol{\xi}' \cdot \boldsymbol{\nabla}) \boldsymbol{\nabla} \cdot \boldsymbol{U} \quad \Rightarrow \quad \boldsymbol{\nabla} \cdot \boldsymbol{\xi}' = 0 \tag{7.10}$$

provided it started that way. In the present case we have

$$\mathrm{D}_t \xi' = u' + \zeta' U_z \quad \text{and} \quad \mathrm{D}_t \zeta' = w'. \tag{7.11}$$

The linear y-vorticity equation is

$$\mathrm{D}_t (u'_z - w'_x) + w' U_{zz} + b'_x = 0, \tag{7.12}$$

which shows a new contribution due to the material advection of particles across the basic-state gradient of vorticity, which is U_{zz}.

We note in passing that for constant shear $U = \lambda z$ (which implies $U_{zz} = 0$) closed-form solutions to the linear initial-value problem exist that exhibit secular growth in the vertical wavenumber much as in the example of a sheared-over disturbance given by (7.3). In fact, for constant shear there are also special solutions to the *nonlinear* initial-value problem for plane waves.

7.1.1 Wave activity measures with shear

The shear $U(z)$ does not upset the basic-state symmetries with respect to x and t and therefore we expect to recover conservation laws for pseudomomentum and pseudoenergy. On the other hand, we will find that wave energy is not conserved in the presence of shear.

The Eulerian pseudomomentum is derived as before, i.e., by multiplying (7.12) with b'/N^2 and averaging. This leads to

$$\mathrm{D}_t \frac{1}{N^2}\overline{b'(u'_z - w'_x)} + (\overline{u'w'})_z + \overline{b'w'}\frac{U_{zz}}{N^2} = 0 \tag{7.13}$$

and therefore to

$$\tilde{\mathsf{p}}_t + (\overline{u'w'})_z = 0 \quad \text{with} \quad \tilde{\mathsf{p}} = \frac{1}{N^2}\overline{b'(u'_z - w'_x)} - \overline{b'^2}\frac{U_{zz}}{2N^4}. \tag{7.14}$$

The physical meaning of the new term becomes more apparent if we use $b' = -\zeta' N^2$ to obtain

$$\tilde{\mathsf{p}} = -\overline{\zeta'(u'_z - w'_x)} - \overline{\zeta'^2}\frac{U_{zz}}{2}. \tag{7.15}$$

If we recall the geometric interpretation of $\tilde{\mathsf{p}}$ in terms of the circulation theorem in §6.3.1, then it is easy to show that the new term makes sure that $\overline{u}_2 - \tilde{\mathsf{p}}$ continues to be proportional to the $O(a^2)$ changes in the circulation $\Gamma(\mathcal{C})$.[1] This ensures that $\tilde{\mathsf{p}}$ enters the zonal mean-flow response in the same way as before.

The equation for the usual wave energy with shear is

$$\bar{E} = \tfrac{1}{2}(\overline{|\boldsymbol{u}'|^2} + \overline{b'^2}/N^2) \quad \Rightarrow \quad \bar{E}_t + (\overline{P'w'})_z = -\overline{u'w'}U_z. \tag{7.16}$$

The source term on the right-hand side shows how energy can be converted between the mean and disturbance parts of the total energy that are displayed in (6.102). The conservation law for the frame-dependent Eulerian

[1] The proof adds a basic vorticity term to the integral in (6.39) using the one-term Taylor expansion $U_z(z) = U_z(z_0) + (z - z_0)U_{zz}(z_0)$ near $z = z_0$.

pseudoenergy with density $\tilde{\mathsf{e}}$, say, follows from (7.14) and (7.16) as

$$\tilde{\mathsf{e}} = \bar{E} + \tilde{\mathsf{p}}U \quad \Rightarrow \quad \tilde{\mathsf{e}}_t + (\overline{P'w'} + \overline{u'w'}U)_z = 0. \tag{7.17}$$

7.1.2 Energy changes for a sheared wavetrain

Wave energy is not conserved, even for zero dissipation, and this can lead to unfamiliar situations, which even appear paradoxical at first sight. To illustrate this we consider a lee wave problem with constant N and variable $U(z)$. To be specific, we assume that $U(0) > 0$ and thereafter $U(z)$ varies smoothly across some shear zone $z \in [0, H]$ such that $0 < U(z) \leq U(0)$ and $U(z) = \frac{1}{2}U(0)$ if $z \geq H$. For a steady non-dissipative wavetrain the conservation of pseudomomentum and pseudoenergy then yields

$$(\overline{u'w'})_z = 0 \quad \text{and} \quad (\overline{P'w'} + \overline{u'w'}U)_z = 0. \tag{7.18}$$

Moreover, the pseudoenergy flux is zero for a zero-frequency lee wave[2] and therefore the wave energy flux from (7.16) is

$$\overline{P'w'} = -\overline{u'w'}\,U \quad \Rightarrow \quad \overline{P'w'}(z) \propto U(z). \tag{7.19}$$

Hence the wave energy flux *decreases* if $U(z) < U(0)$, which makes it obvious that wave energy is lost. In particular, the energy flux in $z > H$ is only half of the flux at $z = 0$. This implies a corresponding destruction of wave energy as the wavetrain passes through the shear flow, which is indeed consistent with the sign of the source term in (7.16). At first sight, it would then appear natural to expect that the mean flow should continually be altered in the sheared regions where $U_z \neq 0$ in order to account for this loss of wave energy. However, the constancy of $\overline{u'w'}$ for a steady non-dissipative wavetrain implies that there is in fact *no* mean-flow acceleration anywhere, including in the shear zone. Indeed, this fact is also obvious from Kelvin's circulation theorem.

As in §6.5.4, the resolution of this paradox can be found by considering the causal initial-value problem, in which the flow starts from an undisturbed rest state at $t = 0$. At any altitude z there is then the transient mean-flow response $U \to U + a^2\tilde{\mathsf{p}}$ as the vertically propagating wave front arrives, which implies a change in the kinetic energy of the mean flow equal to $a^2U(z)\tilde{\mathsf{p}}$. It is this change in mean-flow kinetic energy that is available for the wave energy, and as $U(z)$ decreases less such energy is available, consistent with the reduced wave energy flux in (7.19).

[2] This follows from multiplying the steady (7.7a) by the disturbance stream function ψ' (defined such that $(u', w') = (\psi'_z, -\psi'_x)$) and averaging.

Once again, this example shows that energy budgets are very subtle in wave–mean interaction theory and that careful attention needs to be paid to all parts of the problem, including boundary conditions and wave fronts.

7.1.3 Rayleigh and Fjørtoft theorems for shear instability

The combination of pseudomomentum and pseudoenergy conservation provides an accessible and insightful route to some important instability theorems for shear flows. To begin with, the pseudomomentum equation implies *Rayleigh's instability theorem* for unstratified shear flows. To illustrate this we set stratification to zero in this section only, which corresponds to the limit $b' \to 0$ and $N \to 0$ whilst $b'/N^2 = -\zeta' = O(1)$. The buoyancy equation then drops out of the system, which reduces to a single vortical degree of freedom described by initial conditions for $u'_z - w'_x$, say.

The evolution of disturbance vorticity is then entirely due to advection of particles across the basic vorticity gradient, which means that (7.12) can be integrated to yield

$$u'_z - w'_x = -\zeta' U_{zz}. \tag{7.20}$$

Using (7.20) in (7.15) yields

$$\tilde{\mathsf{p}} = -\overline{\zeta'(u'_z - w'_x)} - \overline{\zeta'^2}\frac{U_{zz}}{2} = +\frac{1}{2}U_{zz}\,\overline{\zeta'^2}. \tag{7.21}$$

The support of $\tilde{\mathsf{p}}$ is confined to the support of the basic vorticity gradient U_{zz} and the sign of $\tilde{\mathsf{p}}$ is equal to the sign of U_{zz}.

Rayleigh's theorem follows once (7.21) is integrated over a bounded domain $z \in [0, H]$, say, with flat impermeable walls at the boundary. This implies that the flux $\overline{u'w'} = 0$ at the boundaries and therefore the total pseudomomentum is conserved, i.e.,

$$\mathcal{P} = \int_0^H \tilde{\mathsf{p}}\,dz = \int_0^H \frac{1}{2}U_{zz}\,\overline{\zeta'^2}\,dz \quad \Rightarrow \quad \frac{d\mathcal{P}}{dt} = 0. \tag{7.22}$$

This conservation law implies two immediate facts for exponentially unstable normal modes, which are solutions of the form (real parts understood) $\zeta'(x, z, t) = \zeta'(x, z, 0)\exp(-i\omega t)$ where ω has a positive imaginary part. In this case $\overline{\zeta'^2} \propto \tilde{\mathsf{p}}$ grows exponentially without changing shape, and therefore $\mathcal{P}$ grows exponentially as well. This contradicts the conservation of $\mathcal{P}$ unless $\mathcal{P} = 0$ and therefore an unstable normal mode must have zero total pseudomomentum.

The second fact follows from combining (7.22) with $\mathcal{P} = 0$, which for non-zero ζ' is possible only if U_{zz} changes sign somewhere in the domain.

This is Rayleigh's theorem: a necessary condition for the existence of an exponentially growing normal mode on a shear flow is that the basic vorticity gradient must change sign somewhere in the domain.

Simple explicit solutions can be computed in the case where U_z is piecewise constant, which corresponds to a delta-function structure in U_{zz}. Evanescent *vorticity edge waves* are then supported at these delta function locations, with a sign-definite pseudomomentum density given by (7.21) interpreted as a delta function. Rayleigh's theorem then implies that there must be counter-propagating edge waves in the domain, and any instability can then fruitfully be analysed in terms of the interplay between these counter-propagating waves and how they form a compound disturbance with zero net pseudomomentum.

A sharpening of Rayleigh's theorem is provided by *Fjørtoft's theorem*, which exploits the conservation of pseudoenergy in (7.17) in the form

$$\frac{d}{dt}\int_0^H \left(\bar{E} + \frac{1}{2}U_{zz}U\,\overline{\zeta'^2}\right)\,dz = 0. \tag{7.23}$$

For an unstable normal mode the total pseudoenergy is zero by the same argument that applied to the total pseudomomentum $\mathcal{P}$, so (7.23) already implies that U_{zz} and U must take opposite signs somewhere in the domain, which is Fjørtoft's theorem. At first sight, this seems to be a frame-dependent statement, which would be in conflict with the obvious Galilean invariance of the physical problem. However, adding any constant to U simply adds a term proportional to $\mathcal{P}$ to the total pseudoenergy, but $\mathcal{P} = 0$ anyway! Hence one can in fact replace U by $U-c$ with an arbitrary c in (7.23), which greatly widens the impact of Fjørtoft's theorem. For example, picking a c outside the range of U means that $U-c$ is sign-definite, which shows that Fjørtoft's theorem in fact implies Rayleigh's theorem. Conversely, for monotone $U(z)$ with a single inflexion point at $z = z_I$, say, another convenient choice is $c = U(z_I)$, which can be shown to rule out the instability of some shear profiles even though they pass Rayleigh's theorem.

7.1.4 Ray tracing in a shear flow

For simplicity we restrict to constant N from now on. In ray tracing the shear enters only via the Doppler-shifting term $U(z)k$ in the dispersion relation and the remaining basic-state symmetries in x and t imply that k and ω are constant along a ray. At large R we may use (6.19) for the intrinsic frequency

$\hat{\omega}$:

$$\hat{\omega} = \omega - Uk \quad \Rightarrow \quad \frac{d\hat{\omega}}{dt} = -\frac{dU}{dt}k = -w_g U_z k. \tag{7.24}$$

Hence

$$\hat{\omega}(z) = \hat{\omega}(z_0) - k(U(z) - U(z_0)), \tag{7.25}$$

where $z(t)$ is the altitude along the ray and $z_0 = z(0)$. We note in passing how (7.24) leads to the conservation of wave action. The wave energy equation for ray tracing is

$$\bar{E}_t + (\bar{E}w_g)_z = -\overline{u'w'}U_z = -\frac{k\bar{E}}{\hat{\omega}}w_g U_z = \frac{\bar{E}}{\hat{\omega}}\frac{d\hat{\omega}}{dt} \tag{7.26}$$

and this can be re-written as

$$\frac{\partial}{\partial t}\left(\frac{\bar{E}}{\hat{\omega}}\right) + \frac{\partial}{\partial z}\left(\frac{\bar{E}}{\hat{\omega}}w_g\right) = \frac{d}{dt}\left(\frac{\bar{E}}{\hat{\omega}}\right) + \left(\frac{\bar{E}}{\hat{\omega}}\right)\boldsymbol{\nabla}\cdot\boldsymbol{c}_g = 0. \tag{7.27}$$

Clearly, changes in $\hat{\omega}$ go together with changes in $\bar{E}$.

Now, it is possible that Doppler-shifting may move $|\hat{\omega}|$ towards the bounds $|\hat{\omega}| = N$ and $|\hat{\omega}| = 0$ that follow from the intrinsic dispersion relation (6.19). More specifically, the vertical wavenumber m along a ray has to evolve such that (7.25) remains consistent with the dispersion relation (6.19), which implies that

$$|m| \to 0 \quad \text{if} \quad |\hat{\omega}| \to N \quad \text{and} \quad |m| \to \infty \quad \text{if} \quad |\hat{\omega}| \to 0. \tag{7.28}$$

The first situation leads to a reflection level and the second to a critical level. Physically, reflection corresponds to a local wave structure in which the intrinsic particle motion is nearly vertical, as in a buoyancy oscillation. On the other hand, near a critical level the particle motion is nearly horizontal, as in a shear flow. Both levels are caustics and one has to investigate carefully whether ray tracing remains a valid approximation to linear theory there and also whether linear theory remains a valid approximation to nonlinear reality. The second question is especially important in practice.

For example, for a lee wave with $\omega = 0$, reflection at some level $z = z_r$ occurs if $U(z_r) = N/|k|$, which is the upper speed limit for the generation of propagating waves. The vertical group velocity w_g goes to zero as $z \to z_r$, but the ray-tracing equations

$$\frac{dz}{dt} = w_g = \text{sgn}(m)\frac{kU^2}{N^2}\sqrt{N^2 - k^2U^2} \quad \text{and} \quad \frac{dm}{dt} = -kU_z \tag{7.29}$$

show that $m(t)$ continues to evolve (if $U_z(z_r) \neq 0$) and simply changes sign. This means that w_g changes sign as well and the wave is reflected. It is a standard textbook exercise to show that ray tracing breaks down near z_r, but that it can be repaired locally by fitting an Airy function centred at $z = z_r$. Thus, at a reflection level ray tracing fails locally, but linear theory remains valid.

The situation is quite different for a critical level $z = z_c$, which for a lee wave corresponds to a zero-wind line $U(z_c) = 0$. In this case $m(t)$ retains its sign and grows without bound as $z \to z_c$, and the time to reach z_c is infinite. Clearly, this situation requires more analysis.

7.2 Critical layers

The generic definition of a *critical level* follows from ray tracing: it is a level $z = z_c$ where the dispersion relation implies that $|m| \to \infty$ and therefore $w_g \to 0$. We will see that a wave incident on a critical level from below, say, undergoes a non-uniform transition in a region below $z = z_c$, which is called the *critical layer*. Basically, the robust result is that the wave is absorbed in the critical layer, which therefore acts like a wave dissipation site. This has consequences for the mean-flow response: the critical layer acts like a sponge for pseudomomentum, which leads to concomitant growth in the mean-flow response. This leads to the standard paradigm for wave drag parametrizations in the atmosphere.

We will first look at the footprint of critical layers in ray tracing, and at the question of whether ray tracing may remain valid all the way up to the critical level. This is followed by a discussion of why the steady non-dissipative linear wave problem is ill-posed in the presence of a critical level, and how this can be remedied by looking at the causal initial-value problem. Then, the sponge-like nature of a critical layer is discussed in the context of so-called singular wave absorption and nonlinear wave breaking and finally the critical-layer structure of the mean-flow response and its parametrization in numerical models is described.

7.2.1 Validity of ray tracing in critical layers

We investigate the behaviour of (7.29) in a critical layer for the lee wave configuration studied before. We assume that $U_z(z_c) \neq 0$, which is the generic case, and then it is clear that $m(t)$ grows secularly in time as $z \to z_c$, and as a function of z it is given by

$$m(z) = \operatorname{sgn}(k)\sqrt{\frac{N^2}{U(z)^2} - k^2} \approx \operatorname{sgn}(k)\frac{N}{|U(z)|} \tag{7.30}$$

in the hydrostatic approximation, which is accurate in the critical layer. Furthermore, as $z \approx z_c$ we have

$$\frac{dz}{dt} = w_g \approx \frac{|k|U^2}{N} \propto (z_c - z)^2 \quad \Rightarrow \quad t \propto (z_c - z)^{-1} \tag{7.31}$$

along rays. Therefore, it takes an infinite time for the ray to reach the critical level. This is unlike the reflection level, for which the same argument shows that the ray reaches it in finite time.

Because the ray takes an infinite time to reach the critical level the implication is that conserved wave properties such as the zonal pseudomomentum must accumulate in the critical layer just below $z = z_c$. This raises questions about the possibility to reach a steady state, and also about the validity of both ray tracing and of linear theory.

Whether or not ray tracing remains valid in a critical layer depends on the details of the intrinsic dispersion relation and hence differs from wave problem to wave problem. There is a simple heuristic argument that suggests the surprising fact that ray tracing does indeed remain valid near critical levels for internal gravity waves. The argument is based on the requirement that the remaining distance to the critical layer $|z_c - z|$ along a ray must be large compared to the local wavelength $\propto 1/m$. This means that

$$|m(z)\,(z_c - z)| \gg 1 \tag{7.32}$$

must hold uniformly as $z \to z_c$. In the present case (7.30) implies that $|m| \propto 1/|U| \propto |z_c - z|^{-1}$ near z_c and therefore (7.32) can indeed be satisfied by internal waves. It follows that there are infinitely many oscillations below z_c, specifically, the wave phase diverges logarithmically as $z \to z_c$.

Clearly, this argument uses the specifics of the dispersion relation and therefore it can fail for other wave types. In particular, if the dispersion relation is such that $|m| \propto |\hat{\omega} - \hat{\omega}_c|^{-\sigma}$ asymptotically as $\hat{\omega}$ tends to some $\hat{\omega}_c$ at the critical level, then (7.32) will hold if $U_z(z_c) \neq 0$ and $\sigma \geq 1$. This simple argument is based on $|\hat{\omega} - \hat{\omega}_c| \propto |z_c - z|$ from (7.25), and in the present lee-wave case $\hat{\omega}_c = 0$ and $\sigma = 1$. On the other hand, if Coriolis forces are important then the intrinsic dispersion relation for internal waves is changed to (cf. §8.2.2)

$$\hat{\omega}^2 = N^2 \frac{k^2}{k^2 + m^2} + f^2 \frac{m^2}{k^2 + m^2} \tag{7.33}$$

where $f \ll N$ is the Coriolis parameter. Now the critical layer arises at $\hat{\omega}_c = f$, which leads to $\sigma = 1/2$ and therefore ray tracing is not valid. A similar result applies to horizontal critical layers for Rossby waves.

It is important to bear in mind that the validity of ray tracing as a uniformly accurate approximation to linear theory in a critical layer may only mask the overall breakdown of linear theory to nonlinear reality there. We investigate this further by looking at the linear theory for a critical layer.

7.2.2 Failure of steady linear theory for critical layers

Introducing a stream function ψ' with $(u', w') = (\psi'_z, -\psi'_x)$ in (7.12) yields

$$\mathrm{D}_t\mathrm{D}_t(\psi'_{xx} + \psi'_{zz}) - U_{zz}\mathrm{D}_t\psi'_x + N^2\psi'_{xx} = 0. \tag{7.34}$$

We consider a normal mode in x and t such that $\psi' = \hat{\psi}(z)\exp(i[kx - \omega t])$ and solve for $\hat{\psi}(z)$. In fact, it turns out to be convenient to use the speed $c = \omega/k$ as a parameter for the wavetrain.[3] This is the phase speed as observed at a fixed altitude z and in terms of c we have $\psi' = \hat{\psi}(z)\exp(ik[x-ct])$. Also, we clearly have $\mathrm{D}_t = ik(U(z) - c)$ and thus (7.34) becomes the celebrated *Taylor–Goldstein equation*

$$\frac{\mathrm{d}^2\hat{\psi}}{\mathrm{d}z^2} + \left\{\frac{N^2}{(U-c)^2} - \frac{U_{zz}}{U-c} - k^2\right\}\hat{\psi} = 0. \tag{7.35}$$

For constant U the quantity in curly brackets is m^2 as given by the internal wave dispersion relation in terms of k and $\hat{\omega} = -k(U - c)$. The second term in the curly brackets is related to the vorticity waves mentioned before; it has no essential dependence on stratification and in the envisaged regime of large Richardson number it is usually small compared to the first term. Moreover, this term plays no essential role at a critical layer. For simplicity, we will hence neglect this term and consider

$$\frac{\mathrm{d}^2\hat{\psi}}{\mathrm{d}z^2} + M^2(z)\,\hat{\psi} = 0 \quad \text{with} \quad M(z) = \sqrt{\frac{N^2}{(U(z)-c)^2} - k^2} \tag{7.36}$$

as the relevant modal equation. We are not interested in reflection and therefore we assume $M^2 > 0$ throughout. On the other hand, we are interested in the possibility that $M(z_c) \to \infty$ at some critical level $z = z_c$, which corresponds to $U(z_c) = c$, so the background flow equals the phase velocity of the internal wave at the critical level.

For definiteness, we assume a finite domain $z \in [0, H]$ in which U is constant over a finite interval near each boundary and then undergoes a smooth transition in the interior between its boundary values $U(0)$ and $U(H)$. The assumption of constant U near the boundaries allows a simple

[3] Lee waves correspond to the special case $c = 0$.

formulation for the boundary conditions for $\hat{\psi}$ based on plane waves. For ease of comparison with the lee-wave problem (in which $U > 0$ and $c = 0$) we will assume that $U(0) > c \geq 0$; this means that the intrinsic horizontal phase speed $c - U$ is negative at the ground, but we make no assumption about its sign at $z = H$.

We assume that the waves are generated at $z = 0$ and radiate out of the domain at $z = H$. This implies a Dirichlet condition for $\hat{\psi}$ at $z = 0$ and a radiation condition at $z = H$, i.e.,

$$\hat{\psi}(0) = 1 \quad \text{and} \quad \hat{\psi}_z(H) = iM(H)\text{sgn}(k(U(H) - c))\hat{\psi}(H). \tag{7.37}$$

The sign choice in the second relation assures a positive vertical group velocity for plane internal waves with $\hat{\omega} = -k(U - c)$. This radiation condition implies that $\hat{\psi}^*(H)\hat{\psi}_z(H)$ is purely imaginary, which for *smooth* $\hat{\psi}$ (one continuous derivative is sufficient) ensures that the boundary-value problem has a unique solution. The standard uniqueness proof considers $\hat{\psi}$ as the difference between two solutions (which therefore satisfies a homogeneous version of (7.37a)) and exploits the constancy of the Wronskian (cf. §2.1.5)

$$W = \hat{\psi}^*\hat{\psi}_z - \hat{\psi}\hat{\psi}_z^* = 2i\Im(\hat{\psi}^*\hat{\psi}_z). \tag{7.38}$$

The physical content of constant W becomes apparent upon consideration of the momentum flux

$$\overline{u'w'} = \frac{ik}{4}W. \tag{7.39}$$

Now, if $\hat{\psi}(0) = 0$ then $W = 0$ and therefore $\hat{\psi}^*\hat{\psi}_z$ must be real, which leads to a contradiction with the upper boundary condition unless $\hat{\psi} = 0$ everywhere; this proves uniqueness for smooth solutions.

There is no uniqueness proof for non-smooth $\hat{\psi}$, which turns out to be relevant for critical layers. Indeed, the critical level is a singular point of (7.36) because the coefficient of the highest derivative vanishes there. We assume that $U_z(z_c) \neq 0$ and then $z = z_c$ is a so-called 'regular' singular point. We can investigate the structure of the solution near the critical level by approximating the basic flow near $z = z_c$ as

$$U \approx c + \lambda s \quad \text{where} \quad s = z - z_c \tag{7.40}$$

and the constant λ measures the shear at the critical level such that the local Richardson number $R_c = N^2/\lambda^2$. The modal equation (7.36) for small s is then approximated by

$$s^2\hat{\psi}_{ss} + R_c\hat{\psi} = 0, \tag{7.41}$$

which has power law solutions

$$\hat{\psi} = As^{\alpha_1} + Bs^{\alpha_2} \quad \text{with} \quad \alpha_{1,2} = \frac{1}{2} \pm i\sqrt{R_c - \frac{1}{4}}. \tag{7.42}$$

Note the occurrence of the famous Richardson number threshold 1/4 familiar from the theory of Kelvin–Helmholtz instability for stratified shear flows. We have assumed large values of R from the outset, so the relevant exponents $\alpha_{1,2}$ are complex numbers. This local solution has magnitude proportional to $\sqrt{|z - z_0|}$ and hence $|\hat{\psi}(z_c)| \propto w' = 0$ whilst $|\hat{\psi}_z(z_c)| \propto u'$ diverges. In addition, the phase of $\hat{\psi}$ is proportional to $\sqrt{R_c - 0.25}\ln s$ and therefore diverges logarithmically as $z \to z_c$, which is consistent with the predictions from ray tracing in §7.2.1.

Thus, the solution exhibits unbounded u' at the critical level, which is clearly unphysical and which also violates the small-amplitude approximation that led to the linear equation (7.36) in the first place. Moreover, the solution is not continuously differentiable at the critical level $z = z_c$ and this goes hand-in-hand with the loss of a unique solution to the boundary-value problem. This loss of a unique solution concerns the entire wave field and not just the neighbourhood of the critical level. For instance, the value of $\overline{u'w'}$ and hence of the drag D is undetermined at this stage.

7.2.3 Causal linear theory for critical layers

To obtain a well-posed model for the linear critical layer requires adding either dissipation or time dependence to the wavetrain dynamics. We take the view that it is most important to look at how a wave field came to be in order to understand what it is today, so we add time dependence first and consider dissipation later, in §7.2.4. We add time dependence by letting $\omega \to \omega + i\epsilon$ where ϵ a small positive real number. The resulting wave field as $\epsilon \to 0$ is the *causal solution* to (7.36) and the structure of this evolving wave field will allow us to understand the flow in the critical layer.

The phase speed c is modified to $c + i\epsilon/k$ and this means c is now a complex number. There is now no real level s at which $U(s) = c$, i.e., the singularity has moved to the complex location $s = i\epsilon/(k\lambda)$. This leads to a well-posed equation with a unique smooth solution. Specifically, the local power-law solution based on (7.41) and (7.42) remains valid, but with s replaced by $s - i\epsilon/(k\lambda)$.

We look in detail at the local structure of an upward-propagating lee wave, which for $k > 0$, $U(0) > 0$ and $c = 0$ corresponds to α_2 in (7.42) in order to satisfy the radiation condition of downward phase propagation in the region below the critical level. No attempt is made to match this local solution to an appropriate lee wave pattern far below the critical level, so

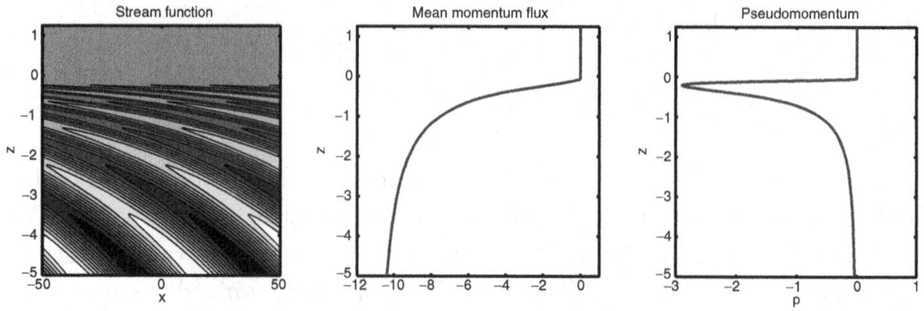

Figure 7.1 Causal internal wave critical layer centred at $z = 0$ with parameters $c = 0$, $N = 0.01$, $\lambda = -0.001$, $k = 2\pi/50$ and $\epsilon = 0.0025|\lambda|$ in units of kilometres and seconds. Left: stream function in arbitrary units, showing phase lines tilted by negative shear and confinement of wavetrain below the critical level. Middle/right: momentum flux $\overline{u'w'}$ and pseudomomentum as functions of altitude, showing momentum flux divergence in the critical layer and the concomitant appearance of a narrow peak of pseudomomentum there.

the wave amplitudes are meaningless, though easily determined via $\overline{u'w'}$. The corresponding stream function is illustrated in Figure 7.1 together with profiles of

$$\overline{u'w'} = \frac{1}{2}\Re(ik\hat{\psi}^*\psi_z) \quad \text{and} \quad \tilde{\mathsf{p}} = \frac{1}{2}\Re\left(\frac{-N^2|\hat{\psi}|^2}{(U - i\epsilon/k)^3}\right). \tag{7.43}$$

The latter is the hydrostatic approximation to $\tilde{\mathsf{p}}$.

The panels in Figure 7.1 show a remarkable structure of the wave field. First, the tilting of the phases lines by the negative shear $U_z = \lambda < 0$ is obvious as is the confinement of the incident wave field below the critical level. Second, the momentum flux far below is at a constant negative level appropriate for a lee wave, but then the flux drops off sharply to zero in the critical layer. This implies a convergence of negative zonal momentum in this critical layer. Third, the pseudomomentum starts with a small negative value far below, but then exhibits a narrow large-amplitude peak in the critical layer before dropping off to zero.

As far as this incident wave structure is concerned there is no apparent wave transmission across the critical level. A more detailed analysis (which includes appropriate matching of the local solution (7.42) to the far-field wave structure away from the critical level) reveals that a very weak transmission is possible, but that the relative amplitude of the momentum flux of the transmitted wave is the very small

$$\exp\left(-2\pi\sqrt{R_c - \frac{1}{4}}\right). \tag{7.44}$$

This drops quickly to zero for large R_c (e.g. $R_c = 100$ in Figure 7.1), so in the atmosphere internal wave transmission is neglected at critical levels.

What does the causal solution teach us about the failure to find a steady-state solution? In a regular problem the limit of the causal solution as $\epsilon \to 0$ yields the steady solution with the correct radiation condition. In the present singular problem this limit leads to a highly non-uniform convergence of the solution. In particular, if we define the critical layer as the region below the critical level in which the momentum flux changes significantly, then the width of this layer decreases as $\epsilon \to 0$, but the momentum flux still goes to zero as $z \to z_c$. For this same reason the peak in $\tilde{\mathsf{p}}$ contained in this layer diverges as $\epsilon \to 0$.

We are led to the conclusion that the occurrence of a critical layer presents a singular perturbation to the linear problem without a critical layer. Specifically, instead of a steady state, the long-time limit of the causal problem yields secular growth of $\tilde{\mathsf{p}}$ in the critical layer: the causal critical layer acts as a peculiar wave absorber, in which the integral of $\tilde{\mathsf{p}}$ is conserved via accumulation of pseudomomentum in the critical layer.

The notion of the causal critical layer as a wave absorber is also supported by the budget for the intrinsic wave energy $\bar{E}$ in (7.16), in which the shear-related source term $-\overline{u'w'}U_z$ is negative for a lee wave encountering a critical layer as in Figure 7.1.

7.2.4 *Singular wave absorption by dissipation*

Ray tracing suggests that a group-velocity ray spends an infinite time approaching the critical level. It is then obvious that any amount of dissipation per unit time, regardless of how small, will lead to the complete attenuation of the wave in the critical layer, simply because the dissipation has an infinite time to act. This is the hallmark of *singular wave absorption*. For example, we can now expect that a small amount of radiative damping will lead to a well-posed dissipative version of a steady critical layer in which the waves are completely absorbed. This consideration of a dissipation complements the analysis of the causal problem in §7.2.3.

In the simple case of weak radiative damping with damping rate $\alpha > 0$ we can use $\tilde{\mathsf{p}} w_g = \overline{u'w'}$ as well as the steady pseudomomentum law

$$(\overline{u'w'})_z = -\alpha \tilde{\mathsf{p}} \tag{7.45}$$

in order to derive (for hydrostatic waves)

$$\overline{u'w'}(z) = \overline{u'w'}(0) \exp\left(-\int_0^z \frac{\alpha d\tilde{z}}{w_g(\tilde{z})}\right) \tag{7.46}$$

$$= \overline{u'w'}(0) \exp\left(-\frac{\alpha N}{|k|}\int_0^z \frac{d\tilde{z}}{(U(\tilde{z})-c)^2}\right). \tag{7.47}$$

This links the momentum flux at altitude z to its starting value at $z = 0$. The flux decays monotonically with altitude and the attenuation integral diverges at the critical level, which shows that $\overline{u'w'}(z_c) = 0$.

The steady pseudomomentum has a similar structure except that there is an additional factor $1/w_g$, i.e.,

$$\tilde{\mathsf{p}}(z) = \tilde{\mathsf{p}}(0)\frac{(U(0)-c)^2}{(U(\tilde{z})-c)^2} \exp\left(-\frac{\alpha N}{|k|}\int_0^z \frac{d\tilde{z}}{(U(\tilde{z})-c)^2}\right). \tag{7.48}$$

This factor leads to transient growth of $\tilde{\mathsf{p}}(z)$ before the inevitable decay to zero as $z \to z_c$. Inspection of (7.48) shows that the maximum of $\tilde{\mathsf{p}}$ occurs at a distance $O(\alpha)$ from the critical level and that the size of this maximum is $O(\alpha^{-2})$. Overall, the momentum flux and $\tilde{\mathsf{p}}$ have the same qualitative structure as in the case of the causal critical layer in Figure 7.1, which suggests the consistency of causal and dissipative critical layers.

The leading-order mean-flow acceleration $\overline{u}_t$ in the critical layer of a steady dissipative wavetrain is given by

$$\overline{u}_t = -(\overline{u'w'})_z = +\alpha\tilde{\mathsf{p}} \tag{7.49}$$

and therefore equals (7.48) times α. This implies that $\overline{u}_t$ shares the vertical structure of $\tilde{\mathsf{p}}$, which includes in particular the narrow sharp peak just below the critical level. The maximum of $\overline{u}_t$ at the peak is $O(\alpha^{-1})$. Clearly, this is a strong wave–mean interaction, with a secularly growing mean-flow response in the critical layer.

7.2.5 Strongly nonlinear critical layers

So far we have looked at the linear theory of the critical layer but, as noted before, there is good reason to doubt that the linear solution presents a valid approximation to the true, nonlinear behaviour of the fluid in the critical layer. Indeed, there is overwhelming observational and numerical evidence that the flow inside critical layers is almost always *strongly* nonlinear and indeed often turbulent. This is the hallmark of nonlinear *wave breaking*. This puts a lot of theoretical work on weakly nonlinear critical layers into perspective, because weakly nonlinear theory can never capture the fully nonlinear character of wave breaking.

It also puts attempts at direct numerical simulation of critical layers into perspective, which would require resolving not only the space–time scales of the waves, but also of the much smaller scales of the three-dimensional turbulence, all the way down to the viscous cut-off scale.

Based on this, our theoretical task is to understand the nonlinearity in the critical layer in order to select the robust features of the linear critical layer model, i.e., those features that remain meaningful in the presence of strong nonlinearity within the critical layer. This leads to a reinforcement of the importance of exact conservation laws.

To begin with, we consider the local wave amplitude as predicted by linear theory in order to understand how linear theory can predict its own breakdown. From what was said before, it is clear that linear wave amplitudes are greatly magnified within the critical layer and are bounded only by the size of the dissipation. For example, it is easy to show that the non-dimensional disturbance shear variance $a^2\overline{u_z'^2}/N^2 \propto |a^2\tilde{\mathsf{p}}/(U-c)| = O(a^2\alpha^{-3})$ in the dissipative critical layer. The same scaling statement holds for the variance $a^2\overline{\zeta_z'^2}$ of the overturning amplitude $a\zeta_z'$, which exceeds unity when the fluid is overturning and thereby signals the onset of convective instability. Stable linear waves therefore require delicate scaling arrangements such that $a \ll \alpha$ in the appropriate non-dimensional sense. Typically, nature does not observe such delicate scaling arrangements and therefore the linear solution is typically unstable in a critical layer, which leads to wave breaking.

The generic reason for wave breaking in the critical layer can be gleaned from a plot of the total streamlines, i.e., basic flow plus wave disturbance, in the lee wave problem. In the critical layer these streamlines form vertically compressed closed loops; these loops are usually referred to as *Kelvin's cat's eyes*. Inside the closed loops the fluid is curling up and stratification surfaces are wound up like spaghetti on a fork. This curling-up grossly exceeds the kinematics that can be described by linear theory, which is confined to describing small undulations relative to a flat rest configuration. Indeed, the irreversible deformation of material lines that in linear theory would merely undulate reversibly is a good practical criterion for nonlinear wave breaking.

The resulting flow is highly unstable to further perturbations. For example, the curling-up of stratification surfaces implies that heavier fluid comes to lie on top of lighter fluid, which is a configuration that is unstable to three-dimensional convection and subsequent breakdown into three-dimensional turbulence. There can also be Kelvin–Helmholtz instabilities due to the strong shear inside the critical layer, which again leads to three-dimensional turbulence. This turbulence mixes the fluid, erodes the background shear and partly destroys the stable stratification inside the critical layer.

This whole process can be observed now and then in rows of billow clouds in the sky, which demarcate the location of a critical layer in its early stages. Within twenty minutes or so the billows break down and mix irreversibly. The appearance of the clouds is coincidental and simply makes the layer visible. Indeed, the wave-breaking process occurs without clouds as well and in this case it is known as *clear-air turbulence*, which is hard to avoid by airplanes because it is not associated with visible convective clouds.

Despite the breakdown of linear theory we can still draw some robust conclusions about the mean-flow response in a critical layer. This is because our arguments were based on the exact conservation laws for mass and momentum. It is only in the detailed evaluation of different terms such as $\overline{u'w'}$ in these conservation laws that we used the linear solutions. Specifically, sufficiently far away from the critical layer we expect linear theory to be quantitatively accurate.

For example, as long as the critical layer absorbs incoming waves without significant reflection or transmission then there must be a concomitant mean-flow acceleration due to the convergence of the wave-induced momentum flux $\overline{u'w'}$. The sign of the acceleration equals the sign of the incoming pseudomomentum, which can be diagnosed below the critical layer via

$$\operatorname{sgn}(\overline{u}_t) = \operatorname{sgn}(\tilde{\mathsf{p}}) = \operatorname{sgn}(k/\hat{\omega}) = \operatorname{sgn}(c - U). \tag{7.50}$$

This shows that during wave absorption the mean flow is always accelerated towards the phase speed c of the incoming wave.

The caveat here is that we cannot establish from within linear theory whether a mature critical layer acts as a wave absorber or reflector. For internal gravity waves the available evidence seems to be that absorption is dominant even for very long times. On the other hand, for a Rossby wave critical layer in the horizontal it is possible to go some way towards a nonlinear model for the critical layer region based on the material invariance of PV; this will be pursued in §9.3.2 and leads to the robust conclusion that a Rossby-wave critical layer may act as an absorber in the early stages, but must act as a reflector in the later stages of its life cycle.

7.2.6 Numerical simulations and drag parametrization

Even if the nonlinear breakdown of waves into three-dimensional turbulence is neglected, the linear critical-layer structure alone poses a non-trivial problem for the direct numerical simulation of the waves and of the wave–mean interactions. This is because the divergence of the vertical wavenumber in the critical layer implies that the wave field will reach the finite vertical grid-scale of a numerical model. At this grid-scale the model does not approximate accurately the governing PDEs, and whether or not the simulated dynamics then resembles the theoretical predictions depends crucially on the type of numerical scheme that is being used.

In a stable discretization, the cascade towards the grid scale will lead to increased numerical dissipation, which reinforces the radiative damping rate α, say, and which may lead to singular numerical absorption of waves even if the explicit $\alpha = 0$. However, the absorption of the incoming waves in the numerical model does not guarantee a physically correct mean-flow response. Indeed, this will depend on whether the momentum equation (7.49) is discretized in a manner that guarantees local conservation of momentum at the grid scale. This would be the case, for instance, in a *finite-volume scheme*, which is based on the flux-divergence form of the governing conservation laws and therefore conservative by construction.

In a *wave drag parametrization scheme*, on the other hand, the entire wave field is modelled by a theoretical procedure based essentially on the kind of linear wave dynamics we have discussed here. The upshot of such a procedure is that much coarser space–time grids based on the evolution of the basic flow only can be used, which is much less costly than using a fine space–time grid based on the evolution of the waves. We will describe a typical example of such a parametrization scheme in the next section.

7.2.7 Saturation parametrization of critical layers

A popular parametrization scheme is based on the idea of *wave saturation*, which combines linear wave theory with a nonlinear threshold for the wave amplitude that models the saturation of the incoming wave due to instabilities and wave breaking. The linear theory is based on steady and non-dissipative ray tracing, which is simple and does not require introducing adjustable parameters, and the nonlinear saturation mechanism is then used to regularize the linear solution in the critical layer. This yields explicit expressions for the saturated momentum flux, which can then be fed back into the numerical model for the evolution of the basic flow.

For example, consider the ray tracing solution to the lee wave problem (7.36) and (7.37) in the hydrostatic approximation:

$$\hat{\psi}(z) = \sqrt{\frac{M(0)}{M(z)}} \exp\left(i \operatorname{sgn}(k) \int_0^z M(\tilde{z})\, d\tilde{z} \right) \quad \text{with} \quad M(z) = \frac{N}{U(z)}. \qquad (7.51)$$

The saturation scheme is based on a stability threshold such as a minimal Richardson number based on the root-mean-square shear, i.e.,

$$\frac{\overline{u_z'^2}}{N^2} = \frac{M^4|\hat{\psi}|^2}{2N^2} \leq \frac{1}{R_*} \tag{7.52}$$

for a suitable constant R_*. Here we have absorbed the non-dimensional wave amplitude a in the definition of u'. A precise value of R_* could be fixed by solving a relevant stability problem, but in practice this is a tunable parameter.

The saturation scheme uses (7.51) as the wave solution up to the saturation level $z = z_s < z_c$, say, where the amplitude in (7.51) reaches the threshold (7.52). This demarcates the begin of the critical layer in the saturation scheme. For $z > z_s$ the amplitude is now saturated, i.e., it is pinned at the saturation threshold such that

$$z_s < z < z_c\colon \quad |\hat{\psi}| \propto M^{-2} \propto U^2 \tag{7.53}$$

for a suitable constant of proportionality. Under the poorly justified assumption that the phase of $\hat{\psi}$ is still given by the WKB expression in (7.51), this rapid decay of $|\hat{\psi}|$ leads to a vanishing u' and a bounded u'_z at the critical level $z = z_c$.

Moreover, the saturated momentum flux follows as

$$z_s < z < z_c\colon \quad \overline{u'w'} \propto M\,|\hat{\psi}|^2 \propto M^{-3} \propto U^3 \propto (z_c - z)^3 \tag{7.54}$$

and therefore

$$z_s < z < z_c\colon \quad \overline{u}_{2t} \propto (z_c - z)^2 \tag{7.55}$$

near the critical level. This simple power law structure can be compared with the much more complex structure in the case of the dissipative critical layer.

Many variants of this simple parametrization scheme are possible. For instance, another saturation scheme reduces the momentum flux to zero directly at the saturation level $z = z_s$; this models the immediate breakdown of the wave once it becomes unstable. This is motivated by some observations of unstable internal wavepackets in the atmosphere, which feature such a rapid breakdown of the wave structure. Other models fall somewhere between these extremes of gradual saturation and immediate destruction. A drawback of all saturation schemes is the discontinuity of $(\overline{u'w'})_z$, and therefore of the wave drag, at the saturation level, which for reasons of numerical stability needs to be smoothed out. In practice, most saturation schemes yield similar results once their parameters have been tuned for optimal performance.

The practical value of any parametrization scheme based on wave saturation lies in its conceptual and mathematical simplicity. Such schemes can also be used in situations without critical layers, at least if they allow the approximate use of ray tracing in the critical layer. For example, the basic density decay with altitude that leads to the wave amplitude growth discussed in §6.7.2 provides another environment in which saturation is relevant for internal waves and for the parametrization of wave drag. Another physical example (discussed in §13.2) is the breaking of incoming surface waves at a beach.

7.3 Joint evolution of waves and the mean shear flow

So far we have treated the basic flow as given and fixed in time. However, in a strong wave–mean interaction it is possible that slow, wave-induced changes in the basic flow may grow to significant amplitude in the sense that the wave structure itself is affected by these changes. In other words, the waves change the very environment on which the propagate. This leads to a joint evolution of both waves and the prevailing mean flow, and to new dynamical possibilities.

For example, in a lee wave problem with critical layers the location of the zero-wind line may change significantly due to wave-induced mean-flow changes, and this will affect the wave structure because of the concomitant shift in the critical levels, which then changes the subsequent locations of strong mean-flow changes and so on.

This joint evolution of waves and mean shear flows over long time scales can lead to very interesting dynamical features, the most prominent of which is the spontaneous occurrence of oscillations in the zonal mean flow. This is believed on strong grounds to be the dynamical explanation for one of the most remarkable features of the equatorial stratosphere: the famous quasi-biennial oscillation (QBO) of the zonal winds there.

From a mathematical point of view, the new element is that the regular perturbation method that underpinned the mathematical set-up described in §5 is not valid over the long, amplitude-dependent times that allow significant changes of the basic flow to accrue. Instead, a suitable singular perturbation method should be used, which allows for the joint evolution of both waves and the mean flow on multiple time scales. To the best of my knowledge, this mathematical program has not been executed fully in situations of practical interest, but by outlining this program it is at least possible to understand clearly what assumptions have to be made in order to justify the use of the commonly used equations in this context.

We will look at the mathematical structure of the multi-scale problem first and then turn to simple examples of wave-induced significant mean-flow changes involving one or several wave modes. Finally, we consider a simple model for the equatorial QBO.

7.3.1 Multi-scale expansion in wave amplitude

The point of departure is the regular perturbation expansion in small wave amplitude $a \ll 1$ for the abstract flow vector $U(t)$ (not to be confused with the zonal mean flow) governed by (5.1) that was described in abstract terms in §5. Formally, the validity of solutions is restricted to times $t = O(1)$, which implies that $U \approx U_0$ throughout. In other words, it is part of the regular perturbation set-up that the changes in U are small and ignorable for the structure of the linear waves.

Now, in a singular perturbation expansion we allow for multiple time scales such that to second order in wave amplitude

$$U(t, a) = U_0(t_0, t_1, t_2) + aU_1(t_0, t_1, t_2) + a^2U_2(t_0, t_1, t_2), \tag{7.56}$$

say. Here the time variables are defined by

$$t_0 = t, \quad t_1 = at, \quad t_2 = a^2t \tag{7.57}$$

and therefore the time derivative is expanded as

$$\frac{\partial}{\partial t} = \frac{\partial}{\partial t_0} + a\frac{\partial}{\partial t_1} + a^2\frac{\partial}{\partial t_2} \tag{7.58}$$

when acting on (7.56). The crucial feature of the multi-scale expansion is that the time variables in (7.57) can be treated as *independent* variables in the intermediate steps of the expansion procedure. Consistency with the single-time reality is achieved by enforcing (7.57) at the end of the computation.

Substitution in (5.1) and collecting terms leads to

$$O(1){:} \quad \frac{\partial U_0}{\partial t_0} + \mathcal{L}(U_0) + \mathcal{B}(U_0, U_0) = 0 \tag{7.59}$$

$$O(a){:} \quad \frac{\partial U_1}{\partial t_0} + \mathcal{L}(U_1) + \mathcal{B}(U_0, U_1) + \mathcal{B}(U_1, U_0) = -\frac{\partial U_0}{\partial t_1} \tag{7.60}$$

$$O(a^2){:} \quad \frac{\partial U_2}{\partial t_0} + \mathcal{L}(U_2) + \mathcal{B}(U_0, U_2) + \mathcal{B}(U_2, U_0) = -\frac{\partial U_1}{\partial t_1} - \frac{\partial U_0}{\partial t_2} - \mathcal{B}(U_1, U_1). \tag{7.61}$$

The new terms on the right-hand side are to be determined by demanding that the expansion (7.56) remains well-ordered for times $t_0 = O(a^{-2})$, i.e., $t_1 = O(a^{-1})$ and $t_2 = O(1)$. This translates to demanding sub-linear growth in time for the various terms in (7.56), which implies that the right-hand side terms in the hierarchy above must not project onto resonant modes of the linear operator apparent on the left-hand side of (7.60) and (7.61). In practice, success in a multi-scale expansion depends on the appropriate choices of the time scales as well as on overcoming the technical difficulties associated with exploiting the sub-linear growth conditions.

Proceeding formally, we view U_0 as the zonal mean flow, which is a trivial steady solution of the governing equation (7.59) and therefore we need to consider its slow dependencies $U_0(t_1, t_2)$ only. It follows that the right-hand side of (7.60) does not depend on t_0, which corresponds to a steady forcing on this time scale. This can resonate with a zero-frequency mode of the linear operator on the left-hand side, and therefore generate a secularly growing $U_1 \propto t_0$, unless the right-hand side does not project onto this mode. We assume that this projection is zero only if the right-hand side is zero, which means we need to consider $U_0(t_2)$ only.

The same argument applied to (7.61) leads to the requirement that the right-hand side vanishes when time-averaged over a long interval in t_0 at fixed t_1 and t_2, i.e., the requirement is that there should be no time-averaged[4] component of the right-hand side. This leads to

$$\frac{\partial U_0}{\partial t_2} + \frac{\partial \langle U_1 \rangle_0}{\partial t_1} = - \langle \mathcal{B}(U_1, U_1) \rangle_0 \tag{7.62}$$

where $\langle \ldots \rangle_0$ denotes time-averaging over t_0. A fundamental problem appears at this stage because it is not clear what the balance between these three terms is.

For instance, in a problem with a trivial basic state $U_0 = 0$ the second and third term describe the weakly nonlinear wave–wave interactions that occur between $O(a)$ waves over the long time scale $t_1 = at$. This is a standard fluid dynamics problem, which typically leads to the consideration of *resonant triads* formed by plane waves that satisfy resonance conditions in wavenumber and frequency. Such resonant triads certainly exist for internal waves, for example, they include the so-called *parametric sub-harmonic instability*, which is believed to play an important role in the ocean. Such resonant wave–wave interactions cannot be transformed away and they may

[4] In single-time reality, the notion of time-averaging over t_0 at fixed t_1 and t_2 corresponds to time-averaging over a time interval $T(a)$ that is large compared to t_0 but small compared to t_1 and t_2 in the limit $a \to 0$. For example, in the present case a time interval $T(a) \propto a^{-1/2}$ would work.

lead to significant changes in the wave field U_1 over the first slow time scale t_1.

Thus, at this stage one has to assume that these wave–wave interactions can be ignored. It is then possible to time-average (7.62) over the first slow time t_1 whilst demanding that $\langle U_1 \rangle_0$ grows sub-linearly in t_1, which leads to

$$\frac{\partial U_0}{\partial t_2} = - \langle \mathcal{B}(U_1, U_1) \rangle_{01} . \tag{7.63}$$

The clumsy notation indicates that the forcing term has been time-averaged over both t_0 and t_1. Physically, this time-average covers many oscillation periods of the waves in t_0 and also many weakly nonlinear interaction cycles in t_1. This equation together with the homogeneous part of (7.60) is the sought-after abstract system for the joint evolution of U_0 and U_1 over very long time scales.

Now, if this abstract system is translated to the concrete Boussinesq system, then we obtain equations in which the zonal mean flow evolves slowly in response to the long-time average of the momentum-flux divergence whilst the structure of the internal waves continually adjusts to the slowly evolving structure of the mean flow. In a nutshell, the wave field is at all times in equilibrium with the slowly evolving mean flow. This is a very natural result, but the caveat about the neglected wave–wave interactions should be noted clearly.

Is it possible to overcome this caveat and take wave–wave interactions into account in a situation of practical interest? To my knowledge, the theory of wave–wave interaction is only well-developed for uniform basic states in which the waves are globally given in terms of orthogonal plane waves. However, in the presence of a shear flow it is virtually impossible to compute these interactions rigorously because there is no simple set of wave shapes that can serve as a complete expansion set. This situation is made even worse by the presence of critical layers, which are of great practical importance.

Despite this negative outlook there are three heuristic reasons why wave–wave interaction may nevertheless be negligible in practice, at least for wave drag problems. The first reason is that wave–wave interactions are very sensitive to intermittency in the linear wave fields, because the persistent overlap of the waves in space and time is essential for the workings of the resonant interactions. For example, a gusty lee wave situation in which wavepackets are generated intermittently in time will presumably invoke much weaker wave–wave interactions than a situation with a steady wavetrain. On the other hand, the long-time averaged forcing term in (7.63) is not affected by intermittency.

Second, in an upward-propagating spectrum of internal waves the time available for wave–wave interactions may in fact be limited by the total vertical propagation time from wave source to wave dissipation site, as determined from the vertical group velocity. In this case wave-wave interactions cannot unfold over the full time of integration and are therefore limited in scope and amplitude.

Finally, a third reason is that resonant wave–wave interactions conserve certain quantities such as wave energy and also pseudomomentum. In fact, the conservation of pseudomomentum under such interactions can even be made precise using a fully nonlinear definition of pseudomomentum, as in GLM theory. This suggests that wave–wave interactions may have no impact on the total pseudomomentum flux and therefore on the zonal mean-flow acceleration. Of course, this neglects changes in the vertical structure of a dissipative wavetrain (such as a shift in the vertical damping rate), which can be caused by wave–wave interactions.

With these caveats and heuristic arguments in mind, we will now look at the consequences of the proposed model for the long-term joint evolution of waves and mean flows.

7.3.2 Examples of joint wave–mean dynamics

We simplify the notation compared to the last section by omitting any special notation for slow times and small-amplitude expansions. Thus, we write $\overline{u}(z,t)$ and view t as t_2. We explicitly want to allow for critical layers, and we use dissipation based on radiative damping for the wave structure in these layers; other models such as saturation would also be possible, of course. To obtain a simple well-posed model system we add some vertical diffusion of zonal mean momentum over the long time scales of interest; this is also relevant for laboratory experiments.

We consider first the case of a single upward-propagating internal wave with horizontal wavenumber k, phase speed $c > 0$, and ground-level momentum flux $\overline{u'w'}(0)$. There is no feedback between $\overline{u}$ and these wave parameters, i.e., we consider this wave as forced near the ground by mechanisms unrelated to the evolving shear flow. The vertical structure of the wave is affected by the shear flow, of course.

The resulting mean-flow equation is

$$\begin{aligned}\overline{u}_t - (\nu\overline{u}_z)_z &= -(\overline{u'w'})_z \\ &= \overline{u'w'}(0)\,\frac{\alpha N}{|k|(\overline{u}-c)^2}\exp\left(-\frac{\alpha N}{|k|}\int_0^z \frac{d\tilde{z}}{(\overline{u}(\tilde{z},t)-c)^2}\right). \qquad (7.64)\end{aligned}$$

Here $\nu(z) > 0$ is the kinematic diffusivity, which may depend on altitude in order to mimic boundary layer effects near the ground. The boundary conditions for $\overline{u}$ are chosen as

$$\overline{u}(0,t) = 0 \quad \text{and} \quad \overline{u}_z(H,t) = 0, \tag{7.65}$$

for some upper altitude H, say. The mean-flow equation is a nonlinear integro–differential equation for $\overline{u}(z,t)$, which does not fit any standard theory for PDEs. This makes it important to understand qualitatively how the various terms interact.

To begin with, the lower boundary condition implies

$$\text{sgn}(\overline{u'w'}(0)) = \text{sgn}(\tilde{\mathsf{p}}(0)) = \text{sgn}(c - \overline{u}(0,t)) = \text{sgn}(c) \tag{7.66}$$

and therefore the wave-induced force in (7.64) always points in the direction of c. Further, if we assume that the typical situation contains at least one critical level where $\overline{u} = c$ then $0 < \overline{u} < c$ in the lower region between the ground $z = 0$ and the first critical level $z = z_c(t)$. The waves are absorbed in the critical layer and therefore the right-hand side of (7.64) is zero in the upper region $z > z_c$. For slowly evolving mean flows we therefore expect the diffusive equilibrium $\overline{u} = c$ in $z_c < z < H$.

Now, the crucial point is that in the critical layer below $z = z_c$ the mean flow $\overline{u} < c$ is dragged towards the critical value $\overline{u} = c$. Specifically, in the shear zone approaching the critical layer we have $\overline{u}_z > 0$ and the refraction caused by this shear will move $|m|$ to higher values, which reduces w_g and $|\overline{u} - c|$ along rays and therefore increases the radiative damping rate per unit altitude. This leads to the momentum flux convergence in the critical layer that pushes $\overline{u}$ towards c.

In other words, the dynamics described by (7.64) naturally leads to a *descent* of the time-dependent critical level $z_c(t)$. This is the key dynamical feature in this setting. The speed of this critical-layer descent is proportional to the momentum flux $\overline{u'w'}(0)$, but its absolute value depends on the mean-flow structure just below the critical level. Finally, for sufficiently small ν a steady state can be reached in which $\overline{u} = c$ everywhere except in a viscous boundary layer near the ground in which very strong shears arrest the critical-layer descent and $\overline{u}$ reduces to zero at the ground. This lower boundary layer is the least robust and least realistic feature of the solution.

We now consider the very interesting case of multiple waves, say two waves A and B with corresponding parameter values (k_A, k_B) and so on. These waves do not interact at the linear level and if $|k_A| \neq |k_B|$ then by orthogonality the total momentum fluxes are also simply the sum of the individual fluxes, i.e.,

$$\overline{u'w'} = \overline{u'_A w'_A} + \overline{u'_B w'_B}. \tag{7.67}$$

However, the special case $|k_A| = |k_B|$ is important and then (7.67) is augmented by cross-correlation terms that fluctuate in time and altitude according to the 'beat' modulations $\omega_A \pm \omega_B$ and $m_A \pm m_B$. Still, on the implied slow time scale these $O(1)$ flux oscillations average to zero and therefore (7.67) can be used for all kinds of waves.

For multiple waves the mean-flow equation (7.64) therefore contains a sum of similar terms on the right-hand side, one for each wave. Of particular interest is the case of a standing wave in x near the ground, which corresponds to the symmetrical set-up

$$|k_A| = |k_B| = |k|, \quad -c_B = c_A = c > 0, \tag{7.68}$$

$$\text{and} \quad -\overline{u'_B w'_B}(0) = \overline{u'_A w'_A}(0) = \overline{u'w'}(0) > 0. \tag{7.69}$$

Thus wave A propagates with speed $c > 0$ to the right and wave B propagates with the same speed and amplitude to the left. At the ground the wave therefore has the standing-wave form $\cos(kx)\exp(-ickt)$. However, the refraction by the mean shear acts differently on the two waves and therefore in the interior of the domain the wave pattern will not be that of a standing wave.

Specifically, wave A has a critical level where $\overline{u} = c$ whilst for wave B the critical level occurs where $\overline{u} = -c$. This difference leads to very interesting dynamical affects in the mean-flow equation (7.64), whose right-hand side is

$$\overline{u'w'}(0)\left\{\frac{\alpha N}{|k|(\overline{u}-c)^2}\, e^{-\int_0^z \frac{\alpha N d\tilde{z}}{|k|(\overline{u}-c)^2}} - \frac{\alpha N}{|k|(\overline{u}+c)^2}\, e^{-\int_0^z \frac{\alpha N d\tilde{z}}{|k|(\overline{u}+c)^2}}\right\} \tag{7.70}$$

now. If the flow starts from rest $\overline{u}(z,0) = 0$ then the right-hand side is zero by symmetry. However, this is an unstable equilibrium, as any small perturbation in $\overline{u}(z,0)$ will act differently on the two waves.[5]

For example, we can consider an initial condition given by the typical situation for wave A in isolation, i.e., $0 < \overline{u} < c$ in a lower region whilst $\overline{u} = c$ in an upper region above the critical level at $z = z_{cA}$, say. Now, wave B is not affected by $z = z_{cA}$ and passes right through this level. In addition, the positive shear $\overline{u}_z > 0$ in the lower region refracts wave B to lower values of $|m_B|$ and therefore to higher values of w_g, as indicated by the concomitant increase in $\overline{u} + c$ with altitude. Thus the radiative damping per unit altitude

[5] Here saturation models differ from dissipation models because in a saturation model there would be no wave-induced force until $|\overline{u}|$ exceeds c somewhere in the domain.

is *reduced* for wave B in the shear zone associated with the approach to the critical layer for wave A.

In a nutshell, a critical layer for wave A is particularly well suited for transmission of wave B, and vice versa. This transmission leads to a negative zonal force in the upper region $z > z_{cA}$, because in this region only wave B is present. Consequently, the mean flow in the upper region will accelerate towards $-c$, and once a critical level at z_{cB}, say, has been established where $\overline{u} = -c$, then this critical level will descend in the same fashion as described before. Of course, the first critical level z_{cA} also descends by the same mechanism.

Thus, the present consideration of two equal-and-opposite waves has led to the conclusion that the mean flow develops an oscillatory profile that descends in the vertical. This remarkable mean-flow pattern appears spontaneously, i.e., it appears from arbitrary non-zero initial conditions for the mean flow in the present model.

It is not easy to guess the dynamics of the system once the first critical level at $z = z_{cA}$ reaches the boundary layer near the ground. A steady state is now not possible because the second critical level at $z = z_{cB}$ keeps on descending. This crowding of the flow structure near the ground also exceeds the simple modelling assumptions that went into (7.64) and (7.70). If one assumes that for dynamical reasons outside the present model a critical level is destroyed once it has descendent below a certain altitude, then it follows at once that wave A is now free to propagate past the critical level $z = z_{cB}$, and therefore the two waves have switched their role. Now it is the turn of wave A to accelerate the fluid above $z = z_{cB}$ towards the speed c and therefore the cycle repeats. This leads to the remarkable prediction of a recurring cycle of mean-flow oscillations that descend in the vertical.

Finally, it is noteworthy that the fundamental dynamical mechanisms of preferential wave transmission and dissipation in a shear flow that are at work here will function also in the presence of near-critical layers, in which $|\overline{u} - c|$ becomes small but never zero. This underpins the robustness of these results, which makes them accessible to relatively simple laboratory experiments.

7.3.3 The quasi-biennial oscillation

It has been possible to observe the spontaneous occurrence and subsequent long-term persistence of such a mean-flow oscillation in a celebrated laboratory experiment. Much later, this remarkable experimental result has also been verified by direct numerical simulations. The experiment used an upright cylindrical tank filled with stratified water. Internal waves were generated at one end of the cylinder by means of a flexible membrane, which oscillated in a low-mode standing-wave pattern in the azimuthal direction. The mean flow was made visible by suspended neutrally buoyant particles whose motion was captured by a camera. There was no radiative damping in this experiment, but molecular viscosity always plays a role on laboratory scales. In fact, the wave damping rate per unit time due to molecular viscosity is proportional to the wavenumber squared and therefore refraction in a critical layer increases the wave damping rate per unit altitude even more strongly than in the case of radiative damping. Starting from rest and running this experiment for a long time showed the robust appearance of counter-propagating bands of azimuthal mean flows and their descent towards the wave source.

The experiment was motivated by theoretical efforts aimed at understanding the quasi-biennial oscillation (QBO) in the equatorial stratosphere. This is a descending oscillation pattern of the zonal mean winds between about 18–30 km altitude, with a descent speed of about 1km per month. The observational track record for these oscillations goes back more than 50 years and the average period of these well-observed oscillations is about 27–28 months, which is manifestly *not* a sub-harmonic of the annual cycle.

The QBO is one of the most predictable large-scale circulation patterns on Earth. Despite this apparent non-randomness, the dynamical workings of the QBO remained mysterious until an explanation based on equatorial waves and wave–mean interaction theory was proposed. The real waves that contribute to the momentum flux in the QBO are not straightforward internal waves, but rather they are a composite of several different wave types that occur in the equatorial region. Still, the fundamental dynamical processes that lead to the QBO are believed to be captured by the simple model described in the last section.

In addition, this model also leads to a ready explanation for the equatorial confinement of the three-dimensional QBO, which does not extend beyond about ±15 degrees latitude. This is due to radiative damping, which in the stratosphere is characterized by damping time scales for large-scale temperature anomalies of about 20 days. Away from the equator, a shear flow is subject to horizontal Coriolis forces which in turn lead to horizontal temperature gradients. Therefore, an off-equatorial multi-year oscillation in the mean shear is indirectly linked to a temperature oscillation and therefore subject to strong radiative damping, and hence cannot be maintained.

On and near the equator, however, the horizontal Coriolis forces are very weak, which allows a shear-only oscillation to persist. Overall, the latitudinal extent of the QBO is consistent with an estimate based on the oscillation period, the Coriolis forces and realistic damping rates.

Even though the basic model for the QBO in terms of waves and wave–mean interactions is broadly accepted, there are a number of observational facts that are hard to pin down in a simple model. These include an observed asymmetry in strength between eastward and westward phases of the QBO as well as the problem of obtaining a quantitative match for the observed descent speed of the QBO once equatorial mean vertical motions due to three-dimensional circulation patterns unrelated to the QBO are taken into account. From a theoretical point of view another intriguing observational fact that is not directly explained by the present theory is the apparent regularity of the observed period. This is because in the basic model the mean-flow forcing is proportional to the ground-based wave amplitude squared, which in the absence of viscous mean-flow stresses directly leads to an oscillation period that is inversely proportional to the wave amplitude squared. This model suggests a strong sensitivity of the period on wave amplitudes.

Hence, the observational record suggests either quite repeatable wave amplitudes in the equatorial regions over many years, or perhaps it suggests the possibility that a nonlinear thresholding mechanism such as saturation might produce a QBO that is insensitive to increasing wave amplitudes near the ground.

Finally, obtaining a realistic QBO in a numerical model has been an important challenge in the development of global atmospheric models for climate research. Even with parametrized wave drag it has been difficult to obtain a realistic QBO because of the very strong numerical mean-flow damping due to coarse vertical resolution in the models. Present-day general circulation models are beginning to exhibit spontaneous QBOs due to their directly resolved wave and mean-flow dynamics.

It will be an important milestone, not too far away now, when a subtle multi-scale feature such as the QBO can be routinely simulated in our global numerical models.

7.4 Notes on the literature

Two classic papers on wave propagation through shear are Bretherton and Garrett (1968), which deals with wave action conservation, and Booker and Bretherton (1967), which deals with internal wave critical layers. A recent review article on critical layers in atmospheric flows is Haynes (2003). Drag parametrizations based on wave saturation were pioneered the late 1960s; a central reference is Lindzen (1981). The inability to obtain a QBO in a zonally-symmetric model without waves unless a mysterious zonal force on the mean flow is added was demonstrated in the pioneering study of Wallace and Holton (1968). A wave-based explanation for the QBO was then suggested in Lindzen and Holton (1968) using special wave types relevant for equatorial regions; the inevitability of QBO-type oscillations for generic wave situations was stressed in Plumb (1977) and the famous laboratory experiment on QBO-type oscillations is described in Plumb and McEwan (1978). A review article of recent QBO research is Baldwin et al. (2001).

7.5 Exercises

1. *Sheared-over disturbances with stratification.* Consider (7.34) in the case of a uniform shear flow $U = \lambda z$ and assume a disturbance of the form

$$\psi'(x, z, t) = f(t)g(x - \lambda z t) \tag{7.71}$$

in terms of two functions $f(\cdot)$ and $g(\cdot)$. Show that a solution for arbitrary $g(\cdot)$ is obtained only if $f(t)$ satisfies the ODE

$$\frac{d^2}{dt^2}\left[\left(1 + \lambda^2 t^2\right) f(t)\right] + N^2 f(t) = 0. \tag{7.72}$$

Find b' and verify that this is in fact a solution of the nonlinear equations as well, because both Jacobians $\partial(\psi', \nabla^2\psi') = \partial(\psi', b') = 0$. For large t show that $f(t) \sim t^\mu$ where

$$\mu = -\frac{3}{2} \pm i\sqrt{R - \frac{1}{4}} \quad \text{and} \quad R = \frac{N^2}{\lambda^2}. \tag{7.73}$$

Even though ψ' decays in time if $R > 1/4$, show that both u'_z and b'_z grow without bound, which shows the inevitable breakdown of the solution due to some combination of shear and convective instability.

2. *Oscillatory mean-flow dynamics without critical layers.* Consider the wave-induced mean-flow evolution given by the first part of (7.64) together with the momentum flux given in (7.70). Write a simple finite difference code for $\overline{u}(z, t)$ based on these equations and investigate the possibility that for strong enough mean-flow viscosity ν a critical layer may never form, yet mean-flow oscillations can nevertheless arise. This is because the preferential transmission and dissipation of waves by the shear flow is robustly at work even if $\overline{u}$ is below critical layer strength.

8
Three-dimensional rotating flow

Building on the vertical slice model, both the extension to three-dimensional flow and the addition of Coriolis forces are important and non-trivial steps in classic wave–mean interaction theory. Specifically, in three dimensions we recover the generic $2+1$ structure of the linear problem, in which we have two gravity-wave modes and one balanced or vortical mode controlled by the PV distribution. This leads to the generic importance of the zero-frequency PV mode for strong interactions.

The Coriolis forces lead to source terms in the horizontal momentum budgets and therefore to differences between Lagrangian and Eulerian fluxes of horizontal momentum. As noted before, this leads to the important definition of the vertical Eliassen–Palm flux or form stress in the context of wave drag computations. Once again, the zonal pseudomomentum plays a crucial role in formulating the interaction theory.

We briefly recall the governing equations and then look at the modifications of the linear dynamics, including that of pseudomomentum and its flux. This is followed by rotating three-dimensional lee waves and by a discussion of how to simplify the considerably more complicated mean-flow equations in this case. The key concept here is to focus on the vortical mode of the mean-flow response. Finally, because its intrinsic importance in idealized modelling, we discuss the vertical slice model with rotation.

8.1 Rotating Boussinesq equations on an f-plane

We consider frame rotation in the traditional approximation, in which only the locally vertical component of the Coriolis vector is retained (cf. §4.2.2) and we work on an f-plane, in which this component is treated as constant.

The Coriolis force enters the Boussinesq equations via

$$\frac{D\boldsymbol{u}}{Dt} + f\hat{\boldsymbol{z}} \times \boldsymbol{u} + \boldsymbol{\nabla} P = b\hat{\boldsymbol{z}} \tag{8.1}$$

where $\boldsymbol{u}$ is now the velocity relative to the rotating Earth and the vertical unit vector $\hat{\boldsymbol{z}}$ points in the local 'upward' direction, which corresponds to the radial outward direction in spherical geometry. The frame rotation has introduced a second frequency scale into the Boussinesq equations, but there is still no intrinsic length scale.

In local Cartesian coordinates the f-plane Boussinesq equations are

$$\frac{Du}{Dt} - fv + P_x = 0, \quad \frac{Dv}{Dt} + fu + P_y = 0, \quad \frac{Dw}{Dt} + P_z = b, \tag{8.2}$$

$$\frac{Db}{Dt} + N^2 w = 0 \quad \text{and} \quad u_x + v_y + w_z = 0. \tag{8.3}$$

The exact PV is given by

$$q = (\boldsymbol{\nabla} \times \boldsymbol{u} + f\hat{\boldsymbol{z}}) \cdot \boldsymbol{\nabla}(b + N^2 z) \quad \text{such that} \quad \frac{Dq}{Dt} = 0. \tag{8.4}$$

For simplicity, we are restricting attention to constant N throughout. In comparison with rotating shallow water, we can note that rotation in the Boussinesq equations defines a meaningful aspect ratio f/N whereas in shallow water it defined a meaningful horizontal length scale $L_D = \sqrt{gH}/f$.

8.2 Linear structure

The linear equations for small disturbances relative to a rest state are

$$u'_t - fv' + P'_x = 0, \quad v'_t + fu' + P'_y = 0, \quad w'_t + P'_z = b', \tag{8.5}$$

$$b'_t + N^2 w' = 0 \quad \text{and} \quad u'_x + v'_y + w'_z = 0. \tag{8.6}$$

The initial-value problem is posed by specifying three independent initial fields, e.g., the buoyancy and two components of the velocity, the third being computable from $\boldsymbol{\nabla} \cdot \boldsymbol{u}' = 0$.

8.2.1 Balanced vortical mode and Rossby adjustment

The structure of this zero-frequency mode is found from (8.5) by setting the time derivatives to zero. This yields

$$P'_x = fv', \quad P'_y = -fu', \quad P'_z = b', \quad N^2 w' = 0 \quad \text{and} \quad u'_x + v'_y = 0. \tag{8.7}$$

Thus the vortical mode is in geostrophic and hydrostatic balance, i.e., the horizontal pressure gradient balances the Coriolis force and the vertical pressure gradient balances the buoyancy force. Also, the vertical velocity is zero and the horizontal flow is non-divergent. This allows introducing a balanced stream function

$$\psi' = P'/f \quad \text{such that} \quad u' = -\psi'_y, \quad v' = +\psi'_x \quad \text{and} \quad b' = f\psi'_z. \tag{8.8}$$

The crucial new feature compared to the non-rotating case is that the balanced pressure field establishes a non-trivial link between the horizontal velocities and the buoyancy. Specifically, the horizontal gradient of b' is now linked to the vertical shear via

$$(b'_x, b'_y) = f(v'_z, -u'_z). \tag{8.9}$$

This is called the *thermal wind relation* in meteorology, and it has many implications for the structure of slowly evolving weather systems. For example, in the atmosphere a positive buoyancy disturbance $b' > 0$ can be broadly associated with higher temperatures and vice versa. Generally, there is a meridional temperature gradient in the troposphere from higher temperatures near the equator to lower temperatures at higher latitudes, which corresponds to $b'_y < 0$ in the northern hemisphere, say. The thermal wind relation then suggests a positive wind shear $u'_z > 0$ at mid-latitudes. This is compatible with the observed zonal *jet stream*, which features prograde velocities of around 100 m/s at the top of the troposphere, at around 10 km altitude.

The balanced mode is controlled by the linear PV based on (8.4), which after scaling can be written as

$$q' = v'_x - u'_y + \frac{f}{N^2} b'_z \quad \text{such that} \quad q'_t = 0. \tag{8.10}$$

The third term equals $-f\zeta'_z$ and it quantifies the stretching of planetary vorticity in a rotating frame. For example, if $q' = 0$ then $v'_x - u'_y = f\zeta'_z$ and vertical stretching $\zeta'_z > 0$ implies creation of relative vorticity with the same sign as f; this is the familiar 'ballerina effect'. The relations (8.10) hold in a general flow situation, balanced or not, and therefore $q'(\boldsymbol{x})$ is known from the initial conditions. This allows an easy computation of the asymptotic balanced end state of the linear initial-value problem, because in a steady balanced flow q' and ψ' are linked by the Poisson equation

$$\left(\frac{\partial^2}{\partial x^2} + \frac{\partial^2}{\partial y^2} + \frac{f^2}{N^2}\frac{\partial^2}{\partial z^2}\right)\psi' = q'. \tag{8.11}$$

For a given q' and suitable boundary conditions this elliptical operator can be inverted to find ψ' and therefore the structure of the vortical mode. For example, in an unbounded domain with decay conditions at infinity the solution in terms of a scaled Green's function is

$$\psi'(x,y,z) = -\frac{N}{4\pi f}\int \frac{q'(x',y',z')\,dx'dy'dz'}{\sqrt{(x-x')^2+(y-y')^2+(z-z')^2N^2/f^2}}. \quad (8.12)$$

Simpler still, in a periodic domain the Fourier solution is

$$\hat{\psi} = \frac{-\hat{q}}{k^2+l^2+m^2f^2/N^2}. \quad (8.13)$$

This shows that the aspect ratio of the balanced flow is the latitude-dependent f/N, which is also called *Prandtl's ratio.* In geophysics $N^2 \gg f^2$ almost always holds[1] and therefore typical balanced flow structures are shallow, pancake-like structures with much larger horizontal than vertical extent.

The preferred aspect ratio also enters the computation of q' itself. For example, consider a situation in which b_0' is the only non-zero initial field and therefore $q' = fb_{0z}'/N^2$. The final balanced buoyancy field is $b' = f\psi_z'$ and in a periodic domain (8.13) then implies the projection

$$\hat{b} = \frac{m^2f^2/N^2}{k^2+l^2+m^2f^2/N^2}\,\hat{b}_0. \quad (8.14)$$

This shows that the balanced b' will be close to the initial b_0' only if the aspect ratio of b_0' is much shallower than the preferred ratio f/N. If b_0' is much taller, then most of it will have been projected onto wavelike modes, which can propagate away and leave the balanced flow behind.

For example, in the classical *Rossby adjustment problem* we consider the initial-value problem for the system (8.5) in R^3 with compact and smooth initial conditions localized at the origin of the coordinate system, say. The general solution then consists of time-dependent waves that propagate away from the origin with non-zero group velocity[2] plus a steady, non-propagating vortical mode. The solution at every fixed $\boldsymbol{x}$ converges to a balanced end state as $t \to \infty$, although this convergence is pointwise rather than uniform because of the propagating waves that are found at larger and larger distances as $t \to \infty$. The balanced end state can be computed directly from (8.11) without having to compute the wave solution.

It is clear that two different initial conditions will give rise to the same

[1] In the atmosphere the magnitude of N/f is 100 at mid-latitudes and in the ocean it ranges from about 100 in the upper ocean to about 10 in the deep ocean.

[2] There is a set of measure zero in wavenumber space on which the group velocity is zero and although for compact and smooth initial conditions this set is negligible as $t \to \infty$, in practice waves near this set are obviously the slowest to depart.

final balanced state ψ' if their PV distributions are identical. There will be differences in the emitted waves, of course. An analogous statement holds for the inhomogeneous version of the linear equations in which a body force $\boldsymbol{F}' = (F', G', H')$ and diabatic heating R' have been added. As far as the vortical mode is concerned the only quantity that matters is the evolution of q', which is

$$q'_t = G'_x - F'_y + \frac{f}{N^2} R'_z. \tag{8.15}$$

This makes it obvious that different combinations of horizontal force and heating can give rise to the same right-hand side in (8.15) and therefore to the same balanced flow response. This equivalence of different forcing terms will greatly simplify the theory of strong wave–mean interactions below.

8.2.2 Internal inertia–gravity waves

The two remaining modes of the system (8.5) describe internal *inertia–gravity waves*, whose intrinsic dispersion relation for plane waves with wave-number vector $\boldsymbol{k} = (k, l, m)$ is[3]

$$\hat{\omega}^2 = N^2 \frac{k^2 + l^2}{k^2 + l^2 + m^2} + f^2 \frac{m^2}{k^2 + l^2 + m^2}. \tag{8.16}$$

Coriolis forces affect shallow, flat waves more whilst the converse is true for buoyancy forces. In addition to the already familiar limiting high-frequency case of buoyancy oscillations with $m = 0$ and $\hat{\omega} = \pm N$ there are now also low-frequency *inertial oscillations* with $k = l = 0$ and $\hat{\omega} = \pm f$. These oscillations are layerwise uniform gyrations of the entire horizontal fluid plane, rotating clockwise if $f > 0$ and vice versa. Propagating waves are confined to the finite frequency band

$$f^2 < \hat{\omega}^2 < N^2. \tag{8.17}$$

For finite f this equation implies that there is now a frequency gap between the steady balanced mode and propagating inertia–gravity waves. This gap is important in practice, because it severely limits the scope for interaction between internal waves and the balanced flow. It also makes clear that f-plane inertia–gravity waves must have zero PV, i.e.,

$$q' = v'_x - u'_y + \frac{f}{N^2} b'_z = 0 \tag{8.18}$$

[3] More generally, if $\boldsymbol{f}$ is not parallel to $\hat{\boldsymbol{z}}$ then (8.16) is replaced by $\hat{\omega}^2 = N^2 |\boldsymbol{k} \times \hat{\boldsymbol{z}}|^2 / |\boldsymbol{k}|^2 + (\boldsymbol{f} \cdot \boldsymbol{k})^2 / |\boldsymbol{k}|^2$.

for inertia–gravity waves. On a β-plane this is replaced by $q' = -\beta\eta'$, of course.

As in the non-rotating case, $\hat{\omega}$ is zeroth-degree homogeneous in $\boldsymbol{k}$, which leads to the already discussed consequences for the intrinsic group velocity etc. Many equivalent expression for the group velocity are possible; one version is

$$(\hat{u}_g, \hat{v}_g) = \frac{(k,l)}{\sqrt{k^2+l^2}} \frac{N^2 - \hat{\omega}^2}{\hat{\omega}|\boldsymbol{k}|} \sqrt{\frac{\hat{\omega}^2 - f^2}{N^2 - f^2}} \tag{8.19}$$

$$\hat{w}_g = -\mathrm{sgn}(m) \frac{\hat{\omega}^2 - f^2}{\hat{\omega}|\boldsymbol{k}|} \sqrt{\frac{N^2 - \hat{\omega}^2}{N^2 - f^2}}. \tag{8.20}$$

This makes explicit that the group velocity is zero for the two limiting oscillations and that $|\widehat{\boldsymbol{c}}_g| \propto 1/|\boldsymbol{k}|$ for fixed $\hat{\omega}$. Explicitly:

$$|\widehat{\boldsymbol{c}}_g|^2 = \frac{(N^2 - \hat{\omega}^2)(\hat{\omega}^2 - f^2)}{\hat{\omega}^2 |\boldsymbol{k}|^2}. \tag{8.21}$$

Thus $\widehat{\boldsymbol{c}}_g$ is zero for both buoyancy oscillations $\hat{\omega} = N$ and inertial oscillations $\hat{\omega} = f$; this differs from the non-rotating case, because there $\widehat{\boldsymbol{c}}_g$ was non-zero in the shear-flow limit $\hat{\omega} = 0$.

A useful alternative form of (8.16) is

$$\frac{m^2}{k^2 + l^2} = \frac{N^2 - \hat{\omega}^2}{\hat{\omega}^2 - f^2}. \tag{8.22}$$

For given $\hat{\omega}$ the inverse of the ratio in (8.22) is the square of the characteristic slope $\mu = dz/dx$ as discussed in connection with time-periodic modes in (6.28). This makes clear that for fixed $\hat{\omega}$ the characteristic slope flattens, i.e., it becomes more horizontal, as f^2 is increased. Also, (8.22) implies that a critical level in the vertical with $m^2 \to \infty$ will arise at altitudes where the intrinsic frequency is pushed towards its finite lower limit $\hat{\omega}^2 \to f^2$. As noted before, ray tracing as well as linear theory breaks down in the corresponding critical layer.

Due to isotropy in the horizontal plane it suffices to describe the detailed structure of inertia–gravity waves for the special case $\boldsymbol{k} = (k, 0, m)$, in which $\boldsymbol{k}$ lies in the vertical xz-plane. The incompressibility condition again implies that these are transversal waves, now with a motion in the xz-plane that is ninety degrees out of phase with motion in the y-direction. Specifically, we obtain

$$u' = -\frac{m}{k} w', \quad v' = i\frac{fm}{\hat{\omega}k} w', \quad b' = -i\frac{N^2}{\hat{\omega}} w', \quad P' = \frac{\hat{\omega}^2 - N^2}{m\hat{\omega}} w'. \tag{8.23}$$

Thus the two sets (u', w', P') and (b', v') are ninety degrees out of phase. The associated particle trajectories are straight lines in the xz-plane and ellipses with aspect ratio $f/\hat{\omega}$ in the xy-plane. The sense of particle rotation along the horizontal ellipses is clockwise when viewed from above for $f > 0$. Direct observation of such elliptical particle orbits is a useful measurement technique to detect inertia–gravity waves.

Turning to wave properties, it turns out that the mean wave energy $\bar{E}$ does *not* satisfy energy equipartition. Instead, there is equipartion between the kinetic energy in the xz-plane and the sum of potential energy plus the kinetic energy in the y-direction, i.e.,

$$\bar{E} = \frac{1}{2}\overline{|\boldsymbol{u}'|^2} + \frac{1}{2}\frac{\overline{b'^2}}{N^2} = \overline{u'^2} + \overline{w'^2} = \frac{\overline{b'^2}}{N^2} + \overline{v'^2}. \tag{8.24}$$

For high-frequency waves $f/\hat{\omega}$ is small and therefore v'^2 is negligible against b'^2/N^2, which recovers the non-rotating equipartition result. On the other hand, for low-frequency waves the opposite is true and potential energy becomes less important. For example, in the limiting case of an inertial oscillation w' and b' are zero and therefore *all* the wave energy is kinetic energy in the xy-plane. Two useful approximate formulas in this context are

$$\frac{\overline{b'^2}}{N^2} = \left(1 - \frac{f^2}{\hat{\omega}^2}\right)\bar{E} \quad \text{and} \quad \overline{|\boldsymbol{u}'|^2} = \left(1 + \frac{f^2}{\hat{\omega}^2}\right)\bar{E}. \tag{8.25}$$

The Lagrangian version of zonal pseudomomentum p follows as in §6.3.1, but with $\boldsymbol{u}'$ replaced by $\boldsymbol{u}' + \frac{1}{2}f\hat{\boldsymbol{z}} \times \boldsymbol{\xi}'$, which is the absolute disturbance velocity at the displaced particle position. This is discussed in the context of GLM theory later.

In keeping with the Eulerian approach to wave–mean interaction theory in simple geometry, we focus here on the Eulerian version of the zonal pseudomomentum $\tilde{\mathsf{p}}$, which even for non-zero f is still given by $\tilde{\mathsf{p}} = \overline{b'(u'_z - w'_x)}/N^2$. Its general evolution law follows from the linear equations as[4]

$$\frac{\partial \tilde{\mathsf{p}}}{\partial t} + \boldsymbol{\nabla} \cdot \boldsymbol{B} = -\overline{q'v'} \quad \text{where} \quad \boldsymbol{B} = \left(0, \overline{u'v'}, \overline{u'w'} - \frac{f}{N^2}\overline{b'v'}\right). \tag{8.26}$$

Crucially, under the zero-PV constraint (8.18) this is a conservation law for zonal pseudomomentum. The flux $\boldsymbol{B}$ is called the *Eliassen–Palm flux* (EP flux) and its peculiar vertical component is a key feature of wave–mean interaction theory with rotation. We will see below how this component is equal to the form stress on undulating stratification surfaces and to the

4 The pseudomomentum law (8.26) also holds *without* zonal averaging if the zonal component of $\boldsymbol{B}$ is augmented by $\frac{1}{2}(u'^2 - v'^2 - w'^2)$. This is useful for diagnosing pseudomomentum fluxes in problems outside simple geometry.

wave drag on undulating topography. The present definition of $\tilde{\mathsf{p}}$ makes no allowance for Rossby waves, but when this is rectified in §9.2.1 and §9.3.1 it will be seen that the EP-flux remains unchanged.

Now, the EP flux is not restricted to plane internal waves, but in this special case the vertical component is

$$\overline{u'w'} - \frac{f}{N^2}\overline{b'v'} = \overline{u'w'}\left(1 - \frac{f^2}{\hat{\omega}^2}\right). \tag{8.27}$$

Hence the 'bare' Eulerian momentum flux $\overline{u'w'}$ always overestimates the true pseudomomentum flux. For slowly varying wavetrains (8.26) reduces to the generic

$$\tilde{\mathsf{p}} = \frac{k}{\hat{\omega}}\bar{E} \quad \text{and} \quad \frac{\partial \tilde{\mathsf{p}}}{\partial t} + \boldsymbol{\nabla}\cdot(\tilde{\mathsf{p}}\boldsymbol{c}_g) = 0. \tag{8.28}$$

Finally, by isotropy we can define a two-dimensional horizontal pseudomomentum vector in obvious analogy with the zonal pseudomomentum via

$$(\tilde{\mathsf{p}}_x, \tilde{\mathsf{p}}_y) = \left(\overline{b'(u'_z - w'_x)}/N^2, \overline{b'(v'_z - w'_y)}/N^2\right) \tag{8.29}$$

and then (8.26) holds with vertical flux *vector*

$$\overline{\boldsymbol{u}'w'} - \frac{f}{N^2}\overline{b'(v', -u')} = \overline{\boldsymbol{u}'w'}\left(1 - \frac{f^2}{\hat{\omega}^2}\right) \tag{8.30}$$

for plane waves, where $(\tilde{\mathsf{p}}_x, \tilde{\mathsf{p}}_y) = (k, l)\bar{E}/\hat{\omega}$.

8.2.3 Rotating lee waves and mountain drag

We first consider a vertical slice model in the two-dimensional xz-plane. As in the non-rotating case, we suppose a constant zonal basic flow $U > 0$ blowing over topography $h(x)$. With rotation there is now a need for a meridional basic pressure gradient in order to balance the meridional Coriolis force fU. However, this basic pressure gradient does not affect the wave solution. The lee wave computation for a single plane wave then follows as in the non-rotating case, with the key steps that for $U > 0$ the frequency and radiation conditions imply

$$\hat{\omega} = -Uk \quad \text{and} \quad m = \operatorname{sgn}(k)\sqrt{k^2\frac{N^2 - U^2k^2}{U^2k^2 - f^2}} \tag{8.31}$$

for propagating waves. For evanescent waves the square root is taken from a negative argument and the factor $\operatorname{sgn}(k)$ is omitted. There is now a double speed limit

$$\frac{f}{|k|} \leq U \leq \frac{N}{|k|} \tag{8.32}$$

for propagating waves, i.e., in the slow band $k^2U^2 < f^2$ no propagating waves are generated. Also, $|m|$ is increased over its non-rotating value and in fact diverges at the lower cut-off for k. Thus, the vertical wavelength of lee waves can be arbitrarily short in the rotating case, which is a new feature.

The mean pressure-related drag force on the mountain is defined by the Lagrangian-mean momentum flux $D = \overline{\zeta'_x P'}$ as before. Using $\mathrm{D}_t\zeta' = w'$ and $\mathrm{D}_t u' + P'_x = fv'$ this leads to

$$D = \overline{\zeta'_x P'} = -\left(\overline{u'w'} - \frac{f}{N^2}\overline{b'v'}\right) \tag{8.33}$$

for a steady wave field. Thus we have the remarkable result that for a steady wave the pressure-related drag force is equal to minus the vertical Eliassen–Palm flux, i.e., equal to minus the vertical flux of zonal pseudomomentum. This shows once again the importance of the pseudomomentum budget. It can also be shown that it is the pseudomomentum flux, and not $\overline{u'w'}$, that remains constant with altitude in the case of a shear flow $U(z)$.

This raises the question of how there can be a systematic difference between the Eulerian momentum flux $\overline{u'w'}$ and its Lagrangian counterpart $-\overline{\zeta'_x P'}$. The answer lies in the systematic zonal Coriolis forces fv' that are exerted on the fluid in the pockets between the Eulerian line $z = z_0$, say, and the undulating material line that occupies $z = z_0$ in the rest configuration. Indeed, it follows from (8.23) that there is a correlation between ζ' and v' and therefore fluid particles above $z = z_0$ are subject to a zonal Coriolis force of a different sign than fluid particles below $z = z_0$. This explanation is analogous to the vertical force computations with buoyancy in §6.5.2.

For a single plane wave with $h(x) = h_0 \cos kx$ the mean drag is

$$D(k, U) = \frac{h_0^2}{2}\sqrt{U^2k^2 - f^2}\sqrt{N^2 - U^2k^2}. \tag{8.34}$$

This shows that $D(k, U)$ goes to zero at both the upper and lower speed limits for wave propagation. Also, comparison with the non-rotating result (6.78) shows that $D(k, U)$ is in fact diminished by the frame rotation, i.e., increasing f with U and $h(x)$ fixed decreases the mountain drag.[5]

[5] Perhaps this is necessary for the stability of the coupled atmosphere–Earth system, as otherwise a drag-induced positive fluctuation of the Earth's rotation rate would have a positive feedback on the drag, which would lead to a further increase in the rotation rate and so on.

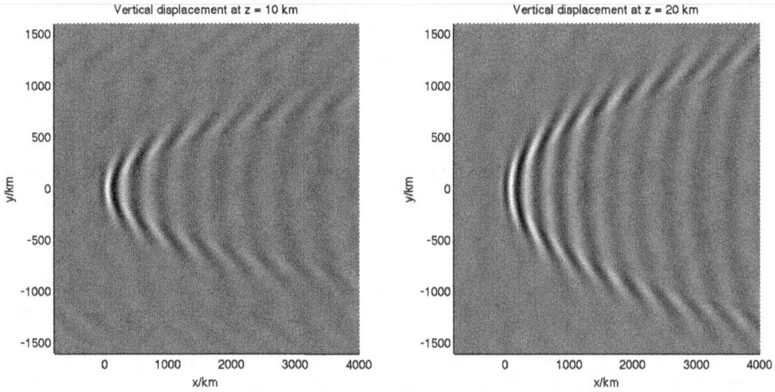

Figure 8.1 Horizontal plan view of three-dimensional rotating lee wave in response to zonal flow $U > 0$ over a Gaussian hill defined by $h = \exp(-(x^2 + y^2)/2R^2)$. Parameter values are $R = 50$km, $U = 10$m/s, $N = 0.01\text{s}^{-1}$ and $f = N/100$. Left: vertical elevation at $z = 10$km. Right: same at $z = 20$km.

Now, for three-dimensional lee waves based on $h(x, y)$ the derivation follows through with obvious changes, e.g., (8.31b) is replaced by

$$m = \text{sgn}(k)\sqrt{(k^2 + l^2)\frac{N^2 - U^2k^2}{U^2k^2 - f^2}} \tag{8.35}$$

for propagating waves. The vertical elevation $\zeta'(x, y, z)$ has the horizontal Fourier representation

$$\hat{\zeta}(k, l, z) = \exp(imz)\hat{h}(k, l), \tag{8.36}$$

which yields a straightforward numerical computation of the wave field as illustrated in Figure 8.1. The most notable new element is the curved bow wave structure in the horizontal plane caused by compact two-dimensional topography.

The mean drag is now a two-dimensional vector

$$(D_x, D_y) = (\overline{\zeta'_x P'}, \overline{\zeta'_y P'}). \tag{8.37}$$

For a single plane wave with $h = h_0 \cos(kx + ly)$ the isotropy of the problem in the horizontal plane together with the fact that the component of the wind blowing perpendicular to (k, l) is irrelevant leads to the result that the drag is parallel to (k, l) and $UD_x > 0$. Specifically, the mean drag is

$$(D_x, D_y) = \text{sgn}(k)\frac{(k, l)}{\sqrt{k^2 + l^2}}\frac{h_0^2}{2}\sqrt{U^2k^2 - f^2}\sqrt{N^2 - U^2k^2} \tag{8.38}$$

for propagating waves and zero for evanescent waves. This equals minus the vertical flux of horizontal pseudomomentum in (8.29) and (8.30). Overall,

(8.38) makes clear that the two-dimensional drag is *not* parallel to the basic wind.

8.3 Mean-flow response and the vortical mode

The mean continuity equation

$$\overline{v}_y + \overline{w}_z = 0 \tag{8.39}$$

now allows for non-trivial circulations in the meridional, yz-plane. Moreover, there are now significantly more terms in the mean-flow equations:

$$\overline{u}_t + (\overline{u}\,\overline{v})_y + (\overline{u}\,\overline{w})_z - f\overline{v} = -(\overline{u'w'})_z - (\overline{u'v'})_y \tag{8.40}$$

$$\overline{v}_t + (\overline{v}^2)_y + (\overline{v}\,\overline{w})_z + f\overline{u} + \overline{P}_y = -(\overline{v'w'})_z - (\overline{v'^2})_y \tag{8.41}$$

$$\overline{w}_t + (\overline{w}\,\overline{v})_y + (\overline{w}^2)_z + \overline{P}_z - \overline{b} = -(\overline{w'^2})_z - (\overline{w'v'})_y \tag{8.42}$$

$$\overline{b}_t + (\overline{b}\overline{v})_y + (\overline{b}\overline{w})_z + N^2\overline{w} = -(\overline{b'w'})_z - (\overline{b'v'})_y. \tag{8.43}$$

The complexity of this exact system can be significantly reduced if we focus on the leading-order mean-flow response and on strong interactions only.

8.3.1 Leading-order response and the TEM equations

We expand

$$\overline{u} = U + a^2\overline{u}_2, \quad \overline{v} = a^2\overline{v}_2, \quad \overline{w} = a^2\overline{w}_2 \quad \text{and} \quad \overline{b} = a^2\overline{b_2} \tag{8.44}$$

with constant U. Dropping expansion subscripts again we obtain (8.39) plus

$$\overline{u}_t - f\overline{v} = -(\overline{u'w'})_z - (\overline{u'v'})_y \tag{8.45}$$

$$\overline{v}_t + f\overline{u} + \overline{P}_y = -(\overline{v'w'})_z - (\overline{v'^2})_y \tag{8.46}$$

$$\overline{w}_t + \overline{P}_z - \overline{b} = -(\overline{w'^2})_z - (\overline{w'v'})_y \tag{8.47}$$

$$\overline{b}_t + N^2\overline{w} = -(\overline{b'w'})_z - (\overline{b'v'})_y. \tag{8.48}$$

On the left we recognize the linear operator from (8.5) acting on zonally symmetric fields and forced by eight wave-induced terms. The mean-flow response consists of a zonally symmetric balanced vortical response plus zonally symmetric inertia–gravity waves. It is not possible anymore to study the zonal mean-flow response $\overline{u}$ in isolation from the other mean fields. Also, at this stage the connection between the mean-flow response and the pseudomomentum evolution is obscured.

One approach at this stage is to absorb various forcing terms in the definition of the mean fields in order to reduce the number of forcing terms that need to be considered. This leads to the *transformed Eulerian-mean* equations (TEM), in the simplest version of which $(\overline{v}, \overline{w})$ are replaced by the so-called *residual meridional circulation* $(\overline{v}^*, \overline{w}^*)$ defined by

$$(\overline{v}^*, \overline{w}^*) = (\overline{v}, \overline{w}) + \left(-\frac{\partial}{\partial z}, +\frac{\partial}{\partial y}\right) \frac{\overline{b'v'}}{N^2}. \tag{8.49}$$

For steady waves these residual velocities are closely related to the Lagrangian-mean velocities defined later in GLM theory. By construction, the residual circulation is still non-divergent and it satisfies slightly simpler equations than (8.45); with some additional effort these equations can then be linked approximately to the pseudomomentum evolution. We will use a slightly different and perhaps more intuitive approach to obtain essentially the same result.

8.3.2 Forcing of mean vortical mode

Here we follow an approach directly based on the zero-frequency vortical mode of the mean-flow response. As discussed before, this mode governs the strong wave–mean interactions in the case of a steady wavetrain and by focusing on this mode we are willing to neglect wavelike parts of the mean-flow response, which presumably can be bounded uniformly at $O(a^2)$. This means that we are looking for the projection of the eight forcing terms in (8.45)–(8.48) onto the single forcing term relevant for the mean vortical mode. At a stroke, this reduces the eight forcing terms to the single linear combination that appears in (8.15) as applied to the mean flow, i.e.,

$$\begin{aligned}\frac{\partial}{\partial t}\left(-\overline{u}_y + \frac{f}{N^2}\overline{b}_z\right) &= (\overline{u'w'})_{zy} + (\overline{u'v'})_{yy} - \frac{f}{N^2}(\overline{b'v'})_{yz} - \frac{f}{N^2}(\overline{b'w'})_{zz} \\ &= (\boldsymbol{\nabla}\cdot\boldsymbol{B})_y - \frac{f}{N^2}(\overline{b'w'})_{zz}.\end{aligned} \tag{8.50}$$

Here $\boldsymbol{B}$ is the EP flux from (8.26). The final term can be re-written as

$$\frac{f}{N^2}\overline{b'w'} = -f\overline{b'b'_t} = -\frac{f}{2}(\overline{b'^2})_t. \tag{8.51}$$

Thus in a steady wavetrain $\overline{b'w'} = 0$ and based on this we can neglect this term in (8.50).

Now, the key step is that the mean vortical mode evolves *as if* the wave-induced forcing was entirely due to a zonal force in the x-momentum equation equal to $-\boldsymbol{\nabla}\cdot\boldsymbol{B}$. We exploit this by replacing (8.45)–(8.48) with the much simpler set

$$\begin{aligned}\overline{u}_t - f\overline{v} = -\boldsymbol{\nabla}\cdot\boldsymbol{B}, &\qquad \overline{v}_t + f\overline{u} + \overline{P}_y = 0, \\ \overline{w}_t + \overline{P}_z - \overline{b} = 0 \quad \text{and} &\quad \overline{b}_t + N^2\overline{w} = 0.\end{aligned} \tag{8.52}$$

To be sure, the mean-flow response as predicted by the simplified set (8.52) will be different at $O(a^2)$ from that predicted by the full set. The difference is described to a first approximation by the transformation (8.49). However, both sets will agree for the evolution of the important vortical mode, i.e., both sets will agree on the evolution of strong interactions. Therein lies the value of this much simpler set of mean-flow equations.

For example, wave forcing and dissipation will add a source term $\mathcal{F}$ on the right-hand side of (8.26) and this leads to the zonal mean-flow equation

$$\overline{u}_t - f\overline{v} = -\boldsymbol{\nabla}\cdot\boldsymbol{B} = \tilde{\mathsf{p}}_t - \mathcal{F}, \tag{8.53}$$

which can be then be used, say, in the case of Newtonian cooling $\mathcal{F} = -\alpha\tilde{\mathsf{p}}$. In this way the results on non-acceleration conditions and the dissipative pseudomomentum rule of the non-rotating two-dimensional case can be transferred to the rotating three-dimensional case.

8.4 Rotating vertical slice model

In the rotating version of the slice model there are two independent spatial variables x and z but three velocity components (u, v, w); this model is also called a 2.5-dimensional model for this reason and it is popular for idealized studies of intermediate complexity.

However, the reduced dimensionality also leads to certain peculiarities in the structure of the equations, which play a role in wave–mean interaction theory. Specifically, although there now is an exactly steady PV-controlled balanced mode, it cannot be forced by wave–mean interactions, because of the lack of y-derivatives. This means that there are no strong wave–mean interactions with steady wavetrains in this system and hence the previous methods based such interactions do not apply.

We first describe the special mathematical structure of the equations and then briefly discuss the issues arising in wave–mean interaction theory.

8.4.1 Stratification and rotation symmetry

The fields are functions of (x, z, t) and the governing equations for constant N and f are

$$u_x + w_z = 0 \quad \text{and} \quad \frac{Du}{Dt} + P_x = fv, \quad \frac{Dv}{Dt} + fu = 0, \tag{8.54}$$

$$\frac{Dw}{Dt} + P_z = b, \quad \frac{Db}{Dt} + N^2 w = 0. \tag{8.55}$$

Notably, the two equations on the right are mathematically equivalent, which implies that in addition to the familiar stratification lines there are now also *rotation lines* such that both

$$S = N^2 z + b \quad \text{and} \quad M = f^2 x + f v \tag{8.56}$$

are materially invariant (M could also be defined as $fx + v$, but the present definition enhances the symmetry between S and M). The existence of such material rotation lines is peculiar to the slice model, i.e., there are no materially invariant lines or surfaces that correspond to M in the three-dimensional case.

In the rest configuration lines of constant S are horizontal and lines of constant M are vertical, so M can be viewed as some form of sideways stratification due to the background rotation. Indeed, stratification seeks to keep the motion horizontal and rotation seeks to keep the motion vertical, as is exemplified by the famous *Taylor columns* in rapidly rotating fluids. With both f and N present there is an interesting competition between these opposing trends.

The similarity between rotation and stratification goes beyond (8.56); in fact, there is a symmetry of the slice model in which the horizontal and vertical coordinates are switched and b is replaced with fv. Thus stratification and rotation have *exactly* the same dynamical effect in the slice model. This also explains the peculiar absence of energy equipartition for plane inertia–gravity waves that was noted in (8.24): the kinetic energy to do with v should more appropriately be counted as part of the potential energy to do with b. Indeed, using the analogy in (8.56), the 'potential energies' due to b and fv have densities b^2/N^2 and $f^2v^2/f^2 = v^2$, respectively.

The potential vorticity in the slice model can also be written in terms of S and M:

$$q = (f\hat{\boldsymbol{z}} + \boldsymbol{\nabla} \times \boldsymbol{u}) \cdot \boldsymbol{\nabla} S = (f + v_x) S_z - v_z S_x = \frac{1}{f} \partial(M, S) \tag{8.57}$$

where the Jacobian $\partial(M, S) = M_x S_z - M_z S_x$. Thus the PV is inversely proportional to the area element of the two-dimensional mesh belonging to the curvilinear coordinate system generated by the family of materially invariant lines M and S. The material invariance of q then follows from the fact that the flow is area-preserving in the xz-plane.

If $q = fN^2$ then the state of minimum energy is the state of rest and the curvilinear system based on M and S reduces to the Cartesian system based on x and z. However, if q is non-uniform then the minimum energy state corresponds to balanced motion with $u = w = 0$ and non-zero b and v. This shows that the balanced states of the slice model are exactly steady.

Finally, the slice model also has a circulation theorem, but its formulation involves the three-dimensional deformation of line elements, so it is not intrinsic to the motion in the xz-plane. This makes the practical use of the circulation theorem somewhat difficult in the slice model.

8.4.2 Wave–mean interactions in the slice model

Formally, the derivation of the simplified set for strong wave–mean interactions in §8.3.2 can also be transferred to the slice model, even though that derivation used y-dependent fields at an intermediate step. Allowing for three-dimensional dependence of fields at intermediate steps, but not in the final result, is a useful gambit that occurs now and then in wave–mean interaction theory. However, it is important to note that the strong mean-flow response apparent in (8.50) was linked to the y-dependence of the EP-flux divergence: in the slice model there can be no y-dependence and therefore there is no strong mean-flow response to a steady wavetrain.

Explicitly, starting from (8.45)–(8.48), we consider the leading-order mean-flow system for the slice system given by $\overline{w} = 0$ and

$$\overline{u}_t - f\overline{v} = -(\overline{u'w'})_z \quad \text{and} \quad \overline{v}_t + f\overline{u} = -(\overline{v'w'})_z. \tag{8.58}$$

The mean vertical velocity is zero because of the continuity equation and the buoyancy has decoupled from the horizontal velocities. The horizontal mean flow in the slice model is horizontally homogeneous by construction and therefore its linear dynamics takes the form of inertial oscillations with frequency $\pm f$. There is no zero-frequency vortical mode and hence strong, resonant interactions are possible only if the wave-induced forcing terms also oscillate at the inertial frequency, which rules out steady wavetrains.

This makes attempts to bring (8.58) into a more standard form involving the EP flux somewhat arbitrary. For example, if we apply the TEM technique by replacing $\overline{v}$ with $\overline{v}^*$ as defined in (8.49) then we obtain

$$\overline{u}_t - f\overline{v}^* = -\left(\overline{u'w'} - \frac{f}{N^2}\overline{b'v'}\right)_z \quad \text{and} \quad \overline{v}^*_t + f\overline{u} = \frac{f}{N^2}(\overline{b'u'})_z. \tag{8.59}$$

This brings in the EP flux and the second flux term $\overline{b'u'}$ is zero for a plane wave, which suggest that it could be negligible. However, the same argument could also have been made for neglecting $\overline{v'w'}$ against $\overline{u'w'}$ in (8.58).

Notably, in §11.2.3 we will obtain a GLM version of (8.59) in which both velocity components are defined in the same way and in which the meridional forcing term is precisely zero whilst the zonal forcing term is precisely the divergence of the vertical flux of Lagrangian zonal pseudomomentum.

8.5 Notes on the literature

The TEM equations were proposed in Andrews and McIntyre (1976b), Andrews and McIntyre (1976a) and Boyd (1976). Important generalizations that are increasingly used in oceanography were given in Andrews and McIntyre (1978b). The extensive use of the TEM equations in GFD is discussed in the textbooks Andrews et al. (1987) and Vallis (2006).

8.6 Exercises

1. Inertial waves and Taylor columns. The dispersion relation (8.16) without stratification (i.e, $N = 0$) is

$$\hat{\omega}^2 = f^2 \frac{m^2}{k^2 + l^2 + m^2} \quad \text{such that} \quad 0 < \hat{\omega}^2 < f^2 \tag{8.60}$$

for propagating waves. These slow waves within a rotating homogeneous fluid are called *inertial waves.* Argue from (8.60) that if such waves are forced at some fixed horizontal scale with a very low frequency $\hat{\omega} \ll f$ then the fluid response will be at a very large vertical scale. In the limit $\hat{\omega} \to 0$ this leads to the celebrated Taylor columns.

2. PV inversion. Find the Green's function $G(\boldsymbol{x}, \boldsymbol{x}')$ used in (8.12) by transforming the defining equation

$$G_{xx} + G_{yy} + \frac{f^2}{N^2} G_{zz} = \delta(x - x')\delta(y - y')\delta(z - z') \tag{8.61}$$

into the standard three-dimensional Poisson equation. Show that the balanced flow around a positive point charge of PV is cyclonic and that the stratification surfaces of constant $b' + N^2 z$ are bending towards the point charge from above and below.

3. Linear hurricane model. Use the previous problem to find the balanced flow corresponding to a three-dimensional point charge in initial buoyancy:

$$u_0' = v_0' = w_0' = 0, \quad b_0' = \delta(x)\delta(y)\delta(z). \tag{8.62}$$

This is a model hurricane formed by localized latent heat release, which increases the buoyancy. Make a sketch of the balanced stratification surfaces in the xz-plane for $y = 0$ and identify cyclonic and anti-cyclonic circulation regions.

9
Rossby waves and balanced dynamics

We now turn to wave–mean interactions involving Rossby waves, the peculiar vorticity waves whose linear dynamics was described briefly in §4.2.2. Unlike acoustic waves or gravity waves, the dynamics of Rossby waves is essentially linked to the layerwise advection of PV, and this gives the mathematical description of Rossby waves and of their interactions with a mean flow a very special character, including the one-way phase propagation of Rossby waves.

The easiest model in which to study this topic is the quasi-geostrophic approximation to the shallow-water equations on a β-plane. However, the results easily generalize to three-dimensional stratified flow.

9.1 Quasi-geostrophic dynamics

We have no interest in gravity waves in this chapter and therefore we use the simplest theoretical approximation that filters these waves whilst retaining the balanced flow structure of Rossby waves and shallow-water vortices. This is accomplished by the quasi-geostrophic approximation to the equations, which is essentially a nonlinear extension of the linear balanced mode. These equations use a single dynamical variable, namely the PV.

Overall, the use of PV and of balanced flow systems based on PV advection and PV inversion (such as the quasi-geostrophic system or its many variants) are key concepts in atmosphere ocean fluid dynamics. For instance, balanced models were an essential component of the first successful numerical weather forecasts. Such a direct quantitative use of balanced models is less important today, but the insights that can be gained from studying such reduced dynamical systems remain as valuable as ever.

9.1.1 Governing equations

Recall from §4.2.1 that in the case of an f-plane shallow-water model with flat topography (i.e., the still water depth H is constant) the linear balanced mode in geostrophic balance is described by a stream function ψ such that

$$u_b = -\psi_y, \quad v_b = +\psi_x, \quad \text{and} \quad h_b = f_0\psi/g. \tag{9.1}$$

Here h_b is the balanced depth disturbance and f_0 is the constant Coriolis parameter. The corresponding linear PV is (up to factors of constant H)

$$q_b = v_{bx} - u_{by} - \frac{f_0}{H}h_b = \nabla^2\psi - \kappa_D^2\psi \tag{9.2}$$

with $\kappa_D = f_0/\sqrt{gH}$, as before. This balanced mode is steady according to linear dynamics. Now, in quasi-geostrophic dynamics we use the exact PV equation

$$q_t + (\boldsymbol{u}\cdot\boldsymbol{\nabla})q = 0 \tag{9.3}$$

and then approximate both q and $\boldsymbol{u}$ in it by their balanced counterparts (9.1) and (9.2). Dropping subscripts, this yields the governing equations

$$q = \nabla^2\psi - \kappa_D^2\psi \quad \text{and} \quad q_t + \partial(\psi, q) = G_x - F_y. \tag{9.4}$$

Here the curl of a body force $\boldsymbol{F} = (F, G)$ has been added. This is a self-consistent dynamical system termed *quasi-geostrophic dynamics*, which has only one degree of freedom because the initial-value problem is complete by specifying only q at the initial time. Note that if $\kappa_D \to 0$ then (9.4) reduces to standard vortex dynamics described by the two-dimensional incompressible flow equations. Even with non-zero κ_D the basic ideas and concepts from two-dimensional vortex dynamics still apply, although the details are different, of course. For example, the velocity field outside a vortex core still decays as $1/r$, but only if r is small compared to the deformation length $1/\kappa_D$. Otherwise, if $\kappa_D r \gg 1$, this algebraic decay is replaced by an exponential decay $\exp(-\kappa_D r)$. This shows that a finite Rossby deformation scale $1/\kappa_D$ leads to an exponential confinement of vortex structures, and therefore to exponentially weak interactions between vortices separated by distances larger than $1/\kappa_D$.

An important aspect of (9.4) is that the stream function ψ, which is proportional to the layer depth disturbance in shallow water, adjusts non-locally and instantaneously to changes in q. This is an inescapable consequence of inverting the elliptic diagnostic equation (9.4a) to find ψ from q. At first sight this is surprising, because the underlying shallow-water equations are hyperbolic, with propagation speeds that are bounded by the shallow-water gravity wave speed. Again, the analogy with incompressible vortex dynamics helps to understand this point: in the incompressible limit the sound speed is essentially taken to be infinite when compared to the fluid speed, i.e., the Mach number goes to zero in the incompressible limit. Similarly, in the quasi-geostrophic equations the speed of gravity waves is taken to be infinite compared to the speed of the quasi-geostrophic fluid motion. Thus, the quasi-geostrophic equations skip over all the gravity-wave-dependent details of how geostrophic balance is approximately maintained in a time-evolving flow, and by doing so the equations necessarily acquire the non-local, action-at-a-distance character that is typical for vortex dynamics.

Implicit in all this is the assumption that ψ and $\boldsymbol{\nabla}\psi$ are small enough that the equations of motion are dominated by the linear terms, with the weaker nonlinear terms then merely facilitating a slow time evolution of the linearly balanced state. For the shallow-water system this assumption implies either a small Froude number $|\boldsymbol{\nabla}\psi|/\sqrt{gH}$ or a small Rossby number $|\nabla^2\psi/f_0|$, or both.

Another aspect of (9.4) that is important in practice is that a unique inversion of (9.4a) generally requires auxiliary conditions on ψ in order to fix the null space of the modified Helmholtz operator $\nabla^2 - \kappa_D^2$. The exceptions to this are the unbounded plane with decay conditions at infinity and the doubly periodic domain (which is often used for numerical simulations), because in these cases the null space is empty. However, in the case of an x-periodic channel with flat impermeable walls at $y = -L/2$ and $y = L/2$, say, the null space consists of the zonally symmetric flows $\psi = A\cosh(\kappa_D y) + B\sinh(\kappa_D y)$, or $\psi = A + By$ in the case $\kappa_D = 0$. The coefficients A and B must then be fixed by two auxiliary conditions, which could be the constancy of net mass and of net absolute zonal momentum (to be discussed in the next section), for example. These conservation laws are inherited from the underlying shallow-water equations and they involve the integrals of ψ and of $-\psi_y - \kappa_D^2 y\psi$ over the channel, which fixes A to be zero and B to be non-zero in proportion to the net absolute zonal momentum in the channel. Alternatively, B could be fixed by monitoring the circulation along the flat material contour $y = -L/2$ at the southern wall, say, which depends only on the zonally symmetric part of the flow. In unforced flows this circulation is constant because $v = 0$ at the wall.

Finally, quasi-geostrophic dynamics is succinctly described in terms of ψ and q, and it is substantially more awkward to find the corresponding quasi-geostrophic equations in terms of the original variables (h, u, v). For instance, quasi-geostrophic dynamics is based on a stream function and yet the values of the stream function may change following a fluid particle, which by $\psi \propto h - H$ implies a weak but non-zero divergence of the velocity field. Hence the non-divergent geostrophic velocity field is generally augmented by a weaker 'ageostrophic' velocity field that supplies the needed flow divergence. This is confusing, especially as the ageostrophic velocity is still balanced in the sense that it can be computed from the PV alone. In contrast, by using ψ and q we can avoid the need for any other flow fields, and hence we will work exclusively with ψ and q.

9.1.2 Conservation properties

The unforced quasi-geostrophic equations have a number of interesting conservation properties. They obviously conserve q on material trajectories, which implies infinitely many integral conservation laws such as the conservation of enstrophy defined as the integral of q^2. More subtly, with suitable boundary conditions the diagnostic relation (9.4a) is a self-adjoint operator, which implies the conservation of total energy in the generic form

$$-\frac{1}{2}\int q\psi \, dx \, dy = +\frac{1}{2}\int \left(|\boldsymbol{\nabla}\psi|^2 + \kappa_D^2|\psi|^2\right) dx \, dy. \tag{9.5}$$

This shows scale-selective contributions of both kinetic and available potential energy such that at small scales kinetic energy dominates and vice versa. The conservation of (9.5) follows directly from the prognostic relation (9.4b) coupled with the self-adjoint nature of (9.4a), i.e.,

$$-\frac{1}{2}\int (q_t\psi + q\psi_t) \, dx \, dy = -\int q_t\psi \, dx \, dy = \frac{1}{2}\int \partial(\psi^2, q) \, dx \, dy = 0. \tag{9.6}$$

Many variants of the quasi-geostrophic equations are possible, and provided the relationship between ψ and q remains self-adjoint they all satisfy a generic energy conservation law of the form (9.5).

Now, in vortex dynamics the conservation of momentum is most easily expressed in terms of *Kelvin's impulse*, which is the first moment of q rotated by ninety degrees. In unforced flow this moment is conserved under appropriate boundary conditions and in forced flows it changes according to the momentum imparted by the external force. Kelvin's impulse will be discussed in detail in §12.4.1 and here we only note its zonal component and the relation of this component to the absolute zonal momentum of the underlying shallow-water equations. Thus, the zonal component of the impulse is

$$I = \int yq \, dx \, dy \quad \Rightarrow \quad \frac{dI}{dt} = \int vq \, dx \, dy + \int y(G_x - F_y) \, dx \, dy. \tag{9.7}$$

In an unbounded domain with compact q and $\boldsymbol{F}$, substitution from (9.4a) and integration by parts leads to

$$\frac{dI}{dt} = \int F\, dx\, dy, \tag{9.8}$$

as promised. This conservation law is related to the conservation of absolute zonal momentum in the full shallow-water system, whose density per unit area is $hu - f_0 hy$ based on the conservation law

$$h\frac{D}{Dt}(u - f_0 y) + \left(\frac{g}{2}h^2\right)_x = hF. \tag{9.9}$$

Clearly, the integral of the $-f_0 hy$ part of the absolute momentum is proportional to the y-position of the centre of mass of the fluid layer. Mass-conserving changes in the layer depth h that change this position will therefore change the absolute zonal momentum. Specifically, for $f_0 > 0$ a southward mass transport increases the absolute zonal momentum and vice versa. The physical interpretation of this follows from considering a mid-latitude tangent plane in the northern hemisphere. On such a tangent plane the southward motion of a fluid particle brings the particle closer to the equator and this goes together with an increase in the distance from the Earth's rotation axis and therefore with an increase of the total angular momentum of the fluid particle. Thus, the absolute zonal momentum is the stand-in for the total angular momentum in a tangent-plane approximation to the spherical Earth.

Returning to the impulse we note that to quasi-geostrophic accuracy the absolute zonal momentum relative to a state of rest has the density $Hu - f_0(h-H)y$, which after division by the constant H integrates to

$$M = \int \left(u - f_0 y\frac{h-H}{H}\right) dx\, dy = \int (-\psi_y - \kappa_D^2 y\psi)\, dx\, dy = I \tag{9.10}$$

after using the decay conditions on ψ. Thus the quasi-geostrophic approximation to the absolute zonal momentum M and the impulse I are equal in an unbounded domain with compact q and $\boldsymbol{F}$.

The situation is slightly different in the practically important case of an x-periodic zonal channel with flat impermeable walls such that $v = 0$ at $y = \pm L/2$, say. Here the net impulse I may differ from the net absolute zonal momentum M, because there are zero-impulse flows that have non-zero absolute zonal momentum. These flows are given by the null space of (9.4a) discussed previously. For example, the absolute zonal momentum of $\psi = B\sinh(\kappa_D y)$ is non-zero, but the flow has zero PV and therefore zero impulse. Specifically, in such a channel (9.8) and (9.10) are replaced by

$$\frac{dI}{dt} = \int F\,dx\,dy - \frac{L}{2}\int (F(x, L/2, t) + F(x, -L/2, t))\,dx \qquad (9.11)$$

and

$$I = M - \frac{L}{2}\int (u(x, L/2, t) + u(x, -L/2, t))\,dx, \qquad (9.12)$$

respectively. Thus, in unforced flows, or in flows with $\overline{F} = 0$, the zonal impulse I is still constant. Combining (9.11) and (9.12) and noting that $v = 0$ at the channel walls yields the evolution of the net absolute zonal momentum as

$$\frac{dM}{dt} = \frac{d}{dt}\int (-\psi_y - \kappa_D^2 y\psi)\,dx\,dy = \int F\,dx\,dy. \qquad (9.13)$$

As noted previously, (9.13) together with mass conservation fixes the null space of (9.4a). For example, the impulsive forcing with $F = F_0\delta(t)$ of an initial state of rest produces a balanced flow with $q = 0$ and $I = 0$, but with an absolute zonal momentum M that is equal to F_0 times the area of the channel. The meridional structure of this flow is then given by the null-space flow

$$\psi = -\frac{F_0}{\kappa_D}\frac{\sinh(\kappa_D y)}{\cosh(\kappa_D L/2)} \quad \text{and} \quad u = F_0\frac{\cosh(\kappa_D y)}{\cosh(\kappa_D L/2)}. \qquad (9.14)$$

Monitoring the circulation along the material wall contour $y = -L/2$, which jumps to F_0 times the channel length during the impulsive forcing, leads to the same result because $u = F_0$ at the walls.

9.1.3 Quasi-geostrophic β-plane

To proceed to the β-plane where $f(y) = f_0 + \beta y$, we simply include the β term in the definition of the PV. Geostrophic balance $f(y)\hat{\boldsymbol{z}} \times \boldsymbol{u} + g\boldsymbol{\nabla} h = 0$ is then compatible with a stream function only at leading order, in which $f \approx f_0$. Based on this, the diagnostic relation (9.4a) is unchanged, i.e., we keep κ_D as based on the constant f_0. This procedure will be consistent only if the meridional extent of the flow domain is restricted such that $f \approx f_0$ uniformly, which is one of the shortcomings of quasi-geostrophic theory. The quasi-geostrophic equations on a β-plane can then be written as

$$q - \beta y = \nabla^2\psi - \kappa_D^2\psi \quad \text{and} \quad q_t + \partial(\psi, q) = G_x - F_y. \qquad (9.15)$$

The conservation laws described before remain valid with obvious modifications; it is easiest, but not essential, to replace q by $q - \beta y$ in (9.5) and (9.7).

9.1.4 Response to effective zonal mean force

An important question in the context of simple geometry is how a zonal mean force $\overline{\boldsymbol{F}} = (\overline{F}, 0)$ changes the mean PV. It follows from (9.15) and simple geometry that

$$\overline{q}_t + \overline{\partial(\psi', q')} = -\overline{F}_y \quad \Rightarrow \quad \overline{q}_t = -(\overline{v'q'} + \overline{F})_y = -(\overline{F_e})_y. \tag{9.16}$$

Thus, a northward mean PV flux $\overline{v'q'} > 0$ is equivalent to a prograde zonal mean force in their impact on the mean PV. Clearly, as far as $\overline{q}$ is concerned, we can combine $\overline{F}$ and $\overline{v'q'}$ into a single effective zonal mean force $\overline{F_e}$ as indicated and this is sufficient to capture the impact of wave-induced fluxes on the mean PV. Of course, how a change in $\overline{q}$ translates into a change in $\overline{u} = -\overline{\psi}_y$ or layer depth $\overline{h} - H = f_0\overline{\psi}/g$ depends on the inversion of (9.15a) for the mean fields, which is nontrivial for finite κ_D. This illustrates a general principle for forced balanced flows: the PV changes in an obvious and local manner dictated by the curl of the force field, whereas the velocity and layer depth adjust non-trivially and non-locally to the changed PV.

To illustrate this we consider a simple example in which a state of rest with $\beta = 0$ is at time $t = 0$ subjected to an impulsive zonal mean force, i.e., to a zonal body force of the form $F = A(y)\delta(t)$. Here $A(y)$ is a smooth localized envelope of the force as illustrated in the first panel of Figure 9.1. The result of the forcing is a changed balanced state, which is still steady, and which follows from inverting the zonally symmetric version of (9.15a):

$$q(y) = \psi_{yy} - \kappa_D^2\psi = -A_y. \tag{9.17}$$

We consider an unbounded fluid in the y-direction as the easiest case. After multiplying (9.17) by y and integrating the zonal impulse I (per unit horizontal length) is clearly equal to the y-integral of $A(y)$, which is the net zonal momentum imparted by the body force. Now, in the non-rotating case f_0 and $\kappa_D = f_0/\sqrt{gH}$ are both zero and hence this yields $u(y) = A(y)$ and $h = H$. The absolute zonal momentum is then entirely due to zonal flow.

However, this picture is drastically changed for non-zero f_0, no matter how small. In other words, going from a non-rotating to a rotating case is a singular perturbation for the balanced flow. To illustrate this both $u(y)$ and $h(y) - H$ are computed numerically and sketched in the second and third panels of Figure 9.1. Specifically, the total y-integral of u is now zero for

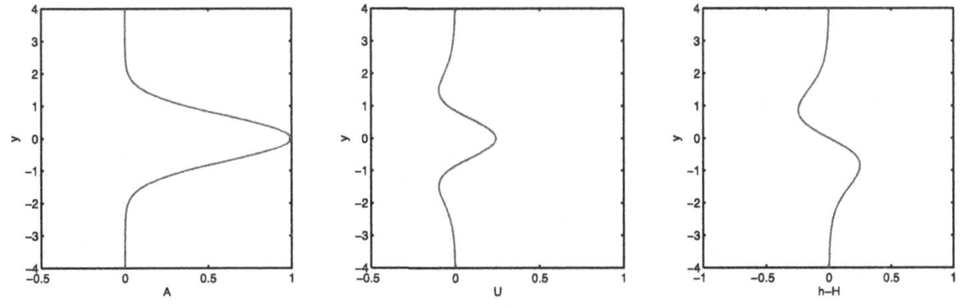

Figure 9.1 Balanced flow response to impulsive zonal forcing in an unbounded domain. Left: meridional structure $A(y)$ of the impulsive zonal force $F = A(y)\,\delta(t)$. This is also the shape of the zonal flow $u(y)$ in the non-rotating case. Middle: zonal flow $u(y)$ in a rotating case with $g = H = 1$ and $f_0 = 2$. This zonal flow has no net zonal momentum. Right: depth disturbance $h(y) - H$ in the same case. This surface elevation corresponds to positive net absolute zonal momentum.

any non-zero κ_D. This follows at once from $u = -\psi_y$ and the decay of $\psi \propto \exp(-\kappa_D|y|)$ as $y \to |\infty|$. Hence the y-integrated absolute zonal momentum must reside *entirely* in the contribution due to the layer depth. The depth distribution in the third panel of Figure 9.1 does indeed correspond to a net southward motion of fluid particles and it is the absolute zonal momentum associated with this motion that precisely accounts for the zonal momentum imparted by the forcing.[1]

It remains to understand how the fluid 'knows' how to establish the peculiar flow structure depicted in Figure 9.1. This is explained by considering a Rossby adjustment problem for the linear shallow-water equations with initial conditions given by the non-rotating solution $u(y) = A(y)$ and $h = H$. Clearly, in this simple zonally symmetric problem the final state of this adjustment problem yields the balanced state of quasi-geostrophic theory. Now, with $f_0 > 0$ the prograde initial jet will experience a Coriolis force that pushes it southward and the ensuing southward fluid motion will lead to a surface elevation at the southern flank of the jet and to a surface depression at the northern flank. This explains the structure of $h - H$. Moreover, the southward motion will then lead to a *retrograde* Coriolis force, which decelerates the jet, as indicated as well. Lowering the Coriolis parameter delays but does not prevent this sequence of events, which explains the singular perturbation for the final balanced state.

[1] In a finite channel it is possible that the total integral of $u = -\psi_y$ is non-zero even for non-zero f_0 and this means that in a finite channel the absolute zonal momentum is partitioned in a non-trivial manner into zonal momentum and layer-depth changes. The details of this partition then depend on the channel width.

9.2 Small amplitude wave–mean interactions

Linearizing (9.15) around a steady state with a basic zonal flow $U(y)$ and basic PV given by $Q = \beta y - U_y$ leads to

$$q' = \nabla^2\psi' - \kappa_D^2\psi' \quad \text{and} \quad \mathrm{D}_t q' + (\beta - U_{yy})v' = 0. \tag{9.18}$$

Consistent with §4.2.2 we find that for constant U plane waves satisfy the famous intrinsic dispersion relation

$$\hat{\omega} = \omega - Uk = -\beta\frac{k}{k^2 + l^2 + \kappa_D^2}. \tag{9.19}$$

Now, the single Rossby-wave frequency branch in (9.19) leads to a sign-definite pseudomomentum for Rossby waves, which is always negative in accordance with the negative intrinsic phase speed $\hat{\omega}/k$. This fact gives wave–mean interaction problems involving Rossby waves their special flavour.

9.2.1 Rossby-wave pseudomomentum

The linear PV equation can be integrated to yield

$$\mathrm{D}_t q' + (\beta - U_{yy})v' = 0 \quad \Rightarrow \quad q' + (\beta - U_{yy})\eta' = 0. \tag{9.20}$$

Multiplication by $\eta' = -q'/(\beta - U_{yy})$ and zonal averaging then leads to an equation for the Eulerian pseudomomentum for Rossby waves:

$$\tilde{\mathsf{p}} = -\frac{1}{2}\frac{\overline{q'^2}}{(\beta - U_{yy})} = -\frac{1}{2}(\beta - U_{yy})\,\overline{\eta'^2} = \frac{1}{2}\overline{\eta' q'} \quad \text{and} \quad \tilde{\mathsf{p}}_t = \overline{v'q'}. \tag{9.21}$$

The first expression for $\tilde{\mathsf{p}}$ extends most easily to forced-dissipative flows. We will see later in §10.4.4 how the small-amplitude pseudomomentum in (9.21) is related to a finite-amplitude definition of the same quantity.

Now, using

$$q' = v'_x - u'_y - \frac{f_0}{H}h' \tag{9.22}$$

the PV flux on the right-hand side of (9.21b) can be re-written as

$$\overline{v'q'} = -(\overline{u'v'})_y. \tag{9.23}$$

This relation (which is exact in quasi-geostrophic theory) is an instance of the Taylor identity, which will be discussed in more detail in §9.3.1. Clearly, by virtue of (9.23), pseudomomentum is conserved and its meridional flux is $\overline{u'v'}$.

9.2.2 Localized forcing and dissipation

A common situation in the atmosphere involves mid-latitude flow instabilities (such as baroclinic instability) that are generating Rossby waves, which subsequently travel to higher and lower latitudes where they are then dissipated. Of course, in the real atmosphere this scenario also involves significant vertical wave propagation, but the basic mechanisms at work can be studied using a two-dimensional model in combination with a localized body force $\boldsymbol{F} = (F, G)$ that effects the wave production and dissipation as needed.

We assume that $U = 0$ and $\overline{F} = 0$, i.e., the localized body force does not add or subtract any net zonal momentum to the fluid, which is consistent with an atmospheric situation in which intrinsic flow instabilities and decay mechanisms are responsible for the wave production and dissipation. For the purpose of wave production such a zero-net-momentum body force is often referred to as a 'stirring' force in the literature; however, one needs to be careful about the precise meaning of the term 'stirring' in any given situation, as the dynamical consequences of the 'stirring' depend rather sensitively on whether or not there is a net momentum input associated with it! Worse still, the word 'stirring' has unwanted associations with a different process, "potential-vorticity mixing" (§9.3.2), that has a role in more realistic versions of this problem.

In our case we allow for no net momentum input and therefore both the net absolute zonal momentum and the net zonal impulse are constant throughout. The pseudomomentum equation (9.21b) changes to

$$\tilde{\mathsf{p}}_t = \overline{v'q'} - \frac{\overline{q' \boldsymbol{\nabla} \times \boldsymbol{F}'}}{\beta} = \overline{v'q'} + \mathcal{F}, \tag{9.24}$$

say. The sign-definiteness of $\tilde{\mathsf{p}} \leq 0$ implies that $\mathcal{F} < 0$ for wave production and $\mathcal{F} > 0$ for wave dissipation. Now, the corresponding effective zonal mean force is

$$\overline{F_e} = \overline{v'q'} = \tilde{\mathsf{p}}_t - \mathcal{F}, \tag{9.25}$$

which makes the dissipative pseudomomentum rule obvious. Without body forcing the relation (9.25) implies that the arrival of a Rossby wave goes hand-in-hand with a negative effective mean force, i.e., the mean flow is subject to a retrograde force at a progressing wave front. Again, the extent to which such retrograde forces lead to retrograde accelerations depends on inverting $\overline{q}$ in order to find $\overline{\psi}$ and therefore $\overline{u} = -\overline{\psi}_y$.

We now consider a definite example with a steady wavetrain that is being generated in a production region $y \in [-D, D]$, say, and hence $\mathcal{F} < 0$ there by construction. Outside the production region $\mathcal{F} = 0$ and the waves are propagating away towards increasing $|y|$. We can now imagine that the waves are subject to momentum-conserving dissipation in some remote dissipation regions where $|y| \gg D$ and $\mathcal{F} > 0$. Hence, for a steady wave field the mean flow will encounter a persistent retrograde effective force in these dissipation regions, which is in accordance with the dissipative pseudomomentum rule. At first sight this leads to a puzzle: how is this persistent retrograde mean-flow forcing in the dissipation regions consistent with the conservation of absolute zonal momentum, and of zonal impulse, as there is no net input of zonal momentum in the production region?

The answer to this puzzle comes from considering the effective mean force in the production region, which is *positive* by construction because $\overline{F_e} = -\mathcal{F}$ there. The flow response to this prograde effective force in the production region is similar to Figure 9.1, which corresponds to positive absolute zonal momentum and impulse. This prograde effective force in the production region precisely balances the retrograde effective force in the dissipation regions and therefore absolute zonal momentum is indeed conserved globally, albeit in a highly non-local way that depends crucially on the wave-induced momentum fluxes.

The nature of the mean-flow response in the production region can be understood in more detail by considering a transient variant of the thought experiment, in which the wave-producing force acts only for a finite period of time. If this period is short compared to the propagation time scale of the Rossby waves (essentially, the width of the forcing region divided by the typical meridional group velocity), then it is possible to distinguish clearly the impact on the impulse of the wave production itself and of the subsequent propagation of the Rossby waves away from the production region.[2]

Now, by (9.25) the effective mean force is zero during the wave production period and hence the production itself leads to no changes in $\overline{q}$ and therefore to no changes in the impulse. However, as the waves are subsequently leaving the production region the material PV contours are flattening out, and the decaying undulations of these contours correspond to an *up-gradient* flux of PV. Basically, this is the time-reverse of the familiar process of a wave front arriving at a quiescent fluid location, in which case growing undulations always correspond to a *down-gradient* flux of PV (or of stratification in the internal-wave case). This up-gradient PV flux due to departing Rossby waves leads to a dipolar change in $\overline{q}$ such that $\overline{q}$ has a positive anomaly at the northern edge of the production region and a negative anomaly at the southern edge, just as in the case of Figure 9.1; recall (9.17). After multi-

[2] Essentially, this is a kind of Rossby adjustment problem in which a zonally asymmetric initial state adjusts to a zonally symmetric steady state by emitting Rossby waves.

plication with y, this dipolar structure clearly corresponds to net positive zonal impulse in the production region.

The wave-induced momentum fluxes in this problem and in its more realistic versions (see §9.3.2) are the key to explaining the old "negative viscosity enigma" of V.P. Starr and E.N. Lorenz.

9.3 Rossby waves and turbulence

The balanced, PV-controlled nature of quasi-geostrophic dynamics allows significant progress to be made into studying the interplay between strongly nonlinear layerwise two-dimensional turbulence and Rossby-wave dynamics. In fact, the situation is much simpler here than in the full Boussinesq equations, say, because in quasi-geostrophic dynamics *both* the turbulence and the waves are controlled by the PV distribution.

We will first discuss the so-called *Taylor identity* for quasi-geostrophic dynamics and then we will look in more detail into the intriguing nonlinear dynamics of waves and turbulent mixing.

9.3.1 The Taylor identity for quasi-geostrophic dynamics

The Taylor identity links the divergence of the EP flux to the meridional flux of PV. It is related to the elementary vector identity

$$\nabla \cdot (\boldsymbol{u}\boldsymbol{u}) = (\boldsymbol{u} \cdot \nabla)\boldsymbol{u} = \nabla \left(\frac{1}{2}|\boldsymbol{u}|^2\right) + (\nabla \times \boldsymbol{u}) \times \boldsymbol{u}, \tag{9.26}$$

where the first step is valid only for incompressible flows such that $\nabla \cdot \boldsymbol{u} = 0$. The zonal component of the zonally averaged (9.26) is

$$\nabla \cdot (\overline{u\boldsymbol{u}}) = (\overline{uv})_y + (\overline{uw})_z = \overline{(u_z - w_x)w} - \overline{(v_x - u_y)v}. \tag{9.27}$$

In horizontally two-dimensional incompressible flow with $w = \overline{v} = 0$ this identity takes the simpler form, known as the Taylor identity,

$$(\overline{u'v'})_y = -\overline{v'q'} \tag{9.28}$$

where $q = v_x - u_y$. As was shown in § 9.2.1, the same expression also holds for two-dimensional quasi-geostrophic dynamics in shallow water with finite κ_D, with q now standing for the quasi-geostrophic PV. We now demonstrate that an analogous expression also holds for the divergence of the EP flux in the three-dimensional quasi-geostrophic equations. In the Boussinesq system these equations are defined by (cf. §8.2.1)

$$u = -\psi_y, \quad v = +\psi_x, \quad w = 0, \quad b = f_0\psi_z \tag{9.29}$$

in conjunction with the quasi-geostrophic PV disturbance

$$q - \beta y = v_x - u_y + \frac{f_0}{N^2} b_z = \nabla^2 \psi + \frac{f_0^2}{N^2} \psi_{zz} \tag{9.30}$$

and the single layerwise evolution equation

$$q_t + \partial(\psi, q) = 0 \tag{9.31}$$

where the Jacobian acts on the horizontal coordinates (x, y) only. It follows readily from (9.29) and (9.30) and simple geometry (which implies $\overline{v} = \overline{\psi_x} = 0$) that the quasi-geostrophic EP flux divergence

$$(\overline{u'v'})_y - \frac{f_0}{N^2} (\overline{b'v'})_z = -\overline{v'q'}. \tag{9.32}$$

This relation is also known as the Taylor identity. For internal gravity waves the vertical component of the EP flux also contained a term $\overline{u'w'}$, but because $w = 0$ in quasi-geostrophic theory this term is absent here. Note that (9.32) made use of $\overline{b'v'_z} = 0$, which holds because $\overline{\psi'_z \psi'_{zx}} = 0$ in simple geometry.

Thus, we have found that the divergence of the EP flux can be equated to the meridional PV flux. In other words, the EP flux divergence and the PV flux are both equal to the effective zonal mean force $\overline{F_e}$ that is designed to capture the impact of the wave dynamics on the mean PV field. We recognize here the same argument that was used to derive the EP flux in §8.3.2.

Overall, the Taylor identity makes explicit that quasi-geostrophic mean-flow changes are linked to meridional fluxes of PV as well as to the divergence of the EP flux. The interplay between these two kinds of fluxes also makes it obvious why, as a rule, persistent mean flow changes induced by dissipating or breaking waves do not conserve zonal momentum locally, as was already exemplified by the two-dimensional problem considered in §9.2.2.

9.3.2 *Turbulent mixing of PV*

The quasi-geostrophic flow is controlled by the PV distribution and therefore it is possible to gain insight into the flow manifestations of turbulent PV mixing. For example, this allows us to discuss the flow configuration in a mature nonlinear critical layer, which was not possible in the internal wave case. We work in the two-dimensional shallow-water model, but again this generalizes easily to three-dimensional flow.

Basically, we take the reasonable point of view that turbulence mixes and homogenizes the distribution of materially invariant fields, including the PV field. The mixing aspect follows almost by definition from the chaotic trajectories of a turbulent flow and the homogenization aspect follows once sufficiently small flow scales are reached such that viscous diffusion becomes relevant. Actually, in the case of PV mixing the homogenization of q by diffusion is secondary because of the scale-selective nature of the PV-inversion operator in (9.22), which implies that small-scale features of q lead to a very weak stream function and therefore to weak balanced flows. In other words, small-scale PV features are dynamically very passive, and this makes actual homogenization less relevant.

Where does the turbulence comes from? The most important case in atmospheric dynamics is the generation of two-dimensional turbulence within horizontal critical layers formed by breaking Rossby waves in the stratosphere; these huge, planetary-scale turbulent regions are also called Rossby-wave 'surf zones' in analogy with breaking waves on a beach.

Now, we know from the causal linear theory of critical layers that such a layer acts as a sink for zonal pseudomomentum and that there is a concomitant mean-flow forcing in the critical layer, which for Rossby waves is necessarily retrograde. This is as far as linear theory can be pushed, but what happens in a mature nonlinear critical layer? As noted before, there is no simple answer to this question in the case of internal waves, but for Rossby waves we can consider the balanced flow structure within a critical layer in which the PV has been completely homogenized by the turbulence. Specifically, we consider a steady flow within a layer of meridional width $2D$ embedded in an infinite domain. The layer is centred at $y = 0$ and within the layer q is constant and equal to its layer-averaged value before the mixing, which is zero in the present example (cf. first panel in Figure 9.2). This means

$$\psi_{yy} - \kappa_D^2 \psi = -\beta y \quad \text{if} \quad |y| \leq D \tag{9.33}$$

and zero otherwise. Typically, the width of the critical layer is comparable with the Rossby deformation scale and therefore $\kappa_D D$ is of order unity. In fact, by virtue of (9.17), such a mixed-layer PV distribution is equivalent to an impulsive force of the kind discussed in §9.1.4, namely

$$\overline{F_e} = A(y)\delta(t) \quad \text{with} \quad A(y) = -\tfrac{1}{2}\beta\left(D^2 - y^2\right) \tag{9.34}$$

inside the mixed layer and $A(y) = 0$ outside. Clearly, this is a *retrograde* force and we can expect a response similar to the one sketched in Figure 9.1, but with the zonal direction reversed. That this is indeed the case is illustrated by the numerical solution in Figure 9.2. As before, the balanced zonal flow has zero net momentum, but there is a non-zero amount of retrograde zonal impulse and of retrograde absolute zonal momentum associated with the mean PV and the balanced layer depth distribution, respectively.

The inevitable decrease in zonal impulse (compared to the unmixed state

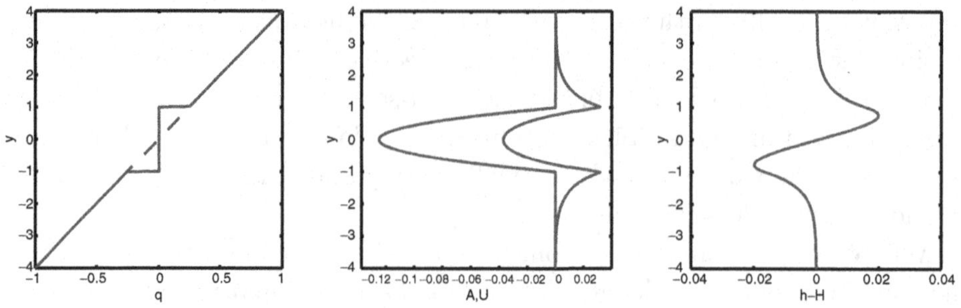

Figure 9.2 Zonal mean flow changes due to PV mixing in unbounded domain. Left: basic PV (dashed) and mixed PV (solid) for critical layer centred at $y = 0$ with $D = 1$ and $\beta = 1/4$. Locally, this mixing corresponds to a retrograde effective force and it lowers the zonal impulse and the absolute zonal momentum of the flow in this region. Middle: the left curve is the equivalent impulsive force structure $A(y)$ in (9.34) and the right curve is the corresponding zonal flow $U(y) = -\psi_y$ for $g = H = 1$ and $f_0 = 2$. Right: corresponding layer depth disturbance, indicating retrograde absolute zonal momentum.

$q = \beta y$) associated with the isolated layer of mixed PV also makes it obvious that turbulent PV mixing must inevitably be associated with retrograde radiation stresses into the layer, or with retrograde external forces, because by (9.8)) such stresses or forces are necessary to provide the retrograde impulse. This is of course consistent with the usual picture of a Rossby-wave critical layer and with the dissipative pseudomomentum rule.

Now, the impulse in this mixed-layer flow is also clearly finite and equal to

$$\int A(y)\, dy = -\frac{2}{3}\beta D^3, \tag{9.35}$$

which suggests that the Rossby-wave critical layer cannot absorb the incoming negative pseudomomentum indefinitely. In other words, it seems reasonable that a well-mixed critical layer must turn from an absorber into a reflector for incoming Rossby waves. This theoretical suggestion is indeed borne out by observations, analytical solutions and numerical simulations of nonlinear Rossby-wave critical layers.

There is an interesting and fundamental kinematic feature of PV mixing that can be read-off from the first panel of Figure 9.2. The turbulent mixing in the critical layer corresponds to a down-gradient, southward flux of PV that raises PV values in the southern half of the critical layer and lowers them in the northern. By necessity, this down-gradient flux leads to an intensification of the PV gradient at the boundaries of the critical layer. Indeed, in the illustrated extreme case of perfect mixing the PV gradient has two delta functions at the boundaries, which in a less extreme case of partial mixing would turn into narrow zones of strongly increased PV gradients.

Thus, a mature critical layer offers a very inhomogeneous environment for Rossby waves, because of the strong PV gradients and the mean-flow shear located at the flanks of the critical layer. Such strong inhomogeneity is well documented for the real winter stratosphere.

This robust kinematic fact contributes to the longevity of mixed PV zones. Basically, the boundaries of such zones invariably feature a strong PV gradient, which gives them a strong Rossby-wave restoring mechanism that lends a peculiar kind of 'Rossby-wave elasticity' to these boundaries such that they respond in an elastic manner to sideways displacements. This elasticity, together with the increased shear, makes these boundaries resilient to disturbances that seek to erode the mixed zone.

Now, it is interesting and of practical relevance to generalize the consideration of a single layer of mixed PV to a case where there are several mixed layers. In fact, it is instructive to consider an extreme scenario, in which the smooth PV gradient due to the β-effect has been replaced entirely by layers of well-mixed PV, which together form a 'PV staircase'.

9.3.3 PV staircases and self-sharpening jets

We consider the extreme case of a 'PV staircase', in which the PV is zonally symmetric, steady and a piecewise constant function of y. Specifically, we consider a segment of the unbounded β-plane of meridional width L in which there are $N+1$ zones of constant PV separated by N PV jumps located at y_n with $n \in \{1, 2, \ldots, N\}$. The total PV contrast of the undisturbed flow across the segment is βL and we assume that this continues to be the case for the staircase flow, i.e, if $\Delta q_n > 0$ is the PV contrast across the nth jump then

$$\sum_{n=1}^{N} \Delta q_n = \beta L. \tag{9.36}$$

This ensures that the staircase could have been created by mixing the undisturbed PV. The equation for the zonal flow is

$$U_{yy} - \kappa_D^2 U = \beta - q_y = \beta - \sum_{n=1}^{N} \Delta q_n \delta(y - y_n). \tag{9.37}$$

This is an extension of the equation governing a single well-mixed critical layer. The simplest regime in which to solve (9.37) occurs if the width of the constant-PV layers is significantly larger than the deformation scale $1/\kappa_D$.

In this regime we obtain

$$U(y) = -\frac{\beta}{\kappa_D^2} + \sum_{n=1}^{N} \frac{\Delta q_n}{2\kappa_D} \exp(-\kappa_D |y - y_n|) \tag{9.38}$$

inside the segment. Thus a constant retrograde flow is augmented by a sequence of prograde jets centred at the jump locations. As expected, the total zonal momentum integrated across the segment is zero, but the absolute zonal momentum and impulse is non-zero, and negative, due to the layer depth effect discussed in the context of Figure 9.2. Clearly, the robust negative impulse of a PV staircase implies that a net retrograde momentum flux into the region (or a net retrograde momentum input by an external force) was required in order to set up the staircase state. For example, if this shallow-water PV staircase is a model for a PV staircase on a stratification surface in three-dimensional flow, then this would correspond to the requirement that there was a wave-induced vertical EP flux that converged at the stratification surface under consideration. Once more, this is the basic message of the Taylor identity (9.32).

Now, in the present two-dimensional staircase flow the meridional PV gradient is concentrated at the jump locations, which makes it obvious that Rossby waves relative to such a staircase basic state will be very different from Rossby waves relative to an undisturbed basic state with constant PV gradient equal to β. For one thing, Rossby waves in the staircase state will be localized at the jump locations and evanescent away from there. We can illustrate this by looking at a single jet centred at $y = 0$, say, and an evanescent linear wave relative to this basic flow. The relevant linearized PV equation is

$$q'_t + U(y)q'_x + Q_y v' = 0 \quad \text{where} \quad Q_y = \Delta q \delta(y) \tag{9.39}$$

is the PV jump at the jet centre. For simplicity we will neglect the β term in (9.38), but it is easy to check that the calculation goes through with the same conclusions if this term is included. Assuming a modal structure $\psi' = \hat{\psi}(y)\exp(ik(x-ct))$ in (9.39), enforcing decay of $\hat{\psi}$ as $|y| \to \infty$ as well as continuity of $\hat{\psi}$ and the appropriate jump condition for $\hat{\psi}_y$ at $y = 0$ then leads to

$$\hat{\psi}(y) = \exp(-|y|(k^2 + \kappa_D^2)^{1/2}) \tag{9.40}$$

$$\text{and} \quad c(k) = U(0)\left(1 - (1 + k^2/\kappa_D^2)^{-1/2}\right). \tag{9.41}$$

Here $U(0) = \Delta q/(2\kappa_D)$ is the maximum jet velocity at $y = 0$. Clearly, $c(k)$ is an increasing function of k and

$$0 \le c(k) \le U(0). \tag{9.42}$$

Comparison with $U(y) = U(0)\exp(-|y|\kappa_D)$ shows that the absolute zonal phase speed of a linear mode relative to a single prograde jet falls between the minimum and the maximum of the jet profile. Therefore every such mode has a pair of critical lines where $U(y) = c$ at some $y = \pm y_c$ in the jet flank. Streamline cat's eyes are formed there and we can expect increased mixing across the flanks of the jet. Now, in the present case this mixing is irrelevant because there is nothing to mix, i.e., the PV gradient is zero in the jet flank by assumption. However, if we imagine that the concentrated PV gradient at $y = 0$ is broadened over some width that includes y_c, then the mixing associated with the critical layer would concentrate the PV gradient once more into the core of the jet. As far as the jet profile is concerned, this robust fluid-dynamical scenario suggests a certain self-sharpening mechanism for these jets, in which any broadening of the PV gradient is quickly counteracted by critical-layer mixing at the jet flanks. As in the single critical-layer case, this is a strong argument for the longevity of PV staircases and of the associated zonal flow structure consisting of sharp prograde jets embedded in broader retrograde flow.

This leads to the interesting suggestion that a smooth PV distribution might, under the right flow conditions, be spontaneously replaced by a non-smooth banded PV distribution, in which piecewise constant well-mixed zones of PV are separated by jumps in the PV. Here the 'right flow conditions' are understood to mean a symbiosis between a sufficient amount of nonlinear turbulence and a sufficient amount of Rossby-wave activity and concomitant radiation stresses in order to allow initiating and sustaining the mixing processes, which, as we have seen via the Taylor identity and the considerations regarding the effective zonal mean force, rely crucially on turbulent mixing inside the layer as well as on wave-induced momentum fluxes outside the layer.

The spontaneous creation of such a layered state from a smooth PV distribution is all the more remarkable because the smooth PV distribution $Q = \beta y$ corresponds to the fluid undergoing solid-body rotation. This points to a fundamental instability of solid-body rotation in stably stratified fluid systems with curved stratification surfaces. Basically, in the presence of suitable radiation stresses, such fluid systems are likely to have a tendency to spontaneously deviate from the solid-body-rotation state, and once they have deviated from it there is no obvious reason why they should return to it.

Apart from Earth's atmospheric flows, this may also apply to flows on planets such as Jupiter and the other gas giants in our solar system, which certainly exhibit strong zonal flows coexisting with high levels of turbulence, and also, sometimes, with long-lived coherent vortex structures such as Jupiter's famous red spot. This is a fascinating speculation, although at present not enough is known about the vertical flow structure on other planets to test this speculation conclusively. The ignorance about this vertical flow structure is all the more important in light of the importance of wave-induced vertical fluxes of zonal momentum, as we have discussed several times.

9.4 Notes on the literature

A central reference on the use of PV diagnostics and the scope of PV dynamics in atmospheric science is Hoskins et al. (1985). There is a very large body of theoretical and numerical work on the topic of balanced flow dynamics, including the interactions between Rossby waves and the zonal-mean flow, and the textbooks Andrews et al. (1987) and Vallis (2006) provide general introductions to these. A mathematical analysis of the quasi-geostrophic approximation is given in Majda (2003). The nature of Rossby-wave critical layers is summarized in the recent review Haynes (2003) and the specific topic of wave absorption or reflection for a mature critical layer is discussed in Killworth and McIntyre (1985). PV staircases and the interplay between turbulence and waves are described theoretically and investigated numerically in Dritschel and McIntyre (2008) and McIntyre (2008).

9.5 Exercises

1. *Generic two-dimensional vortex dynamics.* Consider the generic two-dimensional vortex–stream function dynamics

$$\frac{\partial q}{\partial t} + \partial(\psi, q) = 0 \quad \text{and} \quad \mathcal{L}\psi = q \tag{9.43}$$

in a domain D with $\psi = 0$ on the boundary of D. Here $\mathcal{L}$ is a linear self-adjoint operator. Show that the energy and enstrophy

$$\mathcal{E} = -\frac{1}{2}\int_D \psi q \, dxdy \quad \text{and} \quad \mathcal{Z} = +\frac{1}{2}\int_D q^2 \, dxdy \tag{9.44}$$

are conserved. (Hint: differentiate in time and exploit the self-adjointness of $\mathcal{L}$.) If the right-hand side of (9.43a) is changed to $\nu\mathcal{L}q$ with $\nu > 0$ constant, show that $\mathrm{d}\mathcal{E}/\mathrm{d}t = -2\nu\mathcal{Z}$ and hence energy decreases. Under what condition on $\mathcal{L}$ does enstrophy decrease as well? Let D be the entire plane and restrict $\mathcal{L}$ to have constant real coefficients. Show that if the spectral energy density is $E(k, l, t)$ then the spectral enstrophy density is $-\hat{\mathcal{L}}E$, where $\hat{\mathcal{L}}$ is the symbol of $\mathcal{L}$ in Fourier space and k, l are the wavenumbers in x and y.

2. *Small-amplitude pseudomomentum for gravity and Rossby waves.* Consider the linear Boussinesq equations

$$u'_t - fv' + P'_x = 0,\ v'_t + fu' + P'_y = 0,\ w'_t + P'_z = b' \qquad (9.45)$$

$$b'_t + N^2 w' = 0,\ \ u'_x + v'_y + w'_z = 0 \qquad (9.46)$$

with $f = f_0 + \beta y$ and constant N. Show that the PV disturbance

$$q' \equiv v'_x - u'_y + b'_z f/N^2 \quad \text{satisfies} \quad q'_t + \beta v' = 0. \qquad (9.47)$$

Assuming that the flow is x-periodic and that $\overline{(\ldots)}$ denotes x-averaging, show that the pseudomomentum conservation law

$$(\mathsf{p}_G + \mathsf{p}_R)_t + \boldsymbol{\nabla} \cdot \boldsymbol{F} = 0 \qquad (9.48)$$

holds where

$$\mathsf{p}_G \equiv \frac{1}{N^2}\,\overline{b'(u'_z - w'_x)} \quad \text{and} \quad \mathsf{p}_R \equiv -\frac{1}{2\beta}\,\overline{q'^2} \qquad (9.49)$$

are the pseudomomentum densities for gravity and Rossby waves, respectively, and

$$\boldsymbol{F} \equiv (0, \overline{u'v'}, \overline{u'w'} - \frac{f}{N^2}\,\overline{b'v'}) \qquad (9.50)$$

is the Eliassen–Palm flux.

10
Lagrangian-mean theory

We now embark on a journey into new theoretical territory: Lagrangian-mean theory based on particle-following averaging. This theory allows a sharper and more succinct description of material advection in fluid dynamics, which greatly simplifies the mean description of material invariants such as scalar tracers or, crucially, of vorticity and potential vorticity. Based on this, Lagrangian-mean theory is superior to Eulerian-mean theory in the description of flow dynamics to do with vorticity and circulation. This will be particularly important for wave–mean interactions outside simple geometry, where long-range mean pressure fields greatly complicate the description of the mean flow based on momentum budgets.

On the downside of Lagrangian-mean theory we need to count the increased structural complexity of the theory, the requirement to evolve particle displacements alongside with the usual flow variables, and the potential breakdown of the particle-following flow map under averaging in the case of large-amplitude waves. Thus whether Eulerian or Lagrangian averaging is more efficient can depend on the problem at hand. As before, we take the view that Eulerian and Lagrangian concepts are complementary to each other in the optimal description of all aspects of fluid motion, and this is certainly true in wave–mean interaction theory as well.

We begin with a general discussion of Lagrangian averaging and of small-amplitude Stokes corrections and then give a comprehensive introduction to the so-called generalized Lagrangian-mean theory (GLM), which formally is not restricted to small-amplitude waves. This includes finite-amplitude versions of the conservation laws for wave activities based on the symmetries of the Lagrangian-mean flow. For simplicity, we start out in an inertial frame, but Coriolis forces are added in §10.4, which also includes a discussion of the β-plane.

10.1 Lagrangian and Eulerian averaging

We can discuss the basic goals of Lagrangian averaging by considering a material tracer with a density $\phi(\boldsymbol{x}, t)$ that satisfies

$$\frac{D\phi}{Dt} = \left(\frac{\partial}{\partial t} + \boldsymbol{u} \cdot \boldsymbol{\nabla}\right)\phi = 0. \tag{10.1}$$

Clearly, ϕ is invariant along the actual particle trajectories, which are the integral curves of the velocity field $\boldsymbol{u}(\boldsymbol{x}, t)$. This simple advective structure is lost once Eulerian averages are taken. Indeed, the Eulerian mean of (10.1) is

$$\left(\frac{\partial}{\partial t} + \overline{\boldsymbol{u}} \cdot \boldsymbol{\nabla}\right)\overline{\phi} = -\overline{(\boldsymbol{u}' \cdot \boldsymbol{\nabla})\phi'}, \tag{10.2}$$

which shows that $\overline{\phi}$ is *not* invariant along the Eulerian-mean particle trajectories, which are the integral curves of the Eulerian-mean velocity field $\overline{\boldsymbol{u}}(\boldsymbol{x}, t)$. This equation does not depend on the type of averaging, but it is easiest to consider a time average over a rapidly varying time scale, say.

We now consider the question whether a Lagrangian averaging procedure can be defined such that (10.2) holds without a source term on the right-hand side, i.e., we want to define a Lagrangian-mean averaging operator such that the Lagrangian-mean velocity $\overline{\boldsymbol{u}}^L$ and the Lagrangian-mean material tracer density $\overline{\phi}^L$ satisfy

$$\left(\frac{\partial}{\partial t} + \overline{\boldsymbol{u}}^L \cdot \boldsymbol{\nabla}\right)\overline{\phi}^L = 0. \tag{10.3}$$

One answer comes from the full Lagrangian description of fluid dynamics in terms of a flow map such that $\boldsymbol{x} = \boldsymbol{X}(\boldsymbol{x}_0, t)$ is the current position of the fluid particle that was at $\boldsymbol{x} = \boldsymbol{x}_0$ at the initial time $t = 0$. All fields could then be expressed as functions of $\boldsymbol{x}_0$ and t and in these variables the material time derivative is simply a partial time derivative. It follows that (10.1) and (10.2) are replaced by

$$\phi_t(\boldsymbol{x}_0, t) = 0 \quad \text{and} \quad \overline{\phi}^L_t(\boldsymbol{x}_0, t) = 0, \tag{10.4}$$

provided the averaging is now performed over fixed *initial* positions $\boldsymbol{x}_0$. This clearly preserves the invariance of both ϕ and $\overline{\phi}^L$ along particle trajectories, although in $\boldsymbol{x}_0$-coordinates these are trivial zero-motion trajectories of constant $\boldsymbol{x}_0$. Of course, all the dynamics is stored in the flow map $\boldsymbol{X}(\boldsymbol{x}_0, t)$, which together with its inverse needs to be computed explicitly in order to find the solution.[1]

[1] For example, using the chain rule to evaluate the spatial pressure gradient $\boldsymbol{\nabla}p$ in $(\boldsymbol{x}_0, t)$ coordinates requires the gradient of the inverse flow map.

However, the use of the full flow map is entirely impractical in fluid dynamics, at least if the number of spatial dimensions is larger than one. This is because of the irreducible complexity of chaotic particle trajectories in fluid dynamics, which make the full flow map $\boldsymbol{X}(\boldsymbol{x}_0, t)$ extremely complicated and impossible to use in practice. For example, time-dependent two-dimensional flows exhibit chaotic particle trajectories with the concomitant exponentially fast stretching and folding of small-scale features. Worse still, in three dimensions it is possible to obtain chaotic particle trajectories even for steady flows, as is exemplified by the celebrated ABC flow, and the chaotic particle dynamics is much worse for time-dependent flows. Only in one-dimensional flows is the use of particle labels unproblematic, and such mass-following coordinates are indeed a standard tool in one-dimensional gas dynamics.

We therefore conclude that a practical Lagrangian-mean theory must pursue two goals: on the one hand, it must include enough Lagrangian information such that particle-following averaging can be defined consistently and that simple results for mean material invariants can be obtained. On the other hand, there must not be too much Lagrangian information in order to avoid the disastrous complexities of the full flow map. How to best balance these twin goals is to some extent a matter of personal judgement.

GLM theory is a hybrid theory in which the independent variables remain the usual Eulerian position $\boldsymbol{x}$ and time t whilst the Lagrangian information is stored in an additional dependent field, namely the disturbance-associated displacement field $\boldsymbol{\xi}(\boldsymbol{x}, t)$. The kinematic nature of this field (which will be made precise in §10.2.1) is such that the actual particle trajectories are split into a mean part and a disturbance part, and the working assumption is that the chaotic nature of the trajectories is captured mostly by the mean part, leaving a comparatively well behaved disturbance part that can be studied fruitfully with analytical or numerical tools.

One of the attractive features of GLM theory is that it appears as a natural extension of small-amplitude wave theory, e.g., $\boldsymbol{\xi}$ is a natural finite-amplitude counterpart of the linear particle displacement vector $a\boldsymbol{\xi}'_1$. Also, formally at least, GLM theory extends easily to finite-amplitude disturbances, although in practice its precise use is limited to cases of nearly integrable particle trajectories.[2] Most important in our context is the ease with which a finite-amplitude version of pseudomomentum can be defined in GLM theory.

Against the attractions of GLM theory has to be set the formal nature of

[2] The *formal structure* of GLM theory has also motivated a range of turbulence closure schemes that are collectively called α-models. Like all such closure schemes, α-models contain a mixture of derivation and ad hoc assumptions, so their practical value is evaluated in terms of comparison with experiments.

the definitions concerning $\boldsymbol{\xi}$ and its evolution law, which can lead to breakdown of the theory for finite-amplitude waves. We view this possibility of failure as a necessary consequence of trying to retain some, but not too much, explicit Lagrangian information: we are tickling the Lagrangian dragon of chaotic particle trajectories, and the possibility to predict its own breakdown is not necessarily a bad feature for a theoretical framework operating in such a delicate situation.

10.1.1 Stokes corrections

We use the term *Stokes corrections* (or *Stokes drift* in the special case of velocity) to describe the difference between Lagrangian and Eulerian averages of flow fields. Symbolically, we make the convention

$$\text{Lagrange} = \text{Euler} + \text{Stokes} \tag{10.5}$$

between the various terms. It turns out that Stokes corrections are wave properties, so for small-amplitude waves one can freely move between the Eulerian and the Lagrangian mean-flow response at $O(a^2)$ by adding or subtracting terms that are known from the linear solution. We illustrate this for a plane shallow-water wave relative to a basic state of rest and constant layer depth H (cf. §2.1.4). To $O(a)$ the layer depth and velocity are given by $h = H + ah'_1$ and $u = au'_1$ such that

$$h'_1(x,t) = H\cos k(x-ct) \quad \text{and} \quad u'_1(x,t) = c\cos k(x-ct) \tag{10.6}$$

where $c > 0$, say. The linear particle displacement is

$$\xi'_1 = -\frac{1}{k}\sin k(x-ct) \tag{10.7}$$

and the time-dependent linear particle trajectories are given by

$$x^\xi = x + a\xi'_1(x,t), \tag{10.8}$$

say. In this expression $x^\xi(x,t)$ is the *actual* position (to linear accuracy) at time t of the fluid particle whose *mean* position is x. If we use time-averaging over a wave period for the definition of Eulerian averaging then it is clear that $\overline{x^\xi} = x$.

However, in this longitudinal wave a fluid particle that is moved back and forth by the linear wave is spending slightly more time in the forward-moving phase region than in the backward-moving one. This skews a time-average following the particle, because the conditions prevailing in

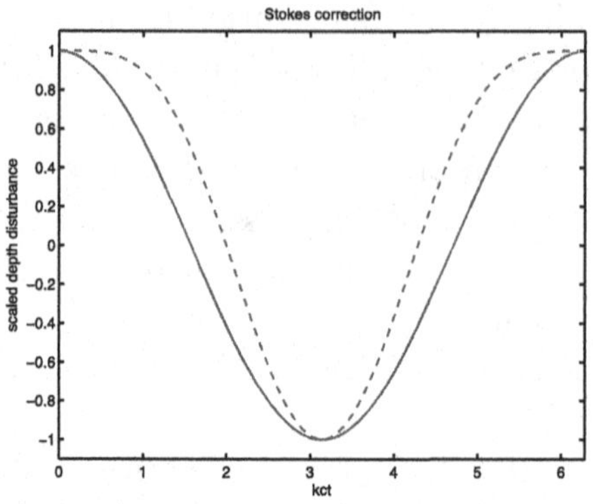

Figure 10.1 The solid line is $h'_1(x,t)$ as a function of time at fixed x and the dashed line is $h'_1(x^\xi,t)$. The wave amplitude has the very large value $a = 0.5$ for ease of illustration. This shows that the particle experiences positive values of h' for longer than negative values, which leads to the positive Stokes correction in (10.12).

the forward-moving phase are sampled more often. For instance, the linear layer depth as seen by a fixed particle is

$$\begin{aligned} h'_1(x^\xi,t) &= h'_1(x,t) + a\xi'_1 h'_{1x}(x,t) + O(a^2) \qquad (10.9)\\ &\approx h'_1(x,t) - ak\xi'_1(x,t)H \sin k(x-ct) = h'_1(x,t) + aH\sin^2 k(x-ct) \end{aligned}$$

to linear accuracy (cf. Figure 10.1). The second term does not average to zero, but to $aH/2$.

If we define Lagrangian averaging for any flow field ϕ via

$$\overline{\phi}^L = \overline{\phi(x^\xi(x,t),t)} \qquad (10.10)$$

and apply this to $h = \phi$ then we obtain

$$\overline{h}^L = H + a\overline{h'_1(x^\xi,t)} + a^2\overline{h}_2 + O(a^3) \approx H + a^2\left(\frac{H}{2} + \overline{h}_2\right). \qquad (10.11)$$

Using the convention (10.5) we can read-off the leading-order Stokes correction as

$$\overline{h}_2^S = \frac{H}{2}, \quad \text{and} \quad \overline{u}_2^S = \frac{c}{2} \qquad (10.12)$$

then follows by the same method.

It is noteworthy that similar Stokes corrections also apply to surface waves in deep water, which are evanescent in the vertical. For instance, the leading-order Stokes correction for the surface height of a plane wave in deep water is $|k|\overline{h'^2}$. This explains the well-observed fact that time-averaged altitude measurements from floating buoys in wavy seas show a slight increase above the Eulerian-mean sea level. On the other hand, the leading-order Stokes corrections are zero in a transversal plane wave, such as an internal gravity wave in the Boussinesq system. This is because the linear particle motion stays within the planes of constant wave phase in this case, so there is no skewing of averages of the kind described here.

One must be careful not to misinterpret results such as (10.12). For example, (10.12b) does *not* imply that particles are drifting forward with mean speed $a^2\overline{u}_2^S$; this would be the case if $a^2\overline{u}_2^L$ were known to be positive. However,

$$\overline{u}_2^L = \overline{u}_2 + \overline{u}_2^S \tag{10.13}$$

and hence this requires knowledge of $\overline{u}_2$. This is the mean-flow response and it is *not* a wave property, so in order to find this quantity one needs to go beyond the linear solution. A simple thought experiment shows that $\overline{u}_2^L$ can be zero even for a progressive plane wave. Here we imagine one-dimensional acoustic waves being generated by an oscillating piston at one end of a narrow tube. The piston oscillates harmonically around $x = 0$ and generates a longitudinal sound wave that propagates towards increasing x. A second movable piston at $x = L$, say, then absorbs the wave without reflection via suitable oscillations. By the gas-dynamical analogy between sound waves and shallow-water waves, the sound wave in the tube has a non-zero positive Stokes drift and yet the fluid as a whole cannot move, i.e., its centre of mass cannot drift within the finite tube. This implies that the Lagrangian-mean velocity is zero, and therefore the Eulerian-mean flow must be equal-and-opposite to the Stokes drift in this thought experiment.

Beyond plane waves we can find the generic expression for the Stokes corrections at $O(a^2)$ by Taylor-expanding (10.10), which leads to

$$\overline{\phi}_2^L - \overline{\phi}_2 = \overline{\phi}_2^S = \overline{\xi_1'\phi_{1x}'} + \frac{1}{2}\overline{\xi_1'^2}\,\overline{\phi}_{0xx}. \tag{10.14}$$

This includes a term due to possible spatial inhomogeneities of the basic state $\phi_0 = \overline{\phi}_0$. Finally, the straightforward multidimensional version of (10.14) without expansion subscripts is

$$\overline{\phi}^S = \overline{(\boldsymbol{\xi}' \cdot \boldsymbol{\nabla})\phi'} + \frac{1}{2}\sum_{i,j} \overline{\xi_i'\xi_j'}\,\overline{\phi}_{ij} + O(a^3). \tag{10.15}$$

For a plane wave $\boldsymbol{\xi}' \cdot \boldsymbol{\nabla} = i\boldsymbol{\xi}' \cdot \boldsymbol{k}$, which is zero for a transversal wave.

10.1.2 Stokes drift, pseudomomentum and bolus velocity

In many situations it is useful to know the relation between the $O(a^2)$ Stokes drift and the pseudomomentum, not least because these false friends are sometimes confused with each other. The required relation is easily computed if the approximations

$$\overline{u}_i^S = \overline{\xi_j u'_{i,j}} \quad \text{and} \quad \mathsf{p}_i = -\overline{\xi_{j,i} u'_j} \tag{10.16}$$

hold, i.e., if the basic shear can be neglected.[3] It then follows that

$$\mathbf{p} = \overline{\boldsymbol{u}}^S + \overline{\boldsymbol{\xi} \times (\boldsymbol{\nabla} \times \boldsymbol{u}')} - \frac{1}{2}\boldsymbol{\nabla}\mathrm{D}_t(\overline{\boldsymbol{\xi} \cdot \boldsymbol{\xi}}) \tag{10.17}$$

holds to the same approximation. Here we used $\overline{\boldsymbol{\xi} \cdot \boldsymbol{u}'} = \frac{1}{2}\mathrm{D}_t(\overline{\boldsymbol{\xi} \cdot \boldsymbol{\xi}})$. There are several interesting aspects of (10.17). For instance, the zonal component of the last term is zero if zonal averaging is used. Otherwise, this term is zero for a steady wave field and $O(\epsilon^2 a^2)$ for a slowly varying wavetrain, where $\epsilon \ll 1$ measures the scale separation, and therefore negligible in those situations.

The second term involves the curl of the disturbance velocity and is therefore zero for irrotational flow; this explains the similarity of $\overline{\boldsymbol{u}}^S$ and $\mathbf{p}$ for acoustic waves, or for surface gravity waves. For internal gravity waves, on the other hand, the leading-order Stokes drift is zero for slowly varying waves and therefore the second term dominates the pseudomomentum. Indeed, for two-dimensional flow in the xz-plane the zonal component of (10.17) is the familiar expression $-\overline{\zeta(u'_z - w'_x)}$.

Another mean velocity field that is sometimes useful for compressible flows (or for flows that appear compressible in the chosen description, such as shallow-water flows, or stratified flows in isentropic coordinates) is the *mass transport velocity* defined by

$$\overline{\boldsymbol{u}}^M = \overline{\boldsymbol{u}} + \frac{1}{\overline{\rho}}\,\overline{\rho' \boldsymbol{u}'} \quad \text{such that} \quad \overline{\rho}\,\overline{\boldsymbol{u}}^M = \overline{\rho \boldsymbol{u}}. \tag{10.18}$$

The disturbance-associated part of $\overline{\boldsymbol{u}}^M$ is occasionally called the *bolus velocity*. It is a wave property and related to the Stokes drift at leading order via

$$\overline{\boldsymbol{u}}^S = \frac{1}{\overline{\rho}}\,\overline{\rho' \boldsymbol{u}'} + \boldsymbol{\nabla} \cdot (\overline{\boldsymbol{u}' \boldsymbol{\xi}}). \tag{10.19}$$

Thus, for slowly varying wavetrains the bolus velocity is approximately equal to the Stokes drift; finally, for compressible irrotational wavetrains both quantities also agree with the pseudomomentum. This often-studied case contributes to the confusion between these three distinct quantities.

[3] Here and later we use component notation with summation over repeated indices understood.

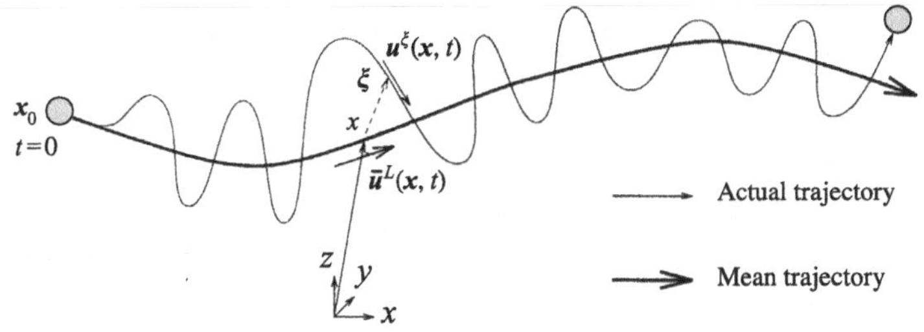

Figure 10.2 Mean and actual trajectory of a particle for a problem with multiple time scales: $\boldsymbol{x} + \boldsymbol{\xi}(\boldsymbol{x}, t)$ is the *actual* position of the fluid particle whose *mean* position is $\boldsymbol{x}$ at (slow) time t. The notation $\boldsymbol{u}^\xi(\boldsymbol{x}, t)$ is shorthand for $\boldsymbol{u}(\boldsymbol{x} + \boldsymbol{\xi}(\boldsymbol{x}, t), t)$.

10.2 Elements of GLM theory

GLM theory is an abstract theory because it applies to arbitrary Eulerian averaging operators: each such Eulerian-mean operator induces a corresponding Lagrangian-mean operator in GLM theory. Still, the kinematics of GLM theory is most easily visualized if the problem has multiple time scales and the averaging operation is defined by time-averaging over the fast time scale (see Figure 10.2). This is the approach that we will use here.

10.2.1 Lifting map and Lagrangian averaging

At the heart of GLM theory is the definition of a disturbance-associated particle displacement field $\boldsymbol{\xi}(\boldsymbol{x}, t)$ that generalizes the linear particle displacements $a\boldsymbol{\xi}_1'(\boldsymbol{x}, t)$ to finite amplitude. Thus, the kinematic definition of $\boldsymbol{\xi}(\boldsymbol{x}, t)$ is that

$$\boldsymbol{x} + \boldsymbol{\xi}(\boldsymbol{x}, t) \tag{10.20}$$

is the *actual* position at time t of the fluid particle whose *mean* position is $\boldsymbol{x}$ (cf. Figure 10.2). It is crucial that $\boldsymbol{\xi}$ is itself evaluated at $\boldsymbol{x}$, because this ensures that the Eulerian mean of (10.20) is simply equal to $\boldsymbol{x}$ and therefore the condition

$$\overline{\boldsymbol{\xi}} = 0 \tag{10.21}$$

holds, so $\boldsymbol{\xi}$ is necessarily a disturbance field.

Another way to look at (10.20) is as the definition of a time-dependent map, which we will call the *lifting map*

$$\boldsymbol{x}^\xi\colon \quad \boldsymbol{x} \to \boldsymbol{x}^\xi = \boldsymbol{x} + \boldsymbol{\xi}(\boldsymbol{x}, t). \tag{10.22}$$

By construction, this nonlinear map lifts mean particle position into actual particle positions. For infinitesimal wave amplitudes this is a near-identity map, but this is not the case in general. It is a basic assumption of GLM theory that this map is smooth and invertible, i.e., each actual particle position corresponds to exactly one mean position and vice versa. As an aside, we note that there is no guarantee that this assumption will hold in all flows, or that it will hold indefinitely for a particular flow even if it was true initially. This is the potential breakdown of nonlinear GLM theory that was mentioned previously.

In essence, GLM theory treats ordinary space (i.e., all the points $\boldsymbol{x}$ in the fluid domain) as a time-dependent reference space for both mean and actual trajectories. Th first reference is established by the trivial map $\boldsymbol{x} \to \boldsymbol{x}$, which associates with $\boldsymbol{x}$ the mean trajectory that touches $\boldsymbol{x}$ at time t. The second reference is established by the lifting map (10.22), which associates with $\boldsymbol{x}$ the actual particle trajectory of the same particle. This dual reference avoids the use of classical Lagrangian particle labels as a reference space.

Before moving on we note that for convex fluid domains the inverse of the lifting map provides a map of the domain into itself, but that this is not necessarily the case otherwise. This is because only in a convex domain do the mean positions necessarily lie inside the domain. For example, if the fluid domain is the exterior of a sphere (as is natural for atmospheric applications), then it is possible that the mean position of a particle lies within the sphere, i.e., underground. To see this it is sufficient to visualize a particle sliding back and forth along a surface geodesic. Clearly, it is the curvature of the fluid boundary that leads to this effect.

Now, the lifting map (10.22) naturally induces a lifting map in function space for arbitrary functions $\phi(\boldsymbol{x}, t)$ such that

$$\phi(\boldsymbol{x}, t) \to \phi^\xi(\boldsymbol{x}, t) = \phi(\boldsymbol{x} + \boldsymbol{\xi}(\boldsymbol{x}, t), t). \tag{10.23}$$

Based on (10.23) and the definition of an Eulerian averaging operator that applies to arbitrary functions of $(\boldsymbol{x}, t)$ we can now define the Lagrangian-mean as the Eulerian mean of the lifted function:

$$\overline{\phi}^L = \overline{\phi^\xi} = \overline{\phi(\boldsymbol{x} + \boldsymbol{\xi}(\boldsymbol{x}, t), t)}. \tag{10.24}$$

The somewhat opulent notation makes explicit that $\boldsymbol{\xi}$ is *always* evaluated at $\boldsymbol{x}$. With this convention understood we will drop the arguments of $\boldsymbol{\xi}$ from now on. The Lagrangian-mean operator involves the nonlinear lifting map, but for a given $\boldsymbol{\xi}$ it is still a linear operator, just as the Eulerian-mean operator.

The definition of the Lagrangian-mean field $\overline{\phi}^L$ suggests a corresponding definition of a Lagrangian disturbance field ϕ^ℓ in analogy with the definition of Eulerian disturbances as $\phi' = \phi - \overline{\phi}$:

$$\phi^\ell(\boldsymbol{x}, t) = \phi^\xi(\boldsymbol{x}, t) - \overline{\phi}^L(\boldsymbol{x}, t) \quad \text{such that} \quad \overline{\phi^\ell} = 0. \tag{10.25}$$

This also implies the Lagrangian version of the quadratic averaging rule

$$\overline{\phi\psi}^L = \overline{\phi}^L\overline{\psi}^L + \overline{\phi^\ell\psi^\ell} \tag{10.26}$$

for arbitrary ϕ and ψ. For small-amplitude disturbances we have the relation

$$\phi^\ell(\boldsymbol{x}, t) = \phi'(\boldsymbol{x}, t) + \boldsymbol{\xi} \cdot \boldsymbol{\nabla}\overline{\phi} + O(a^2) \tag{10.27}$$

between the two kinds of disturbance fields.

10.2.2 The mean material derivative and trajectories

Mean particle positions move with the Lagrangian-mean velocity

$$\overline{\boldsymbol{u}}^L(\boldsymbol{x}, t) = \overline{\boldsymbol{u}^\xi(\boldsymbol{x}, t)} = \overline{\boldsymbol{u}(\boldsymbol{x} + \boldsymbol{\xi}, t)}, \tag{10.28}$$

which induces a mean material derivative via

$$\overline{\mathrm{D}}^L = \partial_t + \overline{\boldsymbol{u}}^L \cdot \boldsymbol{\nabla}. \tag{10.29}$$

The integral curves of $\overline{\boldsymbol{u}}^L$ are mean material trajectories and we want to ensure that if $\boldsymbol{x}(t)$ moves with $\overline{\boldsymbol{u}}^L$ then $\boldsymbol{x}(t) + \boldsymbol{\xi}(\boldsymbol{x}(t), t)$ moves with the actual fluid velocity at the lifted position, which is $\boldsymbol{u}^\xi(\boldsymbol{x}(t), t)$. In other words, if $\boldsymbol{x}(t)$ is a *mean* material trajectory then $\boldsymbol{x}(t) + \boldsymbol{\xi}(\boldsymbol{x}(t), t)$ is an *actual* material trajectory. This leads to an evolution equation for $\boldsymbol{\xi}$ consistent with the kinematic definition in (10.20).

Indeed, applying $\overline{\mathrm{D}}^L$ to (10.20) yields

$$\overline{\mathrm{D}}^L \boldsymbol{x}^\xi = \overline{\boldsymbol{u}}^L + \overline{\mathrm{D}}^L \boldsymbol{\xi} = \boldsymbol{u}^\xi = \overline{\boldsymbol{u}}^L + \boldsymbol{u}^\ell \quad \Rightarrow \quad \overline{\mathrm{D}}^L \boldsymbol{\xi} = \boldsymbol{u}^\ell, \tag{10.30}$$

which is the sought-after evolution law for $\boldsymbol{\xi}$. The small-amplitude version of this law for a basic flow $\boldsymbol{U}(\boldsymbol{x}, t)$ is

$$(\partial_t + \boldsymbol{U} \cdot \boldsymbol{\nabla})\, \boldsymbol{\xi}_1' = \boldsymbol{u}_1' + (\boldsymbol{\xi}_1' \cdot \boldsymbol{\nabla})\boldsymbol{U}. \tag{10.31}$$

This reduces to the earlier relation (7.9) in the case of linear waves on zonal shear flows.

Equation (10.30) contains the relation

$$\overline{\mathrm{D}}^L \boldsymbol{x}^\xi = \boldsymbol{u}^\xi = \left(\frac{D\boldsymbol{x}}{Dt}\right)^\xi, \tag{10.32}$$

which is a special case of the important *GLM identity*

$$\overline{\mathrm{D}}^L \phi^\xi = \left(\frac{D\phi}{Dt}\right)^\xi. \tag{10.33}$$

We will prove this identity in §10.2.3, but for now we notice that its average immediately implies a very simple form of the Lagrangian-mean of a material derivative:

$$\overline{\mathrm{D}}^L \overline{\phi}^L = \overline{\left(\frac{D\phi}{Dt}\right)}^L. \tag{10.34}$$

Here we used that $\overline{\overline{\boldsymbol{u}}^L} = \overline{\boldsymbol{u}}^L$. This key relation shows that GLM theory maintains the advective structure of material invariants because

$$\frac{D\phi}{Dt} = R \quad \text{implies} \quad \overline{\mathrm{D}}^L \overline{\phi}^L = \overline{R}^L. \tag{10.35}$$

If $R = 0$ then ϕ is materially advected by the actual flow and $\overline{\phi}^L$ is materially advected by the Lagrangian-mean flow. Thus GLM theory achieves the main objective for a Lagrangian-mean theory.

The price to pay for the simplicity with which material derivatives can be handled in GLM theory is that Lagrangian averaging does not commute with covariant differentiation with respect to time or space, i.e., $\overline{\boldsymbol{\nabla}\phi}^L \neq \boldsymbol{\nabla}\overline{\phi}^L$ in general. This is a disadvantage compared to Eulerian-mean theory and it stems from the nonlinear lifting map, which does not commute with these derivatives:

$$(\phi^\xi)_t = (\phi_t)^\xi + (\boldsymbol{\nabla}\phi)^\xi \cdot \boldsymbol{\xi}_t \quad \text{and} \quad (\phi^\xi)_{,i} = (\phi_{,j})^\xi(\delta_{ji} + \xi_{j,i}). \tag{10.36}$$

Here the commas denote covariant derivatives and the summation convention is used. The second formula is particularly useful, because it shows that a lifted gradient can be brought back into gradient form by contracting it with $\delta_{ji} + \xi_{j,i}$. The inverse of this lifted gradient map will be given in (10.68) below. With some effort (10.30) and (10.36) can be used for a direct proof of the identity (10.33), but we prefer a more geometric proof in the next section.

10.2.3 Mean mass conservation

The lifting map is a smooth spatial map and we can derive a number of interesting kinematic results from the interplay between this map and standard vector calculus, especially in the form of integral theorems. First, we will consider mean and actual fluid volumes, which will lead to the most useful form of the mass conservation law in GLM theory.

To every fluid volume $\mathcal{D} \subset R^n$ we can associate a lifted volume $\mathcal{D}^\xi \subset R^n$ by applying the lifting map (10.22) to every point $\boldsymbol{x} \in \mathcal{D}$. Thus a point $\boldsymbol{y} \in \mathcal{D}^\xi$ if there is a point $\boldsymbol{x} \in \mathcal{D}$ such that $\boldsymbol{y} = \boldsymbol{x}^\xi$. By the assumed invertibility of the lifting map we can go back and forth between $\mathcal{D}$ to $\mathcal{D}^\xi$ and hence we can write

$$\boldsymbol{x} \in \mathcal{D} \quad \Leftrightarrow \quad \boldsymbol{x}^\xi \in \mathcal{D}^\xi. \tag{10.37}$$

Physically, if $\mathcal{D}^\xi$ is an actual material fluid volume then $\mathcal{D}$ is the volume composed of all the mean positions associated with the particles in $\mathcal{D}^\xi$.

We now consider the mass content of the material volume $\mathcal{D}^\xi$, which by definition is the volume integral

$$M = \int_{\mathcal{D}^\xi} \rho\, dV \tag{10.38}$$

over the fluid density ρ evaluated at all points in $\mathcal{D}^\xi$. By substitution, we can re-write (4.115b) as an integral over the mean material volume $\mathcal{D}$ provided we lift the integrand to the displaced position, i.e.,

$$M = \int_{\mathcal{D}^\xi} \rho\, dV = \int_{\mathcal{D}} (\rho\, dV)^\xi = \int_{\mathcal{D}} \rho^\xi\, (dV)^\xi. \tag{10.39}$$

The key point is that the second integral ranges over the points $\boldsymbol{x} \in \mathcal{D}$ and therefore we must lift ρ and the volume element dV to the actual positions $\boldsymbol{x}^\xi \in \mathcal{D}^\xi$. The lifted volume element is

$$(dV)^\xi = J\, dV \quad \text{where} \quad J = \frac{\partial(\boldsymbol{x}^\xi)}{\partial(\boldsymbol{x})} \tag{10.40}$$

is the Jacobian of the lifting map, which by the assumed invertibility satisfies $J \in (0, \infty)$. For example, if the number of dimensions is $n = 1$ we have

$$J = \frac{\partial(x + \xi)}{\partial(x)} = 1 + \xi_x \tag{10.41}$$

and if $n = 2$ then $\boldsymbol{\xi} = (\xi, \eta)$ and we have

$$J = \frac{\partial(x + \xi, y + \eta)}{\partial(x, y)} = 1 + \boldsymbol{\nabla} \cdot \boldsymbol{\xi} + (\boldsymbol{\nabla}\xi \times \boldsymbol{\nabla}\eta). \tag{10.42}$$

In general, J is a polynomial of degree n in the components of $\boldsymbol{\nabla\xi}$.

Now, (10.39) suggests that we define an *effective mean density* $\tilde{\rho}$ as

$$\tilde{\rho} = \rho^\xi J, \tag{10.43}$$

because then we can write (4.116) in the simple form

$$M = \int_{\mathcal{D}^\xi} \rho\, dV = \int_{\mathcal{D}} \tilde{\rho}\, dV. \tag{10.44}$$

So far we have not used mass conservation, which states that M is constant in time if $\mathcal{D}^\xi$ is a material volume, i.e., if the points in $\mathcal{D}^\xi$ follow the fluid motion. Of course, this is just the basic statement of mass conservation in integral form. What is of interest here is how this statement leads to the local continuity equation. The time derivative of a material volume integral equals the integral over the material derivative of the integrand, and therefore

$$\frac{d}{dt}\int_{\mathcal{D}^\xi} \rho\, dV = \int_{\mathcal{D}^\xi} \frac{D}{Dt}(\rho\, dV) = 0 \tag{10.45}$$

must hold for all $\mathcal{D}^\xi$. This implies the pointwise conservation law $D(\rho\, dV)/Dt = 0$, which by using $D(dV)/Dt = dV\boldsymbol{\nabla}\cdot\boldsymbol{u}$ can be written in the usual forms

$$\frac{D}{Dt}(\rho\, dV) = 0 \quad \Leftrightarrow \quad \frac{D\rho}{Dt} + \rho\boldsymbol{\nabla}\cdot\boldsymbol{u} = 0 \quad \Leftrightarrow \quad \frac{\partial\rho}{\partial t} + \boldsymbol{\nabla}\cdot(\rho\boldsymbol{u}) = 0. \tag{10.46}$$

On the other hand, if the points in $\mathcal{D}^\xi$ move with $\boldsymbol{u}$ then the points in $\mathcal{D}$ move with $\overline{\boldsymbol{u}}^L$. Therefore we can apply precisely the same reasoning to the second half of (10.44), but with ρ and $\boldsymbol{u}$ replaced by $\tilde{\rho}$ and $\overline{\boldsymbol{u}}^L$. The result is

$$\overline{\mathrm{D}}^L(\tilde{\rho}\, dV) = 0 \quad \Leftrightarrow \quad \overline{\mathrm{D}}^L\tilde{\rho} + \tilde{\rho}\boldsymbol{\nabla}\cdot\overline{\boldsymbol{u}}^L = 0 \quad \Leftrightarrow \quad \frac{\partial\tilde{\rho}}{\partial t} + \boldsymbol{\nabla}\cdot\left(\tilde{\rho}\overline{\boldsymbol{u}}^L\right) = 0. \tag{10.47}$$

This shows that the form of the continuity equation has survived under averaging provided we use the effective density $\tilde{\rho}$. Moreover, (10.47) is linear in $\tilde{\rho}$, and $\overline{\boldsymbol{u}}^L$ is a mean field, and therefore (10.47) applies to both the Eulerian-mean and disturbance part of $\tilde{\rho}$ individually. This implies that $\tilde{\rho}$ will remain a mean field if it started out as a mean field. This property of $\tilde{\rho}$ is not obvious from its definition (10.43). Overall, the utility of $\tilde{\rho} = \rho^\xi J$ derives from the fact that it is an almagamate of the physical density and the mathematical lifting map. The drawback is that in general $\tilde{\rho}$ is not equal to either $\overline{\rho}$ or $\overline{\rho}^L$ (see §10.2.4 below).

We can now prove the key identity (10.33) by considering the more general volume integral

$$\int_{\mathcal{D}^\xi} \phi\rho\, dV = \int_{\mathcal{D}} \phi^\xi\tilde{\rho}\, dV, \tag{10.48}$$

which measures the ϕ-content in the material volume $\mathcal{D}^\xi$. Using (10.46), the time derivative of the left-hand integral is

$$\int_{\mathcal{D}^\xi} \frac{D\phi}{Dt} \rho \, dV = \int_{\mathcal{D}} \left(\frac{D\phi}{Dt} \right)^\xi \tilde{\rho} \, dV, \tag{10.49}$$

where we have pulled the equation back to $\mathcal{D}$ in the second step. On the other hand, using (10.47), the time derivative of the right-hand integral in (10.48) yields

$$\int_{\mathcal{D}} (\overline{\mathrm{D}}^L \phi^\xi) \, \tilde{\rho} \, dV \tag{10.50}$$

and as both (10.49) and (10.50) must be equal for all $\mathcal{D}$ we have proven (10.33).

This short geometric proof illustrates the useful method of alternating the lifting map with other operations such as taking the material derivative. We will use the same idea several times in the following sections. With the definition of $\tilde{\rho}$ we have completed the definition of the Lagrangian-mean flow, which consists of the fields $(\tilde{\rho}, \overline{\boldsymbol{u}}^L, \overline{s}^L)$, where $\overline{s}^L$ stands for the Lagrangian-mean of the specific entropy, and also for any other additional material invariants that might be relevant for the thermodynamic fluid state.

10.2.4 Small-amplitude relations for the mass density

We note in passing that (10.43) implies the small-amplitude relation $\rho_1' = -\boldsymbol{\nabla} \cdot (\rho_0 \boldsymbol{\xi}_1)$ that we used repeatedly in linear wave theory. This follows from expanding (10.43) to first order:

$$\rho_0 = (\rho_0 + a\rho_1^\ell)(1 + a\boldsymbol{\nabla} \cdot \boldsymbol{\xi}_1) + O(a^2) \quad \Rightarrow \quad \rho_1^\ell = -\rho_0 \boldsymbol{\nabla} \cdot \boldsymbol{\xi}_1, \tag{10.51}$$

and then using $\rho_1^\ell = \rho_1' + (\boldsymbol{\xi}_1 \cdot \boldsymbol{\nabla})\rho_0$. This relation is also useful to establish a ray tracing expression for the wave property

$$\tilde{\rho} - \overline{\rho}^L = \overline{(J-1)\rho^\xi} \approx \overline{\boldsymbol{\nabla} \cdot \boldsymbol{\xi} \rho^\ell} \approx -\frac{1}{\tilde{\rho}} \overline{\rho^\ell \rho^\ell}. \tag{10.52}$$

This uses that in a slowly varying wavetrain the Jacobian J is dominated by its one-dimensional form $J = 1 + \boldsymbol{\nabla} \cdot \boldsymbol{\xi}$. Moreover, in a slowly varying wavetrain

$$\overline{\boldsymbol{\nabla} \cdot \boldsymbol{\xi} \rho^\ell} \approx -\overline{(\boldsymbol{\xi} \cdot \boldsymbol{\nabla})\rho^\ell} \approx -\overline{\rho}^S \quad \Rightarrow \quad \tilde{\rho} \approx \overline{\rho}. \tag{10.53}$$

Thus in a ray tracing regime the Eulerian-mean density $\overline{\rho}$ equals the effective mean density $\tilde{\rho}$ whilst the Lagrangian-mean density $\overline{\rho}^L$ is increased by a term proportional to the density perturbation variance.

10.2.5 The divergence effect

The effective mean density $\tilde{\rho} = \rho^\xi J$ quantifies the dilation or compression of mean material volume elements. It is an almagamate of the physical density and the mathematical lifting map, and it is possible for such a dilation or compression to take place even if the physical density is constant, as in homogeneous incompressible flow, say, in which $\rho = \rho_0$ throughout. Clearly, this occurs if the lifting map is not volume-conserving, and this *divergence effect* turns out to be an intrinsic property of Lagrangian-mean theory. We will look at the divergence effect at small wave amplitude in order to understand its meaning and to judge its importance for different flow situations. In particular, we will be able to give a physical interpretation of the divergence effect based on the movement of the centre of mass of a mean material volume.

The exact continuity equation (10.47) shows that material changes in $\tilde{\rho}$ are due to non-zero $\nabla \cdot \overline{\boldsymbol{u}}^L$. For incompressible flows $\nabla \cdot \boldsymbol{u} = 0$ and also $\nabla \cdot \overline{\boldsymbol{u}} = \nabla \cdot \boldsymbol{u}' = 0$, but

$$\nabla \cdot \overline{\boldsymbol{u}}^L = \nabla \cdot \overline{\boldsymbol{u}} + \nabla \cdot \overline{\boldsymbol{u}}^S = \nabla \cdot \overline{\boldsymbol{u}}^S \neq 0. \tag{10.54}$$

For small-amplitude waves on top of a resting basic flow the leading-order Stokes drift according to (10.15) is

$$\overline{u}_i^S = \overline{\xi_j u'_{i,j}} + O(a^3) \quad \Rightarrow \quad \nabla \cdot \overline{\boldsymbol{u}}^L = \nabla \cdot \overline{\boldsymbol{u}}^S = \overline{\xi_{j,i} u'_{i,j}} + O(a^3). \tag{10.55}$$

Using the definition of ξ_j and $\nabla \cdot \boldsymbol{\xi} = O(a^2)$ for incompressible flow this can be re-written in the convenient form

$$\nabla \cdot \overline{\boldsymbol{u}}^L = \frac{1}{2} \overline{\mathrm{D}}^L (\overline{\xi_i \xi_j})_{,ij} + O(a^3). \tag{10.56}$$

The corresponding $\tilde{\rho}$ for homogeneous flows follows from (10.47) as

$$\frac{\tilde{\rho} - \rho_0}{\rho_0} = -\frac{1}{2} (\overline{\xi_i \xi_j})_{,ij} + O(a^3). \tag{10.57}$$

This shows that mean dilation effects are related to spatial gradients of the covariance tensor of the material displacements. The appearance of second derivatives of a mean quantity makes clear that the divergence effect is often negligible in situations involving slowly varying wavetrains, because in this case $\tilde{\rho} - \rho = O(\epsilon^2 a^2)$. Conversely, the divergence effect is often important in situations involving evanescent waves.

For example, we consider a two-dimensional fluid domain in a vertical xz-plane with x-periodic flow. We use zonal averaging such that all mean fields are zonally symmetric and hence $\nabla \cdot \overline{\boldsymbol{u}}^L = \overline{w}_z^L$. We consider the upward propagation of a wavepacket into a fluid region that was initially at rest. It

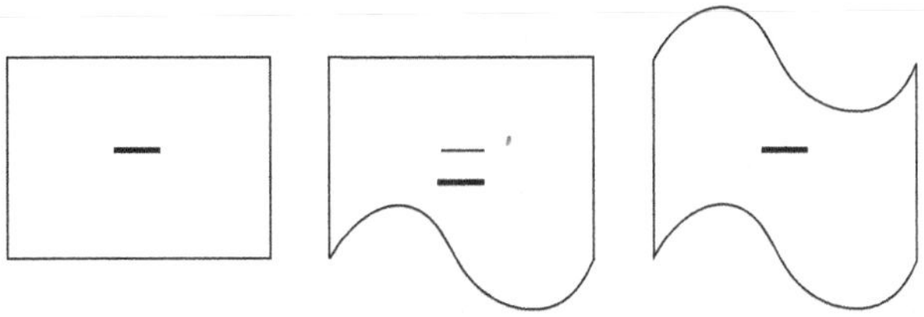

Figure 10.3 Vertical movement of centre of mass during arrival of disturbance from below. Left: undisturbed configuration with rest centre of mass. Middle: arrival of disturbance from below and concomitant lowering of centre of mass. Right: the disturbance now covers the control volume and the centre of mass has returned to its rest position.

then follows from (10.56) that at leading order

$$\overline{w}^L = \frac{1}{2}(\overline{\zeta^2})_{,tz} \quad \text{and} \quad \frac{\tilde{\rho} - \rho_0}{\rho_0} = -\frac{1}{2}(\overline{\zeta^2})_{,zz}. \tag{10.58}$$

Hence (10.58) shows that there will be a transient mean downward motion as the wavepacket arrives at any particular level z, i.e., $\overline{w}^L < 0$ when the wavepacket arrives. This can be understood by considering the centre of mass of a material volume as illustrated in Figure 10.3. The important kinematic observation from Figure 10.3 is that the undulation of the lower edge of the rectangular control volume in the middle panel *lowers* the mean centre of mass of this volume. The mean centre of mass of a material volume $\mathcal{D}^\xi$ is defined by

$$\overline{\left(\int_{\mathcal{D}^\xi} \rho \boldsymbol{x}\, dV\right)} = \int_{\mathcal{D}} \tilde{\rho}\overline{\boldsymbol{x}^\xi}\, dV = \int_{\mathcal{D}} \tilde{\rho}\boldsymbol{x}\, dV, \tag{10.59}$$

which makes it obvious that the mean centre of mass moves with $\overline{\boldsymbol{u}}^L$ and is therefore affected by the divergence effect.

In the present wavepacket example the vertical mean motion due to the divergence effect is transient, but this is not the case for an evanescent wave such as a surface wave on deep water. The relevant displacement sketch is then an upside down version of the middle panel of Figure 10.3, which makes it obvious that in this case the centre of mass has been *raised* near the free surface. This is essentially the same construction that is used to compute the available potential energy of such a disturbed configuration. The effective mean density for deep-water waves is

$$\frac{\tilde{\rho} - \rho_0}{\rho_0} = -\frac{1}{2}(\overline{\zeta^2})_{,zz} = -2k^2(\overline{\zeta^2}) \tag{10.60}$$

where k is the horizontal wavenumber such that the structure of the wave fields is $\zeta(x, z, t) = a \exp|k|z \cos(kx - \omega t)$, say, with z increasing upwards.

Finally, the net vertical raise of the mean free surface is the time integral of (10.58a) at $z = 0$, which is

$$|k|\overline{\zeta^2}(z = 0) = \frac{1}{2}|k|a^2. \tag{10.61}$$

This equals the Stokes correction $\overline{\xi\zeta_x}$ at $z = 0$ for the surface elevation in deep water.

Overall, in practical small-amplitude computations the divergence effect is often ignored unless the situation involves evanescent waves. In this case the strong gradients associated with the trapped wave structure may be relevant at leading order, and the described link between the lowering and raising of the centre of mass then illustrates the importance of the divergence effect for the potential energy budget.

10.2.6 Mean surface elements and conservation laws

After looking at lifted points and volumes, the kinematics of GLM theory is now completed by considering lifted area elements. This will allow us to establish the generic persistence of integral conservation laws under Lagrangian averaging.

Let us denote a differential area element in n dimensions by $\boldsymbol{dA}$ and then its lifted counterpart has components

$$dA_i^\xi = K_{ij} dA_j \quad \Leftrightarrow \quad \boldsymbol{dA}^\xi = \boldsymbol{K} \cdot \boldsymbol{dA}, \tag{10.62}$$

where $\boldsymbol{K}$ is a second-rank tensor and its components $K_{ij} \in R^{n\times n}$ are to be found. Before computing this matrix we note two useful facts about $\boldsymbol{K}$.

First, $\boldsymbol{K}$ has zero (right) divergence. This follows from the integral identity

$$\oint_{\partial\mathcal{D}^\xi} \boldsymbol{dA} = \oint_{\partial\mathcal{D}} \boldsymbol{dA}^\xi = \oint_{\partial\mathcal{D}} \boldsymbol{K} \cdot \boldsymbol{dA} = 0, \tag{10.63}$$

which holds for any closed surface $\partial\mathcal{D}^\xi$ and its mean counterpart $\partial\mathcal{D}$, say. The divergence theorem then implies

$$\int_{\mathcal{D}} \boldsymbol{\nabla} \cdot \boldsymbol{K} dV = 0 \tag{10.64}$$

and this holds for arbitrary $\mathcal{D}$ only if

$$\boldsymbol{\nabla} \cdot \boldsymbol{K} = 0 \quad \Leftrightarrow \quad K_{ij,j} = 0. \tag{10.65}$$

This is the divergence-free property we were after. Second, $\boldsymbol{K}/J$ provides the inverse of the gradient lifting map in (10.36b). This follows from the integral identity

$$\int_{\mathcal{D}^\xi} \boldsymbol{\nabla}\phi\, dV = \oint_{\partial\mathcal{D}^\xi} \phi\, \boldsymbol{dA} \tag{10.66}$$

after pulling back the integrals to the mean volume $\mathcal{D}$ and its boundary $\partial\mathcal{D}$. This yields

$$\int_{\mathcal{D}} (\boldsymbol{\nabla}\phi)^\xi J\, dV = \oint_{\partial\mathcal{D}} \phi^\xi\, \boldsymbol{K}\cdot \boldsymbol{dA} \tag{10.67}$$

and using the divergence theorem together with (10.65) this implies the pointwise relation

$$J(\boldsymbol{\nabla}\phi)^\xi = \boldsymbol{\nabla}\cdot(\phi^\xi \boldsymbol{K}) = \boldsymbol{K}\cdot\boldsymbol{\nabla}(\phi^\xi) \quad \Leftrightarrow \quad J(\phi_{,i})^\xi = K_{ij}(\phi^\xi)_{,j}. \tag{10.68}$$

This is the inverse of (10.36b).

We now compute $\boldsymbol{K}$ by using the fact that volume equals line times area, i.e., a line element $\boldsymbol{dx}$ and an area element $\boldsymbol{dA}$ can be contracted to form the volume element $dV = \boldsymbol{dx}\cdot\boldsymbol{dA}$. The lifted line element $\boldsymbol{dx}^\xi$ is

$$\boldsymbol{dx}^\xi = \boldsymbol{d}(\boldsymbol{x}+\boldsymbol{\xi}) = \boldsymbol{dx} + (\boldsymbol{dx}\cdot\boldsymbol{\nabla})\boldsymbol{\xi}, \tag{10.69}$$

which in index notation corresponds to

$$dx_i^\xi = dx_i + \xi_{i,j}\, dx_j. \tag{10.70}$$

Therefore a lifted volume element can be written as

$$dV^\xi = dx_i^\xi dA_i^\xi = dx_k(\delta_{ik} + \xi_{i,k})K_{im} dA_m, \tag{10.71}$$

where we used (10.70) and (10.62). Alternatively, the lifted volume element can also be written as

$$dV^\xi = JdV = Jdx_k dA_k = dx_k J\delta_{km} dA_m. \tag{10.72}$$

By the quotient rule, demanding that (10.71) equals (10.72) for all choices of $\boldsymbol{dx}$ and $\boldsymbol{dA}$ implies

$$J\delta_{km} = (\delta_{ik} + \xi_{i,k})K_{im}. \tag{10.73}$$

This shows that $\boldsymbol{K}/J$ is the inverse of $(\boldsymbol{I}+\boldsymbol{\nabla\xi})^T$, i.e., K_{ij} are the cofactors[4] of $\delta_{ij}+\xi_{i,j}$. Finally, we apply $\partial/\partial x_m$ to (10.73):

$$J_{,k} = \xi_{i,km}K_{im} + (\delta_{ik}+\xi_{i,k})K_{im,m} = \xi_{i,km}K_{im} \tag{10.74}$$

[4] The cofactors a_{ij} of a matrix A_{ij} are the sub-determinants (with appropriate sign) obtained by removing the ith row and the jth column from the matrix. By Cramer's rule, it follows that $a_{ij}/\det(A_{ij})$ is the inverse of A_{ji}.

after using (10.65). The Jacobian J is a polynomial in the components $\xi_{i,m}$ and applying the chain rule on the left-hand side yields

$$J_{,k} = \frac{\partial J}{\partial \xi_{i,m}} \xi_{i,mk} = \xi_{i,mk} K_{im}. \tag{10.75}$$

The quotient rule now implies the sought-after explicit relation

$$K_{im} = \frac{\partial J}{\partial \xi_{i,m}}. \tag{10.76}$$

For example, if $n = 1$ then we have $J = 1 + \xi_x$ and therefore $K_{11} = 1$. If $n = 2$ then (10.42) leads to

$$K_{ij} = \begin{pmatrix} 1+\eta_y & -\eta_x \\ -\xi_y & 1+\xi_x \end{pmatrix} \quad \Leftrightarrow \quad K_{ij} = \delta_{ij}(1 + \nabla \cdot \boldsymbol{\xi}) - \xi_{j,i}. \tag{10.77}$$

Evidently, the entries K_{ij} are polynomials of degree $n-1$ in the components of $\nabla\boldsymbol{\xi}$ and if $n > 1$ then the area element is scaled and rotated by the lifting map. We note in passing that if $n = 3$ then (10.77) is the first-order approximation to the full K_{ij} in the small-amplitude regime $|\nabla\boldsymbol{\xi}| \ll 1$.

We now have the tools at hand to write down the GLM version of a generic conservation law based on a transport equation of the form

$$\rho \frac{D\phi}{Dt} + \nabla \cdot \boldsymbol{F} = 0, \tag{10.78}$$

which goes together with the integral conservation law

$$\frac{d}{dt} \int_{\mathcal{D}^\xi} \phi \rho dV = \int_{\mathcal{D}^\xi} \frac{D\phi}{Dt} \rho dV = - \oint_{\partial \mathcal{D}^\xi} \boldsymbol{F} \cdot \boldsymbol{dA} \tag{10.79}$$

that holds for arbitrary material volumes $\mathcal{D}^\xi$. The non-advective flux vector $\boldsymbol{F}$ facilitates the exchange of ϕ-content between adjacent volumes, but no ϕ-content is gained or lost in this exchange.

We pull (10.79) back to the mean volume $\mathcal{D}$ and average to obtain the corresponding GLM law

$$\int_{\mathcal{D}} \overline{\mathrm{D}}^L \overline{\phi}^L \tilde{\rho} dV = - \oint_{\partial \mathcal{D}} \overline{(\boldsymbol{F}^\xi \cdot \boldsymbol{K})} \cdot \boldsymbol{dA} \tag{10.80}$$

and its pointwise equivalent

$$\tilde{\rho} \overline{\mathrm{D}}^L \overline{\phi}^L + \nabla \cdot \tilde{\boldsymbol{F}} = 0 \quad \text{where} \quad \tilde{F}_j = \overline{F_i^\xi K_{ij}} \tag{10.81}$$

is the effective mean flux of ϕ-content in GLM theory. Comparing (10.78) and (10.81) we see that the form of the conservation law has not changed.

The same manipulation can also be applied to vector conservation laws such as the momentum equation, where it leads to

$$\rho\frac{Du_i}{Dt} + \Pi_{ij,j} = 0 \quad \Rightarrow \quad \tilde{\rho}\overline{\mathrm{D}}^L\overline{u}_i^L + \tilde{\Pi}_{ij,j} = 0. \tag{10.82}$$

Here the effective mean momentum flux tensor

$$\tilde{\Pi}_{ij} = \overline{\Pi_{ik}^{\xi} K_{kj}} \tag{10.83}$$

is related to the original momentum flux tensor Π_{ij} in obvious analogy with the flux vector relation in (10.81). The relation (10.83) holds for arbitrary momentum flux tensors, so it includes the Navier–Stokes equations, for example. Although Π_{ij} is always symmetric in simple fluids due to conservation of angular momentum, the same need not be true of $\tilde{\Pi}_{ij}$. For example, in the special case of the Euler equations the momentum flux $\Pi_{ij} = \delta_{ij}p$ is isotropic and entirely due to the pressure, whilst the effective mean momentum flux is neither isotropic nor symmetric:

$$\tilde{\Pi}_{ij} = \overline{p^{\xi} K_{ij}}. \tag{10.84}$$

Nevertheless, this tensor accurately describes the mean momentum flux due to pressure forces across undulating material surfaces. It is precisely because these undulations enter into K_{ij} that the effective momentum flux does not inherit the symmetries of the original momentum flux.

There is also an important general property of $\overline{\boldsymbol{u}}^L$ that distinguishes it from $\overline{\boldsymbol{u}}$, namely in GLM theory the mean momentum is entirely contained in the Lagrangian-mean velocity $\overline{\boldsymbol{u}}^L$. This follows directly from

$$\overline{\left(\int_{\mathcal{D}^{\xi}} \rho\boldsymbol{u}\, dV\right)} = \int_{\mathcal{D}} \overline{(\rho^{\xi}\boldsymbol{u}^{\xi}\, dV^{\xi})} = \int_{\mathcal{D}} \tilde{\rho}\overline{(\boldsymbol{u}^{\xi})}\, dV = \int_{\mathcal{D}} \tilde{\rho}\overline{\boldsymbol{u}}^L\, dV. \tag{10.85}$$

This is very different in Eulerian-mean theory, especially for compressible flows, for which the Eulerian-mean momentum density is

$$\overline{\rho\boldsymbol{u}} = \overline{\rho}\,\overline{\boldsymbol{u}} + \overline{\rho'\boldsymbol{u}'}. \tag{10.86}$$

Arguably, the second term can be thought of as a contribution to the mean momentum by the wave field. The same comment applies to the interpretation of (10.86) as the mean mass flux. Of course, that the Eulerian-mean velocity does not account for the entire momentum density or mass flux is related to the loss of the advective form for material invariants in (10.2).

10.2.7 Circulation and pseudomomentum

The ability of GLM theory to follow material advection in a mean sense allows a very convenient formulation of a mean version of Kelvin's circulation theorem. This theorem also provides an exact definition of the GLM pseudomomentum vector.

The circulation Γ around a closed material loop $\mathcal{C}^\xi$ is

$$\Gamma = \oint_{\mathcal{C}^\xi} \boldsymbol{u}(\boldsymbol{x}, t) \cdot \boldsymbol{dx}. \tag{10.87}$$

As written, the material loop $\mathcal{C}^\xi$ is formed by the *actual* positions of a certain set of fluid particles, but we can associate with each such actual position also a *mean* position of the respective particle, and the set of all mean positions then forms another closed loop $\mathcal{C}$, say. Thus $\mathcal{C}^\xi$ is the image of $\mathcal{C}$ under the lifting map:

$$\boldsymbol{x} \in \mathcal{C} \quad \Leftrightarrow \quad \boldsymbol{x} + \boldsymbol{\xi}(\boldsymbol{x}, t) \in \mathcal{C}^\xi. \tag{10.88}$$

This allows re-writing the contour integral in (10.87) in terms of $\mathcal{C}$. The only non-trivial step is the transformation of the line element $\boldsymbol{dx} \to \boldsymbol{dx}^\xi$, which was noted in (10.70). This leads to the identity

$$\Gamma = \oint_{\mathcal{C}^\xi} u_i \, dx_i = \oint_{\mathcal{C}} u_i^\xi \, dx_i^\xi = \oint_{\mathcal{C}} \left(u_i^\xi + \xi_{j,i} u_j^\xi \right) \, dx_i. \tag{10.89}$$

The integration domain is now a mean material loop and therefore we can average (10.89) by simply averaging the bracket multiplying the mean line element dx_i. The first term brings in the Lagrangian-mean velocity and the second term brings in the pseudomomentum:

$$\overline{\Gamma} = \oint_{\mathcal{C}} (\overline{\boldsymbol{u}}^L - \mathsf{p}) \cdot \boldsymbol{dx} \quad \text{where} \quad \mathsf{p}_i = -\overline{\xi_{j,i} u_j^\xi} = -\overline{\xi_{j,i} u_j^\ell} \tag{10.90}$$

is the *exact* GLM definition of the pseudomomentum vector; the minus sign is conventional, of course, and the final form holds because of the averaging. Note that the vector index in p_i corresponds to the derivative index in $\xi_{j,i}$.

This exact kinematic relation shows that the mean circulation is due to a cooperation of $\overline{\boldsymbol{u}}^L$ and p, i.e., both the mean flow and the wave-related pseudomomentum contribute to the mean circulation. Now, if we are in a flow situation where Γ is conserved then its average $\overline{\Gamma}$ is conserved as well, so we have a GLM version of Kelvin's theorem:

$$\frac{d\overline{\Gamma}}{dt} = \frac{d}{dt} \oint_{\mathcal{C}} (\overline{\boldsymbol{u}}^L - \mathsf{p}) \cdot \boldsymbol{dx} = 0. \tag{10.91}$$

Here we observe, and this is the heart of the matter, that the closed loop

$\mathcal{C}$ moves with $\overline{\boldsymbol{u}}^L$ whilst the circulation is formed with a *different* velocity, namely $\overline{\boldsymbol{u}}^L - \mathbf{p}$. This makes it obvious that the mean circulation theorem has a different form compared to the original circulation theorem. This is because different mean velocities are need to move the loop on the one hand and to compute the circulation on the other hand. In the original theorem the fluid velocity $\boldsymbol{u}$ did both jobs, but in the mean theorem we have $\overline{\boldsymbol{u}}^L$ and $\overline{\boldsymbol{u}}^L - \mathbf{p}$, respectively. This points to an irreducible entanglement between disturbances and the mean flow. We will find that the exact conservation law (10.91) provides the most far-reaching consequences of GLM theory.

For example, if the flow is periodic in the Cartesian x-coordinate with finite period L and zonal averaging is used as the definition of the Eulerian-mean operator then a material contour $\mathcal{C}^\xi$ traversing the x-domain qualifies as a closed loop for the purpose of Kelvin's theorem and the corresponding mean contour $\mathcal{C}$ is a straight line parallel to the x-axis. This is the set-up of simple geometry, but without the restriction to small wave amplitude. We now use $\overline{\mathrm{D}}^L \boldsymbol{dx} = \boldsymbol{d}\overline{\boldsymbol{u}}^L = \boldsymbol{dx} \cdot \boldsymbol{\nabla}\overline{\boldsymbol{u}}^L$ in order to re-write (10.91) as

$$\frac{d\overline{\Gamma}}{dt} = \oint_{\mathcal{C}} \overline{\mathrm{D}}^L(\overline{\boldsymbol{u}}^L - \mathbf{p}) \cdot \boldsymbol{dx} - \oint_{\mathcal{C}} \mathbf{p} \cdot \boldsymbol{d}\overline{\boldsymbol{u}}^L = 0 \tag{10.92}$$

$$\Rightarrow \quad \oint_{\mathcal{C}} \overline{\mathrm{D}}^L(\overline{\boldsymbol{u}}^L - \mathbf{p}) \cdot \boldsymbol{dx} = \oint_{\mathcal{C}} \boldsymbol{dx} \cdot \boldsymbol{\nabla}\overline{\boldsymbol{u}}^L \cdot \mathbf{p},$$

where $\mathbf{p}$ contracts with $\overline{\boldsymbol{u}}^L$ and not with $\boldsymbol{\nabla}$. This form is still general, but in simple geometry only dx is non-zero along $\mathcal{C}$ and hence the right-hand side of (10.92) involves only the x-derivative of $\overline{\boldsymbol{u}}^L$, which is zero by the definition of zonal averaging. Likewise, the left-hand side is simply equal to $L\overline{\mathrm{D}}^L(\overline{u}^L - \mathsf{p})$, where p is the x-component of $\mathbf{p}$. Thus we have the exact relation

$$\overline{\mathrm{D}}^L \overline{u}^L = \overline{\mathrm{D}}^L \mathsf{p} \tag{10.93}$$

between the zonally averaged zonal Lagrangian-mean flow and the zonally averaged zonal pseudomomentum in simple geometry. This powerful statement (and its straightforward extension to forced flows) presents the most general expression of non-acceleration conditions for a mean flow in simple geometry.

10.2.8 Why pseudomomentum is conserved

We can now give an elementary yet exact derivation for the integral conservation of zonal pseudomomentum in the case of a zonally symmetric Lagrangian-mean flow. This extends the earlier results for linear waves to arbitrary wave amplitude. As before, we assume that the flow is inviscid such that the circulation theorem holds and that all flow fields are x-periodic with finite period length L. We also assume that the Lagrangian-mean fields $(\tilde{\rho}, \overline{\boldsymbol{u}}^L)$ are zonally symmetric, i.e., they do not depend on x. This may be the case because zonal averaging is used (which is the most relevant case in practice), but this is not necessary.[5] In particular, we do not need to assume that $\mathbf{p}$ is zonally symmetric, so the present set-up is more general than that leading to (10.93). Indeed, following through the steps leading to (10.93) in the present set-up yields the relation

$$L\overline{\mathrm{D}}^L\overline{u}^L = \int_0^L (\overline{\mathrm{D}}^L \mathsf{p})\,dx. \tag{10.94}$$

Now, if $\mathcal{D}$ is a mean material control volume whose zonal extent includes the entire x-range then the time derivative of the zonal pseudomomentum contained in $\mathcal{D}$ is

$$\begin{aligned}\frac{d}{dt}\int_{\mathcal{D}} \tilde{\rho}\mathsf{p}\,dV &= \int_{\mathcal{D}} \tilde{\rho}(\overline{\mathrm{D}}^L\mathsf{p})\,dV = \iint \tilde{\rho}\left(\int_0^L (\overline{\mathrm{D}}^L\mathsf{p})\,dx\right)dy\,dz \qquad (10.95)\\ &= \iint \tilde{\rho}L(\overline{\mathrm{D}}^L\overline{u}^L)\,dy\,dz = \int_{\mathcal{D}} \tilde{\rho}(\overline{\mathrm{D}}^L\overline{u}^L)\,dV = -\oint_{\partial\mathcal{D}} \tilde{\Pi}_{1j}\,dA_j.\end{aligned}$$

This exact result states that zonal pseudomomentum is conserved under the assumptions that we made, which are a combination of mean mass conservation (10.47), Kelvin's mean circulation theorem with mean-flow symmetry (10.94), the zonal symmetry of $\tilde{\rho}$, and the mean momentum conservation (10.82). Indeed, pseudomomentum conservation fails if any of these ingredients is missing. For example, in the presence of viscous stresses mass and momentum are still conserved, but circulation is not; this adds a dissipative term in the second step in (10.95). This is illustrated by the laminar shear flow within a viscous boundary layer over a flat plate: in this case there is a viscous wall stress at the plate and hence mean momentum is exchanged between the fluid and the plate. However, no pseudomomentum is exchanged by the viscous stresses on the flat plate, so in this case the relevant fluxes are different.

It is also noteworthy from (10.95) that the flux of pseudomomentum is closely linked to the flux of mean momentum. Indeed, only the off-diagonal flux components $j \neq 1$ are relevant in the surface integral in (10.95). This

[5] Of course, in nonlinear theory it is virtually impossible for the mean flow to remain zonally symmetric unless wave properties such as pseudomomentum are zonally symmetric as well. Nevertheless, the present set-up is relevant in principle and also in practice for small-amplitude waves, as for their conservation properties it is the symmetry of the $O(1)$ basic state that matters.

proves that the pressure-related off-diagonal fluxes of mean zonal momentum and zonal pseudomomentum are exactly equal in GLM theory (see also (10.124) below). This fact has important implications, for instance it provides an exact underpinning for the earlier result that for small-amplitude lee waves the vertical flux of horizontal pseudomomentum flux equals the mountain drag.

10.2.9 Vorticity and potential vorticity

GLM theory allows capturing a mean version of the three-dimensional vorticity dynamics familiar from the actual fluid equations. Recall from §1.4.2 that the curl of the material acceleration is

$$\nabla \times \left(\frac{D\boldsymbol{u}}{Dt}\right) = \frac{D(\nabla \times \boldsymbol{u})}{Dt} + (\nabla \times \boldsymbol{u})\nabla \cdot \boldsymbol{u} - [(\nabla \times \boldsymbol{u}) \cdot \nabla]\boldsymbol{u}, \qquad (10.96)$$

and that (10.96) is equal to zero if the flow is barotropic. In this case the density-scaled vorticity vector $\nabla \times \boldsymbol{u}/\rho$ is frozen into the fluid.

The analogue of (10.96) in GLM theory can be derived by multiplying the lifted momentum equation by $\delta_{ji} + \xi_{j,i}$ and averaging. This yields a form of the mean momentum equation in which the difference $\overline{\boldsymbol{u}}^L - \mathsf{p}$ appears naturally. Most importantly, this procedure keeps the pressure term in gradient form because

$$\overline{(\delta_{ji} + \xi_{j,i})(p_{,j})^{\xi}} = \overline{(p^{\xi})}_{,i} = \overline{p}^L_{,i}. \qquad (10.97)$$

For barotropic flows the density factor can also be absorbed in the gradient of the specific enthalpy, as usual. Basically, the result of this procedure is

$$\overline{\mathrm{D}}^L(\overline{u}^L_i - \mathsf{p}_i) - \overline{u}^L_{k,i}\mathsf{p}_k + (\overline{\cdots})_{,i} = 0 \qquad (10.98)$$

where the details of the gradient term do not matter for what follows. This is called the circulation form of the GLM momentum equation.[6]

Our main interest here is not in (10.98) itself but in its curl, which is

$$\overline{\mathrm{D}}^L(\nabla \times (\overline{\boldsymbol{u}}^L - \mathsf{p})) + (\nabla \times (\overline{\boldsymbol{u}}^L - \mathsf{p}))\nabla \cdot \overline{\boldsymbol{u}}^L = [(\nabla \times (\overline{\boldsymbol{u}}^L - \mathsf{p})) \cdot \nabla]\overline{\boldsymbol{u}}^L. \qquad (10.99)$$

This is the sought-after Lagrangian-mean vorticity equation; notably, the refraction term $-\overline{u}^L_{k,i}\mathsf{p}_k$ was essential for obtaining it in this form. This equation shows that $\nabla \times (\overline{\boldsymbol{u}}^L - \mathsf{p})/\tilde{\rho}$ is frozen in the flow $\overline{\boldsymbol{u}}^L$, i.e., it is stretched and twisted by $\overline{\boldsymbol{u}}^L$ precisely like a mean material line element.

6 The refraction term $-\overline{u}^L_{k,i}\mathsf{p}_k$ will later be seen to be an essential part of the pseudomomentum dynamics (cf. §10.3.2).

This is an important kinematic aspect of Lagrangian-mean theory, and it appears that a corresponding statement is not available in Eulerian-mean theories. An obvious exact corollary for strictly irrotational flow is that

$$\nabla \times \boldsymbol{u} = 0 \quad \Rightarrow \quad \nabla \times (\overline{\boldsymbol{u}}^L - \mathsf{p}) = 0. \tag{10.100}$$

Now, the material invariance of potential vorticity (PV) in its various forms can be derived from Kelvin's circulation theorem and the same method can be used to derive a surprisingly simple expression for the Lagrangian-mean PV, at least in the absence of diabatic heating effects. Essentially, this amounts to alternating the lifting map with Stokes's integral theorem. For example, in the shallow-water system Stokes's theorem can be used to re-write the circulation in (10.87) as

$$\Gamma = \oint_{\mathcal{C}^\xi} \boldsymbol{u} \cdot \boldsymbol{dx} = \int_{\mathcal{A}^\xi} \nabla \times \boldsymbol{u} \, dx \, dy, \tag{10.101}$$

where $\nabla \times \boldsymbol{u} = v_x - u_y$ and $\mathcal{A}^\xi$ is the material area enclosed by $\mathcal{C}^\xi$, i.e., $\mathcal{C}^\xi = \partial \mathcal{A}^\xi$. The invariance of Γ in the second form in (10.101) for arbitrary $\mathcal{A}^\xi$ implies the material invariance of $\nabla \times \boldsymbol{u} \, dx \, dy$. The area element $dx\,dy$ is not a material invariant in compressible shallow-water flow, but the mass element $h \, dx \, dy$ is. Factorizing with h then leads to

$$\frac{D}{Dt}\left(\frac{\nabla \times \boldsymbol{u}}{h} h \, dx \, dy\right) = 0 \quad \Rightarrow \quad \frac{D}{Dt}\left(\frac{\nabla \times \boldsymbol{u}}{h}\right) = 0, \tag{10.102}$$

which is the definition of PV in shallow water. Mutatis mutandis, the same method applied to the mean circulation in (10.90) yields

$$\overline{q}^L = \frac{\nabla \times (\overline{\boldsymbol{u}}^L - \mathsf{p})}{\tilde{h}} \quad \text{and} \quad \overline{\mathrm{D}}^L \overline{q}^L = 0. \tag{10.103}$$

Here the effective mean layer depth is $\tilde{h} = h^\xi J$, which satisfies the mean continuity equation

$$\overline{\mathrm{D}}^L \tilde{h} + \tilde{h} \nabla \cdot \overline{\boldsymbol{u}}^L = 0. \tag{10.104}$$

Like the mean circulation theorem from which it was derived, the mean PV law (10.103) is an exact statement, which allows making useful general deductions. For instance, in irrotational flows we have $q = 0$ and therefore

$$\overline{q}^L = 0 \quad \Rightarrow \quad \nabla \times \overline{\boldsymbol{u}}^L = \nabla \times \mathsf{p}. \tag{10.105}$$

Of course, even though $\overline{\boldsymbol{u}}^L$ and p have the same curl they can still be different vector fields. This can be either because $\nabla \cdot \overline{\boldsymbol{u}}^L$ is markedly different from $\nabla \cdot \mathsf{p}$, or because $\overline{\boldsymbol{u}}^L$ and p satisfy different boundary conditions, say, at a vibrating wavemaker such as a piston.

If $q \neq 0$ then (10.105) is replaced by

$$\nabla \times \overline{\boldsymbol{u}}^L = \nabla \times \mathbf{p} + \tilde{h}\overline{q}^L. \tag{10.106}$$

This illustrates the scope for further changes in $\nabla \times \overline{\boldsymbol{u}}^L$ due to dilation effects mediated by variable $\tilde{h}$ (i.e., vortex stretching), or due to material advection of different values of $\overline{q}^L$ into the region of interest.

In three-dimensional stratified flow essentially the same construction can be used based on the conserved circulation for material loops within stratification surfaces. This leads to the three-dimensional Lagrangian-mean PV definition

$$\overline{q}^L = \frac{(\nabla \times (\overline{\boldsymbol{u}}^L - \mathbf{p})) \cdot \nabla \overline{s}^L}{\tilde{\rho}} \quad \text{and} \quad \overline{\mathrm{D}}^L \overline{q}^L = 0. \tag{10.107}$$

In both shallow water and three-dimensional flow adding a body force does not upset the functional form of $\overline{q}^L$, although it does change its evolution, of course (see §11.1, for example). However, there is a caveat in the three-dimensional case here that does not arise in the shallow-water equations. Namely, if diabatic heating or cooling is present in the three-dimensional case then the Lagrangian disturbance $s^\ell \neq 0$, which upsets the kinematic construction that leads to the first half of (10.107). Basically, it is part of this construction that actual and mean particles are always on the same isentrope, which can fail in the presence of diabatic heating.

Finally, for the definition of $\overline{q}^L$ in the presence of Coriolis forces see §10.4.1.

10.3 Wave activity conservation in GLM theory

We can now derive finite-amplitude counterparts of the wave activity conservation laws derived in §4.5 for linear waves. As before, continuous symmetries of the mean flow will induce corresponding wave activity conservation laws. In order to discuss these mean-flow symmetries clearly we must be specific about which mean fields actually define the mean flow. We will use $(\tilde{\rho}, \overline{\boldsymbol{u}}^L, \overline{s}^L)$ for this purpose, i.e., the combination of the effective density and the Lagrangian-means of velocity and entropy. In barotropic flows or flows with uniform entropy the smaller set $(\tilde{\rho}, \overline{\boldsymbol{u}}^L)$ is sufficient. These details matter because, in principle, the symmetries of $\tilde{\rho}$ can be different from the symmetries of $\overline{\rho}^L$, say. Of course, if the symmetry under consideration is induced by the averaging operator then all mean fields inherit this symmetry. This case is the most important case in practice for obtaining a robust wave activity conservation law of the form

$$\overline{\mathrm{D}}^L \mathsf{A} + \frac{1}{\tilde{\rho}} \nabla \cdot \boldsymbol{B} = 0 \tag{10.108}$$

for perfect fluid flow. Here A is the wave activity density per unit mass and $\boldsymbol{B}$ is its non-advective flux. For example, a zonally averaged mean flow is zonally symmetric by construction and therefore a suitably defined zonal pseudomomentum *must* be conserved at all times. Other wave activities may or may not be conserved depending on the time-evolving mean-flow symmetry structure. Such wave activities satisfy variants of (10.108) with additional source terms related to mean-flow derivatives with respect to the relevant coordinate, for instance x-derivatives in the case of the zonal pseudomomentum or t-derivatives in the case of pseudoenergy.

Because of these considerations it proves convenient to derive the main GLM wave activity equation using a very general approach based on the dependence of all the flow fields on a smooth parameter α, but without assuming that α is in any way linked to the averaging operator. Specifically, we do not assume that $\overline{\phi}_{,\alpha} = 0$, so the only properties of the averaging operator that we use are pointwise relations such as $\overline{\phi'\psi} = \overline{\phi'\psi'}$.

10.3.1 General wave activity equation

Our point of the departure are the three-dimensional compressible Euler equations:

$$\frac{D\boldsymbol{u}}{Dt} + \frac{\boldsymbol{\nabla} p}{\rho} + \boldsymbol{\nabla}\phi = \boldsymbol{F} \quad \Leftrightarrow \quad \frac{Du_j}{Dt} + \frac{p_{,j}}{\rho} + \phi_{,j} = F_j. \tag{10.109}$$

For ease of later reference, the Lagrangian-mean momentum equation is

$$\overline{\mathrm{D}}^L \overline{u}_i^L + \frac{1}{\tilde{\rho}}(\overline{p^\xi K_{ij}})_{,j} + \overline{(\phi_{,i})}^L = \overline{F_i}^L. \tag{10.110}$$

The external potential ϕ and body force $\boldsymbol{F}$ are quite arbitrary and we could of course absorb $\boldsymbol{\nabla}\phi$ in the definition of $\boldsymbol{F}$. However, we are often interested in keeping separate the effects due to a mean potential $\phi = \overline{\phi}$ (e.g., due to gravity) and those due to localized wave production or dissipation modelled by a force $\boldsymbol{F}$ with a significant disturbance part. For simplicity, we first restrict to barotropic flows; the role of entropy gradients is discussed later in §10.3.3.

Now, we apply the lifting operator to all terms in (10.109b), contract $(10.109\mathrm{b})^\xi$ with $\xi_{j,\alpha}$, and then apply the Eulerian averaging operator. This mirrors the steps used in §4.5, but in a finite-amplitude setting. The material derivative leads to

$$\overline{\xi_{j,\alpha}\left(\frac{Du_j}{Dt}\right)^\xi} = \overline{\xi_{j,\alpha}\overline{\mathrm{D}}^L u_j^\xi} = \overline{\mathrm{D}}^L(\overline{\xi_{j,\alpha}u_j^\ell}) - \overline{u_j^\ell \overline{\mathrm{D}}^L(\xi_{j,\alpha})} \tag{10.111}$$

$$= \overline{\mathrm{D}}^L(\overline{\xi_{j,\alpha}u_j^\ell}) - \overline{u_j^\ell(\overline{\mathrm{D}}^L\xi_j)_{,\alpha}} + \overline{u}_{k,\alpha}^L\overline{\xi_{j,k}u_j^\ell}$$

$$= \overline{\mathrm{D}}^L(\overline{\xi_{j,\alpha}u_j^\ell}) - \overline{u_j^\ell u_{j,\alpha}^\ell} + \overline{u}_{k,\alpha}^L\overline{\xi_{j,k}u_j^\ell}.$$

This expression can be simplified by using the definitions of the GLM wave action and pseudomomentum vector

$$\mathsf{A} = \overline{\xi_{j,\alpha}u_j^\ell} \quad \text{and} \quad \mathsf{p}_i = -\overline{\xi_{j,i}u_j^\ell}, \tag{10.112}$$

which leads to the exact relation

$$\overline{\xi_{j,\alpha}\left(\frac{Du_j}{Dt}\right)^\xi} = \overline{\mathrm{D}}^L\mathsf{A} - \frac{1}{2}(\overline{u_j^\ell u_j^\ell})_{,\alpha} - \overline{u}_{k,\alpha}^L\mathsf{p}_k. \tag{10.113}$$

The appearance of the pseudomomentum vector in this general relation is noteworthy. Next, the pressure gradient term leads to

$$\overline{\xi_{j,\alpha}\frac{(p_{,j})^\xi}{\rho^\xi}} = \overline{\xi_{j,\alpha}\frac{K_{jm}(p^\xi)_{,m}}{\rho^\xi J}} \tag{10.114}$$

$$= \frac{1}{\tilde{\rho}}\left((\overline{\xi_{j,\alpha}K_{jm}p^\xi})_{,m} - \overline{\xi_{j,m\alpha}K_{jm}p^\xi}\right)$$

$$= \frac{1}{\tilde{\rho}}\left((\overline{\xi_{j,\alpha}K_{jm}p^\xi})_{,m} - \overline{J_{,\alpha}p^\xi}\right).$$

This uses $K_{jm,m} = 0$ and the chain rule applied to (10.76). The first term is an explicit flux divergence and it provides the generic wave activity flux

$$B_m = \overline{\xi_{j,\alpha}K_{jm}p^\xi}. \tag{10.115}$$

The final term in (10.114) is more subtle; the task is to show that it is related to α-derivatives of the mean flow. Substituting $J = \tilde{\rho}/\rho^\xi$ leads to

$$-\frac{1}{\tilde{\rho}}\overline{J_{,\alpha}p^\xi} = -\frac{\tilde{\rho}_{,\alpha}}{\tilde{\rho}}\overline{\left(\frac{p}{\rho}\right)}^L + \overline{\frac{p^\xi {\rho^\xi}_{,\alpha}}{{\rho^\xi}^2}}. \tag{10.116}$$

For homogeneous incompressible flow the last term is zero and otherwise it brings in the internal energy per unit mass $\epsilon(\rho)$ defined in (1.25a) via $d\epsilon/d\rho = p/\rho^2$:

$$\overline{\frac{p^\xi {\rho^\xi}_{,\alpha}}{{\rho^\xi}^2}} = \overline{\epsilon_{,\alpha}^\xi} = (\overline{\epsilon}^L)_{,\alpha} \quad \Rightarrow \quad -\frac{1}{\tilde{\rho}}\overline{J_{,\alpha}p^\xi} = -\frac{\tilde{\rho}_{,\alpha}}{\tilde{\rho}}\overline{\left(\frac{p}{\rho}\right)}^L + (\overline{\epsilon}^L)_{,\alpha}. \tag{10.117}$$

Here and below we view $\epsilon(\rho)$ as a given function of the flow field ρ, and not as an independent flow field; this also holds for mean versions such as $\overline{\epsilon}^L = \overline{\epsilon(\tilde{\rho}/J)}$ and so on. We have succeeded in writing the final term in (10.114) as a combination of explicit α-derivatives, but $\overline{\epsilon}^L$ is *not* part of the definition of the mean flow $(\tilde{\rho}, \overline{\boldsymbol{u}}^L)$. Also, $J_{,\alpha} = O(a)$ makes it obvious that (10.117) is a wave property, but the two terms on the right-hand side of (10.117) are not individually wave properties, although their sum must be. This can be rectified by adding and subtracting a mean-flow term in (10.117) such that we obtain the alternative exact expression

$$-\frac{1}{\tilde{\rho}}\overline{J_{,\alpha}p^\xi} = -\frac{\tilde{\rho}_{,\alpha}}{\tilde{\rho}}\left\{\overline{\left(\frac{p}{\rho}\right)}^L - \frac{\tilde{p}}{\tilde{\rho}}\right\} + (\overline{\epsilon}^L - \tilde{\epsilon})_{,\alpha}. \tag{10.118}$$

Here $\tilde{\epsilon} = \epsilon(\tilde{\rho})$ such that $d\tilde{\epsilon} = (\tilde{p}/\tilde{\rho}^2)d\tilde{\rho}$. Using either (10.117) or (10.118), we finally obtain the general wave activity relation

$$\overline{\mathrm{D}}^L A - \frac{1}{2}(\overline{u_j^\ell u_j^\ell})_{,\alpha} - \overline{u}^L_{k,\alpha}\mathsf{p}_k + \frac{1}{\tilde{\rho}}\boldsymbol{\nabla}\cdot\boldsymbol{B} - \frac{\tilde{\rho}_{,\alpha}}{\tilde{\rho}}\overline{\left(\frac{p}{\rho}\right)}^L + (\overline{\epsilon}^L)_{,\alpha} + \overline{\xi_{j,\alpha}(\phi_{,j})^\xi} = \overline{\xi_{j,\alpha}F_j^\ell}. \tag{10.119}$$

This equation extends (4.123) to finite amplitude and arbitrary barotropic flows. The generality of (10.119) is indeed remarkable, because it holds for arbitrary averaging operators and for arbitrary interpretations of the α-derivatives.

In the special case of ensemble averaging with α as the ensemble parameter *any* mean field is independent of α and therefore (10.119) implies the conservation law for wave action, namely

$$\overline{(\ldots)}_{,\alpha} = 0 \quad\Rightarrow\quad \overline{\mathrm{D}}^L A + \frac{1}{\tilde{\rho}}\boldsymbol{\nabla}\cdot\boldsymbol{B} = \overline{\xi_{j,\alpha}F_j^\ell} - \overline{\xi_{j,\alpha}(\phi_{,j})^\xi}. \tag{10.120}$$

Notably, in the important case of a mean potential such that $\phi = \overline{\phi}$ (as in the gravity case $\phi = gz$ with averaging at constant altitude z), the source term due to ϕ is identically zero. This follows because in this case the only dependence of ϕ^ξ on α stems from the α-dependence of $\boldsymbol{\xi}$, so the chain rule yields

$$\phi = \overline{\phi} \quad\Rightarrow\quad (\phi^\xi)_{,\alpha} = \xi_{j,\alpha}(\phi_{,j})^\xi \quad\Rightarrow\quad \overline{\xi_{j,\alpha}(\phi_{,j})^\xi} = (\overline{\phi}^L)_{,\alpha} = 0. \tag{10.121}$$

Clearly, this result relies on the interpretation of α as an ensemble parameter. Conversely, if ϕ has a disturbance part then it can produce or destroy wave action.

10.3.2 Pseudomomentum and pseudoenergy

The evolution laws for the pseudomomentum vector p_i and the pseudoenergy e follow from (10.119) after the substitutions

$$\frac{\partial}{\partial\alpha} = -\frac{\partial}{\partial x_i} \quad\text{and}\quad \frac{\partial}{\partial\alpha} = +\frac{\partial}{\partial t}, \tag{10.122}$$

respectively. In both cases there will be a little finesse: for the pseudomomentum vector this affects the isotropic part of its flux tensor whilst for pseudoenergy this affects its density e. The pseudomomentum version of (10.119) is

$$\overline{\mathrm{D}}^L \mathsf{p}_i + \frac{1}{2}(\overline{u_j^\ell u_j^\ell})_{,i} + \overline{u}^L_{k,i}\mathsf{p}_k + \frac{1}{\tilde{\rho}}B_{im,m} + \frac{\tilde{\rho}_{,i}}{\tilde{\rho}}\overline{\left(\frac{p}{\rho}\right)}^L - (\overline{\epsilon}^L)_{,i} - \overline{\xi_{j,i}(\phi_{,j})^\xi} = -\overline{\xi_{j,i}F_j^\ell}. \tag{10.123}$$

Note the natural appearance of the mean-flow refraction term $\overline{u}^L_{k,i}\mathsf{p}_k$, whose small-amplitude version is familiar from the ray-tracing equations for the wavenumber vector $\boldsymbol{k} \propto \mathbf{p}$. The generic pseudomomentum flux tensor

$$B_{im} = -\overline{\xi_{j,i}K_{jm}p^\xi} = \overline{(K_{im} - \delta_{im}J)p^\xi}. \tag{10.124}$$

The second form uses the identity (10.73) and it shows again that the pressure-related pseudomomentum flux tensor equals the pressure-related mean momentum flux tensor in (10.110) in the off-diagonal terms.

Now, in order to make explicit the source terms of pseudomomentum due to mean-flow gradients it proves convenient to add an isotropic component proportional to δ_{ij} to the definition of the tensor B_{ij}. Specifically, if we add and subtract the gradient of $\tilde{\epsilon} = \epsilon(\tilde{\rho})$ to (10.123) then we can define the total non-advective pseudomomentum flux tensor as

$$B^{\text{tot}}_{im} = B_{im} + \delta_{im}\tilde{\rho}\left[\frac{1}{2}\overline{u_j^\ell u_j^\ell} - (\overline{\epsilon}^L - \tilde{\epsilon})\right]. \tag{10.125}$$

The trick involving $\tilde{\epsilon}$ allows defining this tensor as an explicit wave property; this is the finesse mentioned above. Notably, the square bracket is often negligible at leading order for slowly varying wavetrains. For example, for plane shallow water waves the energy difference $\overline{\epsilon}^L - \epsilon(\tilde{h}) = g\overline{h'^2}/(2H)$, which by equipartition cancels with the mean perturbation kinetic energy.

Using (10.125) the pseudomomentum law takes the form

$$\overline{\mathrm{D}}^L \mathsf{p}_i + \frac{1}{\tilde{\rho}}B^{\text{tot}}_{im,m} = +\overline{\xi_{j,i}(\phi_{,j})^\xi} - \overline{\xi_{j,i}F_j^\ell} \tag{10.126}$$

$$- \overline{u}^L_{k,i}\mathsf{p}_k - \frac{\tilde{\rho}_{,i}}{\tilde{\rho}}\left\{\overline{\left(\frac{p}{\rho}+\epsilon\right)}^L - \left(\frac{\tilde{p}}{\tilde{\rho}}+\tilde{\epsilon}\right) - \frac{1}{2}\overline{u_j^\ell u_j^\ell}\right\},$$

which neatly exhibits the source terms due to forcing and mean-flow gradients. Typically, the curly bracket is non-zero even at leading order for slowly varying wavetrains. For example, in the shallow water case the curly bracket is $g\overline{h'^2}/2 = \bar{E}/2$, and it is with some pleasure that we recognize the consistency between (10.126) and the shallow-water ray-tracing equation (4.150). The potential force term splits into

$$+\overline{\xi_{j,i}(\phi_{,j})^\xi} = -\overline{(\phi_{,i})}^L + (\overline{\phi}^L)_{,i}. \tag{10.127}$$

The first term also appears in (10.110), so for the difference $\overline{\mathrm{D}}^L(\overline{\boldsymbol{u}}^L - \mathsf{p})$ only the second term matters. This is an irrotational term, consistent with the invariance of mean PV in the presence of potential forces.

The derivation of the exact pseudoenergy law follows by replacing α-derivatives by t-derivatives in (10.119). As in linear theory, terms that became extra fluxes in the pseudomomentum law become part of the definition of pseudoenergy here, with the same finesse of subtracting an internal energy term $\tilde{\epsilon}$ in order to obtain an explicit wave property. This leads to the pseudoenergy density[7]

$$\mathsf{e} = \overline{\xi_{j,t}u_j^\ell} - \frac{1}{2}\overline{u_j^\ell u_j^\ell} + (\overline{\epsilon}^L - \tilde{\epsilon}) = \frac{1}{2}\overline{u_j^\ell u_j^\ell} + (\overline{\epsilon}^L - \tilde{\epsilon}) + \overline{u}_k^L \mathsf{p}_k. \tag{10.128}$$

The second step uses $\overline{\mathrm{D}}^L\xi_j = u_j^\ell$. As in linear theory, the pseudoenergy is the sum of an intrinsic, frame-invariant disturbance energy part and a pseudomomentum part that depends on the absolute size of the mean flow. Indeed, the small-amplitude ray tracing version of (10.128) is the familiar

$$\mathsf{e} \approx \bar{E} + \boldsymbol{U}\cdot\frac{\boldsymbol{k}}{\hat{\omega}}\bar{E} = \frac{\hat{\omega} + \boldsymbol{U}\cdot\boldsymbol{k}}{\hat{\omega}}\bar{E} = \frac{\omega}{\hat{\omega}}\bar{E}. \tag{10.129}$$

After some manipulations similar to those leading to (10.126), the exact pseudoenergy law follows as

$$\overline{\mathrm{D}}^L\mathsf{e} + \frac{1}{\tilde{\rho}}C^{\text{tot}}_{m,m} = +\overline{(\phi_{,t})}^L - (\overline{\phi}^L)_{,t} + \overline{\xi_{j,t}F_j^\ell} \tag{10.130}$$
$$+ \overline{u}_{k,t}^L\mathsf{p}_k + \frac{\tilde{\rho}_{,t}}{\tilde{\rho}}\left\{\overline{\left(\frac{p}{\rho}+\epsilon\right)}^L - \left(\frac{\tilde{p}}{\tilde{\rho}}+\tilde{\epsilon}\right) - \frac{1}{2}\overline{u_j^\ell u_j^\ell}\right\},$$

where the total non-advective pseudoenergy flux vector is

$$C^{\text{tot}}_m = \overline{u_j^\ell K_{jm}p^\xi} - \overline{u}_k^L\,\overline{\xi_{j,k}K_{jm}p^\xi} + \tilde{\rho}\overline{u}_m^L\left[\frac{1}{2}\overline{u_j^\ell u_j^\ell} - (\overline{\epsilon}^L - \tilde{\epsilon})\right]. \tag{10.131}$$

[7] Sometimes e is defined by the first term only.

10.3.3 Non-barotropic flows

We now consider the wave activity relations for a non-barotropic flow, in which the pressure is a function of density and also of specific entropy s, say. This implies that the internal energy $\epsilon(\rho, s)$ is also a function of both ρ and s such that

$$d\epsilon = \frac{p}{\rho^2}d\rho + Tds \tag{10.132}$$

where $T(\rho, s)$ is the temperature. If needed, additional thermodynamic variables can be dealt with in analogy to s.

In flows without diabatic heating we have

$$\frac{Ds}{Dt} = 0 \quad \Rightarrow \quad \overline{\mathrm{D}}^L \overline{s}^L = 0 \quad \text{and} \quad \overline{\mathrm{D}}^L s^\ell = 0. \tag{10.133}$$

Subject to initialization with zero Lagrangian disturbance this implies

$$s^\xi = \overline{s}^L, \quad \text{or} \quad s^\ell = 0. \tag{10.134}$$

The Lagrangian-mean flow is now defined by the fields $(\tilde{\rho}, \overline{\boldsymbol{u}}^L, \overline{s}^L)$ and we expect the symmetries of $\overline{s}^L$ to enter the relevant conservation laws on the same footing as the symmetries of $\tilde{\rho}$. Indeed, using (10.132) and (10.134), the Jacobian term in (10.118) can be written as

$$-\frac{1}{\tilde{\rho}}\overline{J_{,\alpha}p^\xi} = -\frac{\tilde{\rho}_{,\alpha}}{\tilde{\rho}}\left\{\overline{\left(\frac{p}{\rho}\right)}^L - \frac{\tilde{p}}{\tilde{\rho}}\right\} + (\overline{\epsilon}^L - \tilde{\epsilon})_{,\alpha} - \overline{s}^L_{,\alpha}(\overline{T}^L - \tilde{T}). \tag{10.135}$$

Here $\overline{\epsilon}^L = \overline{\epsilon(\rho^\xi, s^\xi)} = \overline{\epsilon(\tilde{\rho}/J, \overline{s}^L)}$, and $\tilde{\epsilon} = \epsilon(\tilde{\rho}, \overline{s}^L)$ as well as $\tilde{p}$ and $\tilde{T}$ denote the values of the respective thermodynamics variables if the thermodynamics state is computed at $(\rho, s) = (\tilde{\rho}, \overline{s}^L)$.

The evolution laws for pseudomomentum and pseudoenergy can now be derived as before. This does not lead to changes in their respective densities and fluxes, but it does lead to new source terms due to the space–time gradients of $\overline{s}^L$ stemming from the last term in (10.135).

Finally, in the presence of diabatic heating (10.133) is replaced by

$$\frac{Ds}{Dt} = R \quad \Rightarrow \quad \overline{\mathrm{D}}^L \overline{s}^L = \overline{R}^L \quad \text{and} \quad \overline{\mathrm{D}}^L s^\ell = R^\ell \tag{10.136}$$

for some heating density R. Crucially, this implies that (10.134) does not hold anymore, i.e., $s^\ell \neq 0$. In the present case, (10.135) now involves $\overline{\epsilon}^L = \overline{\epsilon(\rho^\xi, s^\xi)}$ and it also acquires a new term, namely

$$-\overline{s}^L_{,\alpha}(\overline{T}^L - \tilde{T}) \to -\overline{s}^L_{,\alpha}(\overline{T}^L - \tilde{T}) - \overline{T^\ell s^\ell_{,\alpha}}. \tag{10.137}$$

The definitions of $\tilde{T}$ etc. remain unchanged. This new term is linked to the disturbance heating R^ℓ via a time integral. It cannot be written as the α-derivative of a mean quantity and hence it presents an intrinsic source or sink of wave activity. This is analogous to the forcing term $\overline{\xi_{j,\alpha}F_j^\ell}$, but in comparison (10.137) suffers from the implicit, time-integrated dependence on R. As in linear theory, diabatic heating is the Achille's heel of GLM theory.

10.3.4 Angular momentum and pseudomomentum

Angular momentum is important in applications such as atmospheric dynamics, where the angular momentum around the earth's rotation axis plays a significant role for the large-scale circulation of the atmosphere. There are also interesting conceptual points concerning the exact definition of mean angular momentum and of angular pseudomomentum, which is the conserved wave property in the case where the mean flow has an appropriate rotation symmetry. We discuss these issues by starting with the standard definition of angular momentum in fluid dynamics, then proceeding to its Lagrangian-mean form, and finally providing an exact definition of angular pseudomomentum.

Angular momentum is defined in terms of a distance vector $\boldsymbol{r} = \boldsymbol{x} - \boldsymbol{x}_0$ relative to a fixed location $\boldsymbol{x}_0$ such that $Dr_i/Dt = u_i$. The angular momentum density per unit mass is then defined as

$$\boldsymbol{m} = \boldsymbol{r} \times \boldsymbol{u} \quad \Leftrightarrow \quad m_i = \epsilon_{ijk} r_j u_k \tag{10.138}$$

where ϵ_{ijk} is the alternating tensor. In unforced flow with a momentum flux tensor Π_{km} it follows that

$$\frac{Du_i}{Dt} + \frac{1}{\rho}\Pi_{km,m} = 0 \quad \Rightarrow \quad \frac{Dm_i}{Dt} + \frac{1}{\rho}(\epsilon_{ijk} r_j \Pi_{km})_{,m} = \frac{1}{\rho}(\epsilon_{imk}\Pi_{km}) = 0. \tag{10.139}$$

The last term is zero by the symmetry of the momentum flux tensor; this is of course the standard argument in fluid dynamics as to why $\Pi_{km} = \Pi_{mk}$ must hold in the first place.

The generic rules for conservation laws described in §10.2.6 make clear that (10.139) implies a Lagrangian-mean version of angular momentum conservation in the form

$$\overline{\mathrm{D}}^L \overline{m}_i^L + \frac{1}{\tilde{\rho}}\left(\epsilon_{ijk}\, \overline{r_j^\xi \Pi_{km}^\xi K_{mn}}\right)_{,n} = 0. \tag{10.140}$$

Here the Lagrangian-mean density of angular momentum is

$$\overline{m}_i^L = \epsilon_{ijk}\left(r_j \overline{u}_k^L + \overline{\xi_j u_k^\ell}\right), \tag{10.141}$$

which consists of a mean-flow part and a disturbance correlation part. In other words, it is not true that the mean angular momentum is entirely captured by $\overline{\boldsymbol{u}}^L$, although the corresponding statement was true for the mean translational momentum. As $\boldsymbol{u}^\ell = \overline{\mathrm{D}}^L\boldsymbol{\xi}$, the disturbance part measures the rotation of $\boldsymbol{\xi}$ along a mean-flow trajectory, which could be viewed as an intrinsic spin of the fluid motion.

We now turn to the question of how to define an angular pseudomomentum that is conserved if the mean flow has a rotational symmetry. As in §10.2.8, the answer is closely linked to the mean version of Kelvin's circulation theorem. To illustrate the general procedure we look at the simplest case, in which $\Pi_{km} = p\delta_{km}$ and the flow is barotropic such that the circulation theorem holds for all closed material loops. Now, if the Lagrangian-mean flow defined by the fields $(\tilde{\rho}, \overline{\boldsymbol{u}}^L)$ has a rotational symmetry then there is a constant unit vector l_i, say, such that the mean flow is invariant under rotations of the flow field around the rotation axis l_i centred at $\boldsymbol{x}_0$. In this situation a mean material loop $\mathcal{C}$ in the form of a circle around the rotation axis will remain such a circle as the flow evolves. The circulation of $\overline{\boldsymbol{u}}^L - \mathsf{p}$ around $\mathcal{C}$ is conserved and with the assumed circular shape of $\mathcal{C}$ this implies[8]

$$\overline{\mathrm{D}}^L(l_i\epsilon_{ijk}r_j(\overline{u}_k^L - \mathsf{p}_k)) = 0, \tag{10.142}$$

which is analogous to the exact result (10.93) for flows with translational symmetry. With this result in hand we can now define the density of angular pseudomomentum as

$$\hat{m}_i \equiv \epsilon_{ijk}(r_j\mathsf{p}_k + \overline{\xi_j u_k^\ell}) \quad \text{or} \quad \hat{m}_i = \overline{m}_i^L - \epsilon_{ijk}r_j(\overline{u}_k^L - \mathsf{p}_k). \tag{10.143}$$

The key idea is that in a situation with rotational symmetry of the mean flow relative to l_i it follows from (10.142) and (10.143) that

$$l_i\text{-symmetry} \quad \Rightarrow \quad \overline{\mathrm{D}}^L(l_i\hat{m}_i) = \overline{\mathrm{D}}^L(l_i\overline{m}_i^L) \tag{10.144}$$

and hence the corresponding component of angular pseudomomentum is conserved with a flux equal to the flux of angular mean momentum. This recovers the analogous result for translational momentum.

A direct proof of this conservation law is also possible starting from the pseudomomentum law (10.126) and using the rotational mean-flow symme-

[8] For simplicity, we assume that the pseudomomentum field is rotationally symmetric as well. Otherwise one could follow through a derivation in analogy with (10.95) in the case of translational pseudomomentum; this amounts to replacing the azimuthal component of pseudomomentum in (10.142) by its average around the ring $\mathcal{C}$.

try in the Cartesian form

$$l_i\text{-symmetry} \quad \Rightarrow \quad l_i \epsilon_{ijk} r_j \tilde{\rho}_{,k} = 0 \quad \text{and} \quad l_i \epsilon_{ijk} r_j \overline{u}^L_{m,k} = l_i \epsilon_{inm} \overline{u}^L_n. \tag{10.145}$$

The second relation captures the changes of the Cartesian components of a symmetric vector field $\overline{\boldsymbol{u}}^L$ under an infinitesimal rotation; this term cancels with the refraction term $\overline{u}^L_{k,i}\mathsf{p}_k$ in (10.126), which indicates that this term is important for the angular momentum budget of the flow.

10.4 Coriolis forces in GLM theory

As we know, the equations of motion in a frame rotating with constant angular frequency Ω around a fixed rotation axis $\boldsymbol{l}$ differ from the governing equations in an inertial frame by the inclusion of the standard Coriolis and centrifugal force terms (cf. §1.5). Specifically, if $\boldsymbol{\Omega} = \Omega \boldsymbol{l}$ and the Coriolis vector $\boldsymbol{f} = 2\boldsymbol{\Omega}$ then we need to make the changes

$$\frac{D\boldsymbol{u}}{Dt} \to \frac{D\boldsymbol{u}}{Dt} + \boldsymbol{f} \times \boldsymbol{u} \quad \text{and} \quad \phi \to \phi - \frac{1}{2}|\boldsymbol{\Omega} \times \boldsymbol{r}|^2 \tag{10.146}$$

where $\boldsymbol{r} = \boldsymbol{x} - \boldsymbol{x}_0$ measures the distance from a fixed point $\boldsymbol{x}_0$ on the rotation axis such as the Earth's centre. The consequences of (10.146) for GLM theory follow by considering the changes in the fundamental fluid-dynamical conservation laws. This is straightforward in the case of constant $\boldsymbol{f}$, but it requires some additional effort in the case of variable $\boldsymbol{f}$, which is a relevant case for geophysical fluid dynamics based on the traditional tangent-plane approximations.

10.4.1 Rotating circulation and pseudomomentum

Kelvin's circulation theorem remains valid in a rotating frame provided it is based on the absolute circulation, which includes the frame velocity:

$$\Gamma = \oint_C (\boldsymbol{u} + \boldsymbol{\Omega} \times \boldsymbol{r}) \cdot \boldsymbol{dx}. \tag{10.147}$$

The circulation is conserved if the closed loop C moves with $\boldsymbol{u}$ and the usual conditions on the pressure field are satisfied. This is obvious physically from looking at the definition of Γ in an inertial frame, and a direct mathematical proof is straightforward after using $\boldsymbol{dr} = \boldsymbol{dx}$ and the product rule applied to $\boldsymbol{d}(\boldsymbol{r} \times \boldsymbol{u})$, which leads to

$$\frac{d\Gamma}{dt} = \oint_C d[(\boldsymbol{\Omega} \times \boldsymbol{r}) \cdot \boldsymbol{u}] = 0. \tag{10.148}$$

It is noteworthy that in a Cartesian domain with x-periodic boundary conditions the previously used trick to use the circulation theorem on a material line $\mathcal{C}$ that traverses the periodic domain does not work anymore. This is because the frame term $\boldsymbol{\Omega} \times \boldsymbol{r}$ is not x-periodic, which means (10.148) is non-zero. For example, let the flow be x-periodic with finite period length L and let $\mathcal{C}$ be a material contour traversing the domain in the x-direction such that the end points of $\mathcal{C}$ are located at (x, y, z) and $(x+L, y, z)$, respectively. It then follows from (10.148) that

$$\frac{d\Gamma}{dt} = \Omega L v_*, \tag{10.149}$$

where v_* denotes the y-velocity at the end points. This result is confirmed by the special case in which the flow is x-independent and $\mathcal{C}$ is parallel to the x-axis such that $\Gamma = L(u - \Omega y)$. Then (10.149) implies

$$\frac{D}{Dt}L(u - \Omega y) = \Omega L v \quad \Leftrightarrow \quad \frac{Du}{Dt} - 2\Omega v = 0, \tag{10.150}$$

which is correct. An alternative definition of Γ that is indeed conserved in a periodic Cartesian domain is described in § 10.4.4 below.

Now, the definition of pseudomomentum in a rotating frame follows as before from considering the mean form of the circulation integral and collecting all disturbance parts contributing to $\overline{\Gamma}$ in the definition of p. This leads to the conservation of

$$\overline{\Gamma} = \oint (\overline{\boldsymbol{u}}^L - \mathsf{p} + \boldsymbol{\Omega} \times \boldsymbol{r}) \cdot \boldsymbol{dx} \tag{10.151}$$

provided that $\mathcal{C}$ moves with $\overline{\boldsymbol{u}}^L$ and that p is defined as

$$\mathsf{p}_i = -\overline{\xi_{j,i} u_j^\ell} - \overline{\xi_{j,i}(\boldsymbol{\Omega} \times \boldsymbol{\xi})_j}, \quad \text{where} \quad (\boldsymbol{\Omega} \times \boldsymbol{\xi})_j = \epsilon_{jkm}\Omega_k \xi_m. \tag{10.152}$$

This is consistent with the absolute definition of circulation in (10.147). Correspondingly, any suitable definition of Lagrangian-mean PV is affected only by the simple change

$$\boldsymbol{\nabla} \times (\overline{\boldsymbol{u}}^L - \mathsf{p}) \to \boldsymbol{\nabla} \times (\overline{\boldsymbol{u}}^L - \mathsf{p}) + \boldsymbol{f}. \tag{10.153}$$

Finally, if the actual flow has a rotational symmetry around the rotation axis $\boldsymbol{l}$ then $\mathcal{C}$ can be chosen as a ring around the rotation axis and then (10.148) implies that

$$l_i\text{-symmetry} \quad \Rightarrow \quad \frac{D}{Dt}\boldsymbol{l} \cdot (\boldsymbol{r} \times (\boldsymbol{u} + \boldsymbol{\Omega} \times \boldsymbol{r})) = 0. \tag{10.154}$$

This is the same construction as in §10.3.4. Similarly, if the mean flow has a rotational symmetry around the rotation axis $\boldsymbol{l}$, then the mean circulation theorem implies the Lagrangian-mean version of (10.154), namely

$$l_i\text{-symmetry} \quad \Rightarrow \quad \overline{\mathrm{D}}^L \boldsymbol{l} \cdot (\boldsymbol{r} \times (\overline{\boldsymbol{u}}^L - \mathsf{p} + \boldsymbol{\Omega} \times \boldsymbol{r})) = 0. \tag{10.155}$$

10.4.2 Wave activity relations

Repeating the earlier derivations in a rotating frame leads to the conclusion that the only change arises in the definition of the GLM wave action (10.112), where the absolute disturbance velocity must be used. In other words, we must make the shift

$$\boldsymbol{u}^\ell \to \boldsymbol{u}^\ell + \boldsymbol{\Omega} \times \boldsymbol{\xi} \tag{10.156}$$

there and in all subsequent definitions of wave activity densities. This is consistent with the pseudomomentum definition given already. Notably, this does not affect the generic wave activity expressions for ray tracing, e.g., after this change

$$\mathsf{p} = \frac{\boldsymbol{k}}{\hat{\omega}}\bar{E} \quad \text{and} \quad \mathsf{e} = \frac{\omega}{\hat{\omega}}\bar{E} \tag{10.157}$$

remain valid for slowly varying wavetrains in a rotating frame.

10.4.3 Angular momentum and pseudomomentum

Usually, in a rotating fluid system the component of angular momentum that is of interest is that around the rotation axis, which has the density

$$m = \boldsymbol{l} \cdot (\boldsymbol{r} \times (\boldsymbol{u} + \boldsymbol{\Omega} \times \boldsymbol{r})) = \boldsymbol{l} \cdot (\boldsymbol{r} \times \boldsymbol{u}) + \Omega r_\perp^2 \tag{10.158}$$

where $r_\perp$ is the distance from the rotation axis. This density satisfies an integral conservation law as before, and it follows that the Lagrangian-mean density defined by

$$\overline{m}^L = \boldsymbol{l} \cdot (\boldsymbol{r} \times \overline{\boldsymbol{u}}^L) + \boldsymbol{l} \cdot (\overline{\boldsymbol{\xi} \times \boldsymbol{u}^\ell}) + \Omega(r_\perp^2 + \overline{|\boldsymbol{\xi}_\perp|^2}) \tag{10.159}$$

is conserved as well. Here $\boldsymbol{\xi}_\perp$ is the displacement vector in the plane normal to $\boldsymbol{l}$. In analogy with (10.143), this motivates defining the angular pseudomomentum as

$$\begin{aligned} \hat{m} &= \boldsymbol{l} \cdot (\boldsymbol{r} \times \mathsf{p}) + \boldsymbol{l} \cdot (\overline{\boldsymbol{\xi} \times \boldsymbol{u}^\ell}) + \Omega \overline{|\boldsymbol{\xi}_\perp|^2}, \\ \text{or} \quad \hat{m} &= \overline{m}^L - \left(\boldsymbol{l} \cdot (\boldsymbol{r} \times (\overline{\boldsymbol{u}}^L - \mathsf{p})) + \Omega r_\perp^2\right). \end{aligned} \tag{10.160}$$

Comparison with (10.155) now implies that if the mean flow has a rotational symmetry around the rotation axis $\boldsymbol{l}$ then

$$l_i\text{-symmetry} \quad \Rightarrow \quad \overline{\mathrm{D}}^L \hat{m} = \overline{\mathrm{D}}^L \overline{m}^L \tag{10.161}$$

and therefore angular pseudomomentum is conserved in this symmetric situation. In addition, (10.161) shows that the non-azimuthal flux of angular pseudomomentum is equal to the non-azimuthal flux of mean angular momentum. Nothing is implied about the azimuthal flux, which has zero divergence in the present situation. This mirrors the off-diagonal flux equality for the translational pseudomomentum and mean momentum.

Despite the straightforward derivation of the exact angular pseudomomentum density there remains a puzzling fact, namely (10.160) does not obviously reduce to the usual $\boldsymbol{l} \cdot (\boldsymbol{r} \times \mathbf{p})$ in the ray tracing regime, i.e., it is not obvious that the other two terms vanish, or that they are separately conserved in the limit of a slowly varying wavetrain.[9]

10.4.4 Gauged pseudomomentum and the β-plane

The previous derivations for the circulation and the pseudomomentum vector in the presence of Coriolis forces were based on the existence of an inertial frame in which the Coriolis force vanishes, and such a frame always exist for constant $\boldsymbol{f}$. This led to the frequent appearance of the absolute velocity $\boldsymbol{u} + \boldsymbol{\Omega} \times \boldsymbol{r}$ in the relevant definitions.

However, there are several reasons why it is convenient to have a more flexible approach to Coriolis forces. For example, it is often desirable to include Coriolis forces in a flat geometry based on local Cartesian xyz-coordinates, or even in a vertical slice or horizontal plane based on xz- or xy-coordinates, respectively. In such domains it is awkward to define a centre of frame rotation from which $\boldsymbol{r}$ is to be measured, because the equations of motion are homogeneous in space (if explicit centrifugal forces are absent, as usual). Moreover, if the domain is x-periodic with finite period length L, then it might also be desirable to have a definition of circulation that allows conservation for material contours that traverse the domain in the x-direction, which is not the case for the absolute circulation as shown in (10.149).

Last but not least, in geophysical fluid dynamics we want to use the β-

9 Some detailed computations in rotating shallow water indicate that the puzzling terms are indeed small, namely of the same order as the error committed by using in cylindrical coordinates the plane-wave dispersion relation based on Cartesian coordinates, but a general result is missing.

plane model, in which the Coriolis vector is

$$\boldsymbol{f} = f\hat{\boldsymbol{z}} \quad \text{with} \quad f(y) = f_0 + \beta y \tag{10.162}$$

with constants f_0 and $\beta > 0$ and y as the 'northward' coordinate. It is clear that now there is no inertial frame in which this location-dependent Coriolis vector vanishes everywhere, so there is no obvious way to deduce a circulation theorem or a pseudomomentum definition from the previous material on inertial frames. This is also a non-trivial change for GLM theory, essentially because for variable $\boldsymbol{f}$ we must anticipate that

$$\overline{\boldsymbol{f} \times \boldsymbol{u}}^L \neq \boldsymbol{f} \times \overline{\boldsymbol{u}}^L. \tag{10.163}$$

Overall, this motivates looking more generally at the representation of $\boldsymbol{f}$ and on the impact on the circulation theorem and the definitions of pseudomomentum.

First, we discuss the class of vector fields $\boldsymbol{f}$ that should be considered. We take the view that the fundamental property of $\boldsymbol{f}$ is that it should add naturally to the vorticity of the flow. Specifically, with $\boldsymbol{f} = 0$ we have

$$\nabla \times \left(\frac{D\boldsymbol{u}}{Dt}\right) = \frac{D(\nabla \times \boldsymbol{u})}{Dt} + (\nabla \times \boldsymbol{u})\nabla \cdot \boldsymbol{u} - [(\nabla \times \boldsymbol{u}) \cdot \nabla]\boldsymbol{u} \tag{10.164}$$

and with $\boldsymbol{f} \neq 0$ we want to have the analogous

$$\begin{aligned} &\nabla \times \left(\frac{D\boldsymbol{u}}{Dt} + \boldsymbol{f} \times \boldsymbol{u}\right) \qquad (10.165) \\ &= \frac{D(\nabla \times \boldsymbol{u} + \boldsymbol{f})}{Dt} + (\nabla \times \boldsymbol{u} + \boldsymbol{f})\nabla \cdot \boldsymbol{u} - [(\nabla \times \boldsymbol{u} + \boldsymbol{f}) \cdot \nabla]\boldsymbol{u}. \end{aligned}$$

In other words, we want the vector field $\nabla \times \boldsymbol{u} + \boldsymbol{f}$ to be stretched and twisted just as ordinary vorticity would be stretched and twisted in the absence of Coriolis forces. The relation (10.165) will hold for any $\boldsymbol{u}$ precisely if

$$\frac{\partial \boldsymbol{f}}{\partial t} = 0 \quad \text{and} \quad \nabla \cdot \boldsymbol{f} = 0. \tag{10.166}$$

Thus the class of $\boldsymbol{f}$ that we want to consider consists of steady non-divergent vector fields. Such fields can be written in terms of a steady vector potential $\boldsymbol{A}(\boldsymbol{x})$ such that

$$\boldsymbol{f} = \nabla \times \boldsymbol{A} \tag{10.167}$$

and therefore the effective vorticity field is $\nabla \times (\boldsymbol{u} + \boldsymbol{A})$. Indeed, it is then straightforward to show that the circulation

$$\Gamma = \oint_C (\boldsymbol{u} + \boldsymbol{A}) \cdot d\boldsymbol{x} \tag{10.168}$$

is conserved under the usual conditions on the pressure field. The proof of this fact comes down to

$$\frac{d\Gamma}{dt} = \oint_C d[\boldsymbol{A} \cdot \boldsymbol{u}] = 0, \tag{10.169}$$

which in the special case $\boldsymbol{A} = \boldsymbol{\Omega} \times \boldsymbol{r}$ with constant $\boldsymbol{\Omega}$ reduces to (10.148). However, even for constant $\boldsymbol{\Omega}$ there is flexibility in choosing $\boldsymbol{A}$. This is because the gradient of any steady potential $\phi(\boldsymbol{x})$ can be added to $\boldsymbol{A}$ without affecting (10.167) or (10.168). Thus the gauge transformation

$$\boldsymbol{A} \to \boldsymbol{A} + \boldsymbol{\nabla}\phi \tag{10.170}$$

allows changing $\boldsymbol{A}$ in order to adjust its symmetry properties, for instance. In particular, we can find a gauged version of $\boldsymbol{A}$ such that the circulation theorem applies not only for closed material contours but also for material contours that traverse an x-periodic Cartesian domain, as envisaged in the discussion of (10.149). Thus we assume again that the end points of $\mathcal{C}$ lie at (x, y, z) and $(x + L, y, z)$ and then we consider (10.169) for this contour. For an x-periodic flow this will be zero if $\boldsymbol{A}$ is itself x-periodic, or simpler still if $\boldsymbol{A}$ is zonally symmetric. For constant $\boldsymbol{f} = f_0\widehat{\boldsymbol{z}}$ this can be achieved by choosing

$$\boldsymbol{A} = (A, 0, 0) \quad \text{with} \quad A = -f_0 y, \tag{10.171}$$

for example. Thus we obtain exact conservation of a circulation defined by

$$\Gamma = \oint_C (u - f_0 y)\, dx + v\, dy + w\, dz \tag{10.172}$$

for closed contours as well as for traversing contours in x-periodic domains. The gauge between the standard choice $\boldsymbol{A} = \frac{1}{2} f_0 \widehat{\boldsymbol{z}} \times \boldsymbol{r}$ and the present $\boldsymbol{A}$ in (10.171) is $\phi = -\frac{1}{2} f_0 xy$, i.e.,

$$\boldsymbol{A} = \frac{f_0}{2}\widehat{\boldsymbol{z}} \times \boldsymbol{r} + \boldsymbol{\nabla}\left(-\frac{f_0}{2} xy\right). \tag{10.173}$$

This gauge transformation does not change Γ for closed contours, but it does for traversing contours and that has been exploited in (10.172).

We can now derive the definition of pseudomomentum by the same steps used previously, i.e., we evaluate the mean circulation based on (10.168) to obtain

$$\overline{\Gamma} = \oint_C (\overline{\boldsymbol{u}}^L + \overline{\boldsymbol{A}}^L - \mathsf{p}) \cdot \boldsymbol{dx} \tag{10.174}$$

provided the pseudomomentum vector is defined as

$$\mathsf{p}_i = -\overline{\xi_{j,i}(u_j^\ell + A_j^\ell)}. \tag{10.175}$$

For constant $\boldsymbol{\Omega}$ and the standard choice $\boldsymbol{A} = \boldsymbol{\Omega} \times \boldsymbol{r}$ this is identical to the earlier definition. However, different gauges affect the definition of p. In fact, we find the group

$$\boldsymbol{A} \to \boldsymbol{A} + \boldsymbol{\nabla}\phi, \qquad \overline{\boldsymbol{A}}^L \to \overline{\boldsymbol{A}}^L + \overline{\boldsymbol{\nabla}\phi}^L, \tag{10.176}$$
$$\mathsf{p} \to \mathsf{p} + \overline{\boldsymbol{\nabla}\phi}^L - \boldsymbol{\nabla}\overline{\phi}^L, \qquad \overline{\boldsymbol{A}}^L - \mathsf{p} \to \overline{\boldsymbol{A}}^L - \mathsf{p} + \boldsymbol{\nabla}\overline{\phi}^L.$$

For example, in the case of (10.171) we find the gauged pseudomomentum

$$\mathsf{p}_i = -\overline{\xi_{j,i} u^\ell_j} + f_0 \overline{\xi_{,i}\eta}. \tag{10.177}$$

This differs from the standard pseudomomentum by $\boldsymbol{\nabla}(\frac{1}{2} f_0 \overline{\xi\eta})$, which is consistent with (10.176) for $\phi = -\frac{1}{2} f_0 xy$. Now, by construction, the zonal component of the gauged pseudomomentum in (10.176) is conserved if the mean flow is zonally symmetric. This can be seen by considering zonal averaging such that all mean fields are x-independent. It then follows from (10.171) and the conservation of (10.174) applied to a straight mean contour $\mathcal{C}$ parallel to the x-axis that

$$\overline{\mathrm{D}}^L(\overline{u}^L + \overline{A}^L - \mathsf{p}) = \overline{\mathrm{D}}^L(\overline{u}^L - f_0 y - \mathsf{p}) = 0 \tag{10.178}$$
$$\Rightarrow \quad \overline{\mathrm{D}}^L \mathsf{p} = \overline{\mathrm{D}}^L \overline{u}^L - f_0 \overline{v}^L = -\frac{1}{\tilde{\rho}} \tilde{\Pi}_{1j,j},$$

which is conserved by virtue of the Lagrangian-mean zonal momentum equation. Again, only the off-diagonal fluxes $j \neq 1$ matter in (10.178).

We now consider the case of the β-plane defined by (10.162). The zonally symmetric $\boldsymbol{A}$ in (10.171) is augmented with a β-term and becomes

$$\boldsymbol{A} = (A, 0, 0) \quad \text{with} \quad A = -f_0 y - \beta \frac{y^2}{2} \tag{10.179}$$

such that

$$\overline{A}^L = -f_0 y - \beta \frac{y^2}{2} - \beta \frac{\overline{\eta^2}}{2} \quad \text{and} \quad A^\ell = -(f_0 + \beta y)\, \eta - \frac{\beta}{2}(\eta^2 - \overline{\eta^2}). \tag{10.180}$$

Note that the term $\overline{\mathrm{D}}^L \overline{A}^L$ in (10.178) becomes

$$\overline{\mathrm{D}}^L \overline{A}^L = -(f_0 + \beta y)\overline{v}^L - \beta \overline{\eta v^\ell} = -\overline{(f_0 + \beta y) v}^L = -\overline{f v}^L, \tag{10.181}$$

which equals the Lagrangian-mean of the zonal Coriolis force. Now, from (10.175) the β-plane pseudomomentum vector follows as

$$\mathsf{p}_i = -\overline{\xi_{j,i} u^\ell_j} + (f_0 + \beta y)\overline{\xi_{,i}\eta} + \frac{\beta}{2}\overline{\xi_{,i}\eta^2}. \tag{10.182}$$

The zonal component of this pseudomomentum vector is exactly conserved if zonal averaging is used. The generality of the exact definition (10.182) is considerable: it includes the usual quadratic terms for the leading-order wave properties but also a cubic correction to those. In addition, it applies to all kinds of waves (e.g., internal waves or Rossby waves) or disturbances without assuming special scaling such as the hydrostatic approximation or quasi-geostrophic balance. It also applies if $f_0 = 0$, which is relevant for equatorial regions.

Mutatis mutandis, the exact (10.182) reduces to the relevant approximate expressions in the appropriate limits of small wave amplitude and so on. As an example we consider the zonal pseudomomentum for two-dimensional incompressible flow in the xy-plane with small-amplitude disturbances. In this case the linear disturbance fields satisfy

$$\xi'_x + \eta'_y = 0 \quad \text{and} \quad v'_x - u'_y = -\beta\eta' \tag{10.183}$$

by incompressibility and potential vorticity conservation. The zonal pseudomomentum of (10.182) can then be written as

$$\mathsf{p} = -\frac{\beta}{2}\overline{\eta'^2} + \frac{\partial}{\partial y}\left(\overline{\eta' u'} - (f_0 + \beta y)\frac{1}{2}\overline{\eta'^2}\right) + O(a^3). \tag{10.184}$$

The first term is the standard small-amplitude expression for the Eulerian pseudomomentum of Rossby waves. The remaining terms are in divergence form and quantify the difference between the Lagrangian and the Eulerian pseudomomentum in this kind of flow. This is analogous to the internal wave pseudomomentum example in §6.3.1.

10.5 Lagrangian-mean gas dynamics and radiation stress

We conclude our description of GLM theory by considering the wave-induced momentum fluxes in compressible flows, which is important in acoustics. Clearly, the effective Lagrangian-mean momentum flux tensor $\tilde{\Pi}_{ij}$ in (10.84) can be decomposed in a number of ways. In the context of compressible gas dynamics it is of particular interest to highlight the *excess* mean momentum flux due to the presence of disturbances, i.e., the part of the momentum flux that exceeds that of the undisturbed fluid in the same mean density configuration. As an example, we consider the case of a barotropic fluid such that

$$p = f(\rho) \tag{10.185}$$

for some function $f(\cdot)$. Correspondingly,

$$p^\xi = f(\rho^\xi) = f(\tilde{\rho}/J) \quad \text{and} \quad \tilde{p} = f(\tilde{\rho}) \tag{10.186}$$

in obvious notation. The actual flow is governed by

$$\frac{D\rho}{Dt} + \rho \nabla \cdot \boldsymbol{u} = 0 \quad p = f(\rho) \quad \text{and} \quad \frac{Du_i}{Dt} + \frac{1}{\rho} p_{,i} = 0. \tag{10.187}$$

On the other hand, using (10.84), the mean flow is governed by

$$\overline{\mathrm{D}}^L \tilde{\rho} + \tilde{\rho} \nabla \cdot \overline{\boldsymbol{u}}^L = 0 \quad \tilde{p} = f(\tilde{\rho}) \quad \text{and} \quad \overline{\mathrm{D}}^L \overline{u}_i^L + \frac{1}{\tilde{\rho}} \tilde{p}_{,i} = -\frac{1}{\tilde{\rho}} S_{ij,j}, \tag{10.188}$$

where the so-called *radiation stress tensor*

$$S_{ij} = \overline{p^\xi K_{ij}} - \tilde{p} \delta_{ij} \tag{10.189}$$

captures the excess mean momentum flux due to the disturbances. This tensor is analogous to minus the Reynolds stress tensor in the standard Eulerian-mean theory of homogeneous incompressible flows.[10]

In the particular case of a polytropic fluid we have $p \propto \rho^\gamma$ for some constant γ. This implies $p^\xi = \tilde{p} J^{-\gamma}$ and therefore

$$S_{ij} = \tilde{p}(\overline{K_{ij} J^{-\gamma}} - \delta_{ij}). \tag{10.190}$$

In one-dimensional flows the only non-zero component of (10.190) is

$$\tilde{p}(\overline{J^{-\gamma}} - 1) = \tilde{p}(\overline{(1+\xi_{,x})^{-\gamma}} - 1) = \tilde{p} \frac{\gamma(\gamma+1)}{2} \overline{(\xi_{,x})^2} + O(a^3). \tag{10.191}$$

By Jensen's theorem for convex functions, the one-dimensional excess pressure is positive if $\gamma > 0$ or $\gamma < -1$, zero at these thresholds, and negative otherwise. We can note in passing a mathematical curiosity, namely that (10.191) is *exactly* zero in the non-trivial special case $\gamma = -1$, i.e., in this case disturbances do not lead to any excess mean momentum flux in one-dimensional flows. In other words, in this case the one-dimensional Lagrangian-mean flow is entirely unaffected by the presence of the waves, a fact that is hardly obvious from the original equations.[11]

10.5.1 Radiation stress and pseudomomentum flux

In GLM theory the mean momentum flux tensor and the pseudomomentum flux tensor are exactly equal in their off-diagonal components, and by

[10] An alternative definition of S_{ij} is $\overline{p^\xi K_{ij}} - \overline{p}^L \delta_{ij}$; however, $\overline{p}^L$ is not part of the fields $(\tilde{\rho}, \overline{\boldsymbol{u}}^L)$ that define the Lagrangian-mean flow. Also, sometimes S_{ij} is defined with the opposite sign in order to conform with the convention that stress equals minus momentum flux.

[11] In Eulerian variables the curious nature of the $\gamma = -1$ system becomes apparent upon writing the one-dimensional equations in characteristic form. It then turns out that in a simple wave problem (i.e., in a problem involving propagation of waves into a region that was initially at rest) the nonlinear characteristic speed is constant, so there is no wave steepening or shock formation in this curious system.

(10.189) this equality extends to the radiation stress as well. However, this does not imply equality in the diagonal parts of these flux tensors. It is a matter of some importance for problems involving wavemakers and wave-induced forces on solid bodies to understand the difference between the diagonal pseudomomentum flux terms, such as the zonal flux of the zonal pseudomomentum, say, and the corresponding mean momentum flux. It appears that there are no general theorems covering these diagonal fluxes, but in the context of small-amplitude slowly varying wavetrains containing acoustic waves some generic results are indeed available.

For example, in two-dimensional[12] polytropic flows we can use (10.77) and obtain

$$S_{ij} = \tilde{p}(\delta_{ij}(\overline{(1+\boldsymbol{\nabla}\cdot\boldsymbol{\xi})J^{-\gamma}} - 1) - \overline{\xi_{j,i}J^{-\gamma}}). \tag{10.192}$$

At small wave amplitude this becomes

$$\begin{aligned} S_{ij} &= \tilde{p}\left(\delta_{ij}\left(-\gamma\overline{\boldsymbol{\nabla}\xi\times\boldsymbol{\nabla}\eta} + \frac{\gamma(\gamma-1)}{2}\overline{(\boldsymbol{\nabla}\cdot\boldsymbol{\xi})^2}\right) + \gamma\overline{\xi_{j,i}\boldsymbol{\nabla}\cdot\boldsymbol{\xi}}\right) \\ &= \rho_0 c_0^2\left(\delta_{ij}\left(-\overline{\boldsymbol{\nabla}\xi\times\boldsymbol{\nabla}\eta} + \frac{\gamma-1}{2}\overline{(\boldsymbol{\nabla}\cdot\boldsymbol{\xi})^2}\right) + \overline{\xi_{j,i}\boldsymbol{\nabla}\cdot\boldsymbol{\xi}}\right) \end{aligned} \tag{10.193}$$

after using (1.29) and the fact that $\tilde{\rho} = \rho_0$ to sufficient accuracy here. For slowly varying waves we can evaluate these fluxes and find that the first term is zero and that the second term can be written as

$$\rho_0 c_0^2 \frac{\gamma-1}{2}\overline{(\boldsymbol{\nabla}\cdot\boldsymbol{\xi})^2}\delta_{ij} = \rho_0 c_0^2 \frac{\gamma-1}{2}\frac{\overline{\rho^\ell\rho^\ell}}{\tilde{\rho}^2}\delta_{ij} = \rho_0 E\frac{\gamma-1}{2}\delta_{ij} \tag{10.194}$$

where E is the mean acoustic energy density per unit mass.[13] Finally, the last term turns out to be the non-advective pseudomomentum flux tensor, i.e.,

$$\rho_0 c_0^2\, \overline{\xi_{j,i}\boldsymbol{\nabla}\cdot\boldsymbol{\xi}} = -\overline{\xi_{m,i}K_{mj}p^\xi} \tag{10.195}$$

for slowly varying waves. The proof follows directly from expanding the pseudomomentum flux to $O(a^2)$ and noting that a remainder term proportional to $\overline{\xi_{j,i}\boldsymbol{\nabla}\cdot\boldsymbol{\xi}} - \overline{\xi_{m,i}\xi_{j,m}}$ is zero for plane acoustic waves.

The net result is that

$$S_{ij} = \rho_0 E\frac{\gamma-1}{2}\delta_{ij} + (\text{pseudomomentum flux})_{ij} \tag{10.196}$$

at leading order for small-amplitude acoustic wavetrains. Notably, as we have ensured that the linear wave speed c_0 is independent of γ, the first

[12] The same result holds in three-dimensional flow as well.
[13] The previously used energy density per unit volume is related by $\bar{E} = \rho_0 E$.

part depends on the nonlinear details of the pressure function. We will see a couple of examples of the radiation stress in §12.2.1 and 12.2.2, which will illustrate whether or not this isotropic part of the radiation stress equals $-\tilde{p}\delta_{ij}$. If that occurs, then the total mean momentum flux (which is $S_{ij} + \tilde{p}\delta_{ij}$) is equal to the pseudomomentum flux, which has obvious ramifications for the wave-induced force on solid bodies and so on.

For future reference, we note that in the shallow-water system (where $\gamma = 2$ and $p = gh^2/2$ with h identified with ρ) (10.196) yields

$$S_{ij} = \delta_{ij}\,\frac{g}{2}\,\overline{h'^2} + \tilde{h}\overline{u_i'u_j'}. \tag{10.197}$$

This shallow-water radiation stress tensor is relevant for wave-driven flows in the surf zone on beaches; this will be discussed in §13.

Finally, for a general barotropic fluid the above derivation can be carried through as well and then the isotropic term (10.194) is replaced by

$$\rho_0 E \frac{\partial \ln c}{\partial \ln \rho}\delta_{ij} \quad \text{where} \quad c = \sqrt{\frac{\partial p}{\partial \rho}} \tag{10.198}$$

is the nonlinear sound speed; for an ideal gas this derivative is taken at fixed entropy.

10.6 Notes on the literature

Wave–mean interaction theories based on Lagrangian averages have been studied by many authors in a number of fields beginning in the 1960s, with particularly important contributions coming from magneto-hydrodynamics and GFD. GLM theory in essentially the form presented here was built on these foundation and is comprehensively described in Andrews and McIntyre (1978a,c), which also contain references to much of the relevant work in other fields. The definitions given here agree with these references almost everywhere, the modest exception being the notation for integration domains and elements. Further fundamental issues relevant to Lagrangian averaging applied to GFD are discussed in McIntyre (1980a,b). The construction of $\overline{q}^L$ is described in Bühler and McIntyre (1998).

Applications of weakly nonlinear theories for wave–wave and wave–mean interactions using GLM theory and other methods are discussed together in the monograph Craik (1985).

The α-models for turbulence closure problems, whose mathematical structure shares features with the GLM equations, were developed by Holm and his co-workers in a series of papers including Foias et al. (2001).

11
Zonally symmetric GLM theory

The application of GLM theory to zonally symmetric mean flows shows that the small-amplitude results about mean-flow acceleration and pseudomomentum dynamics often have natural counterparts in the fully nonlinear finite-amplitude GLM theory. Overall, the point here is not to discover fundamentally new results, but rather to demonstrate the robustness of the small-amplitude results as well as the ease with which the finite-amplitude GLM theory can be applied. In fact, it is sometimes easier to use the exact GLM definitions rather than their approximate counterparts.

11.1 GLM theory for the Boussinesq equations

Applying GLM theory to an approximate set of fluid equations such as the Boussinesq system can be pursued in two different ways: one can either start from the generic GLM equations based on the compressible Euler equations and then seek the relevant approximate expressions by replicating the scaling steps that lead to the Boussinesq equations, say, or one can re-derive the GLM equations starting from scratch based on the approximate set of fluid equations. The latter route is usually both simpler and more instructive.

Thus we consider the three-dimensional Boussinesq equations in Cartesian xyz-coordinates with z pointing upwards:

$$\frac{D\boldsymbol{u}}{Dt} + \boldsymbol{\nabla} P = b\hat{\boldsymbol{z}} + \boldsymbol{F}, \quad \frac{Db}{Dt} + N^2 w = 0 \quad \text{and} \quad \boldsymbol{\nabla} \cdot \boldsymbol{u} = 0. \tag{11.1}$$

For simplicity we assume N is constant such that the materially invariant stratification variable is $S = b + N^2 z$. The circulation theorem then takes the form

$$\frac{d}{dt} \oint_C \boldsymbol{u} \cdot \boldsymbol{dx} = \oint_C \boldsymbol{F} \cdot \boldsymbol{dx} \tag{11.2}$$

for closed material loops C that lie entirely within a surface of constant stratification S. This goes together with the definition of PV as

$$q = (\boldsymbol{\nabla} \times \boldsymbol{u}) \cdot \boldsymbol{\nabla} S \quad \Rightarrow \quad \frac{Dq}{Dt} = (\boldsymbol{\nabla} \times \boldsymbol{F}) \cdot \boldsymbol{\nabla} S. \tag{11.3}$$

The Lagrangian-mean equations corresponding to (11.1) are

$$\overline{\mathrm{D}}^L \overline{\boldsymbol{u}}^L + \frac{1}{\tilde{\rho}} \boldsymbol{\nabla} \cdot (\overline{P^\xi \boldsymbol{K}}) = \overline{b}^L \hat{\boldsymbol{z}} + \overline{\boldsymbol{F}}^L \quad \text{and} \quad \overline{\mathrm{D}}^L \overline{b}^L + N^2 \overline{w}^L = 0. \tag{11.4}$$

This implies that $\overline{S}^L = \overline{b}^L + N^2 z$ is a mean material invariant, i.e., $\overline{\mathrm{D}}^L \overline{S}^L = 0$ follows without any forcing terms due to the disturbances. The effective density $\tilde{\rho}$ is defined by the generic continuity equation

$$\overline{\mathrm{D}}^L \tilde{\rho} + \tilde{\rho} \boldsymbol{\nabla} \cdot \overline{\boldsymbol{u}}^L = 0. \tag{11.5}$$

We define $\tilde{\rho}$ as scaled with the constant Boussinesq density ρ_*, so $\tilde{\rho} = 1$ holds at rest. As discussed in §10.2.5, for incompressible flow the deviation $\tilde{\rho} - 1$ is a wave property and usually ignorable unless one is dealing with evanescent waves.

The mean circulation theorem is

$$\frac{d}{dt} \oint_C (\overline{\boldsymbol{u}}^L - \mathsf{p}) \cdot \boldsymbol{dx} = \oint_C (\overline{\boldsymbol{F}}^L - \boldsymbol{\mathcal{F}}) \cdot \boldsymbol{dx} \tag{11.6}$$

for mean material loops C that lie entirely within a surface of constant mean stratification $\overline{S}^L$. This uses the generic definitions

$$\mathsf{p}_i = -\overline{\xi_{j,i} u_j^\ell} \quad \text{and} \quad \mathcal{F}_i = -\overline{\xi_{j,i} F_j^\ell}. \tag{11.7}$$

Accordingly, the mean PV satisfies the exact equations

$$\overline{q}^L = \frac{1}{\tilde{\rho}} (\boldsymbol{\nabla} \times (\overline{\boldsymbol{u}}^L - \mathsf{p})) \cdot \boldsymbol{\nabla} \overline{S}^L \quad \Rightarrow \quad \overline{\mathrm{D}}^L \overline{q}^L = \frac{1}{\tilde{\rho}} (\boldsymbol{\nabla} \times (\overline{\boldsymbol{F}}^L - \boldsymbol{\mathcal{F}})) \cdot \boldsymbol{\nabla} \overline{S}^L. \tag{11.8}$$

The corresponding circulation form of the GLM momentum equation is (cf. (10.98))

$$\overline{\mathrm{D}}^L (\overline{u}_i^L - \mathsf{p}_i) - \overline{u}_{k,i}^L \mathsf{p}_k + (\overline{\cdots})_{,i} = \overline{b}^L \delta_{i3} + \overline{\zeta_{,i} b^\ell}. \tag{11.9}$$

The last term on the right-hand side can be simplified by noting that

$$\overline{\mathrm{D}}^L b^\ell + N^2 w^\ell = 0 \quad \Rightarrow \quad b^\ell = -N^2 \zeta \tag{11.10}$$

and therefore the last term is an explicit gradient

$$\overline{\zeta_{,i} b^\ell} = -\frac{1}{2} (N^2 \overline{\zeta^2})_{,i}, \tag{11.11}$$

which does not matter for the circulation. Taking the curl of (11.9) yields the Boussinesq version of the GLM vorticity equation (10.99), namely

$$\overline{\mathrm{D}}^L(\nabla \times (\overline{\boldsymbol{u}}^L - \mathsf{p})) + (\nabla \times (\overline{\boldsymbol{u}}^L - \mathsf{p}))\nabla \cdot \overline{\boldsymbol{u}}^L - [(\nabla \times (\overline{\boldsymbol{u}}^L - \mathsf{p}))\cdot\nabla]\overline{\boldsymbol{u}}^L = \nabla \overline{b}^L \times \hat{\boldsymbol{z}}. \tag{11.12}$$

Notably, the only appearance of the disturbance fields in this exact equation is via the pseudomomentum vector p. The evolution law for p follows from the generic derivation that led to (10.124) in §10.3.2, except that there is no need to involve an internal energy function ϵ because $\rho^\xi = \rho_*$ is constant in the Boussinesq system. However, there is an additional gradient term from the buoyancy acceleration on the right-hand side, which is minus the term in (11.11).

So far all the GLM relations hold generically for all kinds of averaging, i.e., so far we have not yet restricted to zonal averaging. Now, in simple geometry based on periodicity in the x-direction we are only interested in the zonal component of pseudomomentum based on zonal averaging such that all mean fields are zonally symmetric. The zonal component of (11.11) is then zero and it follows from (10.124) in §10.3.2 that

$$\overline{\mathrm{D}}^L\mathsf{p} + \frac{1}{\tilde{\rho}}(\overline{P^\xi K_{1m}})_{,m} = \mathcal{F}_1 = \mathcal{F}. \tag{11.13}$$

Only the off-diagonal flux components $m \neq 1$ matter here. The zonal component of the momentum equation is

$$\overline{\mathrm{D}}^L\overline{u}^L + \frac{1}{\tilde{\rho}}(\overline{P^\xi K_{1m}})_{,m} = \overline{F}_1^L = \overline{F}^L \tag{11.14}$$

and their difference yields

$$\overline{\mathrm{D}}^L(\overline{u}^L - \mathsf{p}) = \overline{F}^L - \mathcal{F}, \tag{11.15}$$

which can also be deduced directly from (11.6) for zonal symmetry. These simple yet exact equations illustrate that the buoyancy acceleration has no direct impact on the horizontal mean-flow equations and that the vertical flux of horizontal pseudomomentum equals that of mean momentum. Moreover, the relevant vertical flux is clearly the pressure-induced form stress across undulating material surfaces. For example, in the special case of two-dimensional flow in a vertical xz-slice we have from (10.77) that $K_{13} = -\zeta_{,x}$ and therefore we obtain the simple expression

$$\overline{P^\xi K_{13}} = \overline{-P^\xi \zeta_{,x}}, \tag{11.16}$$

which makes this fact obvious. It is not hard to show that for the compressible Euler equations (11.13), (11.14) and (11.16) hold with P replaced by the full pressure p.

The two-dimensional case is also useful in order to illustrate the relation between the Lagrangian and the Eulerian pseudomomentum at small wave amplitude. For a constant basic flow the latter was defined correct to $O(a^2)$ as

$$\tilde{\mathsf{p}} = \frac{1}{N^2}\overline{b'(u'_z - w'_x)} = -\overline{\zeta(u'_z - w'_x)}, \tag{11.17}$$

which using $\boldsymbol{\nabla}\cdot\boldsymbol{\xi} = O(a^2)$ can be related to p at the same accuracy by

$$\mathsf{p} - \tilde{\mathsf{p}} = (\overline{\zeta u'})_z = \overline{\zeta u'_z} + \overline{\zeta_z u'} = \overline{\zeta u'_z} + \overline{\xi u'_x} = \overline{u}^S. \tag{11.18}$$

This is a typical relation between Lagrangian and Eulerian definitions of zonal pseudomomentum: the difference captures the Stokes correction between the corresponding zonal mean velocities. This ensures that the generic small-amplitude relations

$$\overline{u}^L_t = \mathsf{p}_t \quad \text{and} \quad \overline{u}_t = \tilde{\mathsf{p}}_t \tag{11.19}$$

both hold.

11.1.1 Dissipative pseudomomentum rule

The exact relation (11.15) shows that the zonal mean-flow accelerations due to transient p that were discussed in small-amplitude theory are robust and have a natural counterpart finite-amplitude theory. Moreover, (11.15) with zero forcing is also a finite-amplitude non-acceleration statement for the zonal mean flow in the presence of a steady wave field. This is also clear from a physical point of view based on the circulation conservation.

Now, we would also like to verify that the dissipative pseudomomentum rule holds, i.e., we want to show that for steady but dissipating waves the effective zonal mean force equals the dissipation density of zonal pseudomomentum. This would follow exactly from (11.15) if $\overline{F}^L$ were zero, or it would follow approximately if

$$|\overline{F}^L| \ll |\mathcal{F}|. \tag{11.20}$$

It is possible to demonstrate that (11.20) holds in general (i.e., even for the compressible Euler equations), but only for momentum-conserving forces. Such forces derive from a momentum flux tensor as

$$F_i = \frac{1}{\rho}\Pi_{ij,j} \quad \text{and therefore} \quad \overline{F}^L_i = \frac{1}{\tilde{\rho}}(\overline{\Pi^\xi_{ik}K_{kj}})_{,j}. \tag{11.21}$$

The explicit presence of a mean divergence makes clear that (11.20) will hold for slowly varying wavetrains. This is because $\mathcal{F}$ as defined in (11.7) can not be written in terms of such a mean divergence and therfore the assumed scale separation appears in the relative magnitudes of $\overline{F}^L$ and $\mathcal{F}$.

But even without a scale separation the net integral of $\tilde{\rho}\overline{F}^L$ over a compact dissipation region is zero by the assumed flux form of $\overline{F}^L$. This contrasts with a non-zero integral for $\tilde{\rho}\mathcal{F}$ in the same situation. In this sense (11.20) holds if applied to the integral of the forces over the dissipation region. Overall, we conclude that the dissipative pseudomomentum rule holds for momentum-conserving dissipative forces, such as those arising from the viscous terms in the Navier–Stokes equations. On the other hand, if dissipation is modelled by body forces that are not momentum-conserving (e.g., via Rayleigh friction such that $F = -\alpha u$) then the pseudomomentum rule usually fails.

This point is important for numerical modelling, where small-scale dissipation is sometimes modelled using ad hoc damping terms, especially near domain boundaries. Lack of momentum conservation can then yield mean-flow responses that are physically unrealistic, and may be significantly in error over long integration times.

11.1.2 Pseudomomentum with vertical shear

The GLM equations derived above apply without modifications to shear flows, as no assumptions were made concerning the structure of $\overline{u}^L$. Thus there are no new fundamental results here, but it is of interest to note the small-amplitude version of Lagrangian pseudomomentum p in the presence of an $O(1)$ basic flow $U(z)$ with shear, say. This computation serves as a typical model for similar small-amplitude situations. The important step is that

$$u^\ell = u' + \zeta U_z, \quad v^\ell = v', \quad \text{and} \quad w^\ell = w' \tag{11.22}$$

at leading order, which matters for the definition of p. The other manipulations follow from incompressibility and integration by parts such that at $O(a^2)$ one obtains (using subscripts without comma to denote derivatives here)

$$\begin{aligned} \mathsf{p} &= -\overline{\xi_x u'} - \overline{\eta_x v'} - \overline{\zeta_x w'} - \overline{\xi_x \zeta} U_z \\ &= (\overline{\eta u'})_y + (\overline{\zeta u'})_z + \overline{\eta(v'_x - u'_y)} - \overline{\zeta(u'_z - w'_x)} - \overline{\xi_x \zeta} U_z. \end{aligned} \tag{11.23}$$

The vertical disturbance vorticity $v'_x - u'_y$ is not affected by the linear buoyancy terms and therefore it is non-zero only due to tilting of the basic state vorticity $U_z\widehat{\boldsymbol{y}}$. Indeed, time-integrating the linear vorticity equation yields

$$\mathrm{D}_t(v'_x - u'_y) = U_z w'_y \quad \Rightarrow \quad v'_x - u'_y = U_z\zeta_y. \tag{11.24}$$

Inserting this back in (11.23) leads to

$$\begin{aligned} \mathsf{p} &= (\overline{\eta u'})_y + (\overline{\zeta u'})_z + \overline{\eta\zeta_y}U_z - \overline{\zeta(u'_z - w'_x)} + \overline{(\eta_y + \zeta_z)\zeta}U_z \\ &= (\overline{\eta u'} + \overline{\eta\zeta}U_z)_y + (\overline{\zeta u'})_z - \overline{\zeta(u'_z - w'_x)} + \overline{\zeta_z\zeta}U_z \\ \Rightarrow \quad \mathsf{p} &= (\overline{\eta u^\ell})_y + (\overline{\zeta u'} + \frac{1}{2}\overline{\zeta^2}U_z)_z - \overline{\zeta(u'_z - w'_x)} - \frac{1}{2}\overline{\zeta^2}\,U_{zz}. \end{aligned} \tag{11.25}$$

The last two terms are the essential terms. If present, then stratification enters via the linear evolution equation for the y-vorticity, namely

$$\mathrm{D}_t(u'_z - w'_x + \zeta U_{zz}) = N^2\zeta_x. \tag{11.26}$$

So, without stratification, $u'_z - w'_x = -\zeta U_{zz}$ and the last two terms in (11.25) combine to the purely shear-related pseudomomentum $+\frac{1}{2}\overline{\zeta^2}U_{zz}$ noted before. On the other hand, with stratification there is a competition between the sign-definite last term and the penultimate term, which can take either sign, and which dominates in the usual ray tracing regime.

Finally, using $b' = -\zeta N^2$ we can also write down the corresponding Eulerian pseudomomentum as the sum of the two essential terms:

$$\tilde{\mathsf{p}} = \frac{1}{N^2}\overline{b'(u'_z - w'_x)} - \frac{1}{2N^4}\overline{b'^2}\,U_{zz}. \tag{11.27}$$

Here the appearance of N is somewhat misleading because there is no real dependence on stratification.

The insensitivity with respect to the number of spatial dimensions of the generic GLM expression for pseudomomentum is in contrast with Hamiltonian theories for conserved wave activity measures in Eulerian variables, which tend to rely significantly on intermediate steps that exploit certain aspects of the dimensionality of the problem. Still, at small amplitude it is usually possible to show that the Eulerian pseudomomentum based on Hamiltonian methods and the GLM pseudomomentum agree in their essential terms.

11.2 Rotating Boussinesq equations on an f-plane

We consider the case of constant vertical rotation $\boldsymbol{f} = f_0\widehat{\boldsymbol{z}}$ in Cartesian xyz-coordinates with zonal periodicity, which is the definition of f-plane dynamics. The circulation theorem can be written in different equivalent forms, depending on the gauge used for the vector potential $\boldsymbol{A}$ in (10.167).

For zonal averaging in Cartesian coordinates it is most convenient to use the zonally symmetric potential (10.171), and the corresponding circulation theorem is

$$\frac{d}{dt}\oint_{\mathcal{C}} \boldsymbol{u}\cdot d\boldsymbol{x} - f_0 y\,dx = \oint_{\mathcal{C}} \boldsymbol{F}\cdot d\boldsymbol{x}. \tag{11.28}$$

The rotating PV is independent of the gauge:

$$q = (\boldsymbol{\nabla}\times\boldsymbol{u} + f_0\hat{\boldsymbol{z}})\cdot\boldsymbol{\nabla}S \quad\Rightarrow\quad \frac{Dq}{Dt} = (\boldsymbol{\nabla}\times\boldsymbol{F})\cdot\boldsymbol{\nabla}S. \tag{11.29}$$

The corresponding mean circulation theorem is

$$\frac{d}{dt}\oint_{\mathcal{C}} (\overline{\boldsymbol{u}}^L - \mathsf{p})\cdot d\boldsymbol{x} - f_0 y\,dx = \oint_{\mathcal{C}} (\overline{\boldsymbol{F}}^L - \boldsymbol{\mathcal{F}})\cdot d\boldsymbol{x} \tag{11.30}$$

with the gauged pseudomomentum vector defined as in (10.177). As noted before, this gauged pseudomomentum differs from the standard pseudomomentum (10.152) only by the gradient of a mean field, which implies that the zonal components are equal if zonal averaging is used. The corresponding mean PV is defined by

$$\overline{q}^L = \frac{1}{\tilde{\rho}}(\boldsymbol{\nabla}\times(\overline{\boldsymbol{u}}^L - \mathsf{p}) + f_0\hat{\boldsymbol{z}})\cdot\boldsymbol{\nabla}\overline{S}^L \tag{11.31}$$

and satisfies (11.8b).

The Lagrangian-mean equations (11.4) are unchanged except for the mean Coriolis term $\boldsymbol{f}\times\overline{\boldsymbol{u}}^L$ in the momentum equation; in particular, $\overline{\mathrm{D}}^L\overline{b}^L + N^2\overline{w}^L = 0$ still holds and the zonal components of the zonally averaged momentum and pseudomomentum equations are

$$\overline{\mathrm{D}}^L\overline{u}^L - f_0\overline{v}^L + \frac{1}{\tilde{\rho}}(\overline{P^\xi K_{1m}})_{,m} = \overline{F}^L \quad\text{and}\quad \overline{\mathrm{D}}^L\mathsf{p} + \frac{1}{\tilde{\rho}}(\overline{P^\xi K_{1m}})_{,m} = \mathcal{F}, \tag{11.32}$$

respectively, and therefore

$$\overline{\mathrm{D}}^L\overline{u}^L - f_0\overline{v}^L = \overline{\mathrm{D}}^L\mathsf{p} + \overline{F}^L - \mathcal{F}. \tag{11.33}$$

11.2.1 Residual and Lagrangian-mean circulations

In GLM theory there are no wave-forcing terms in the exact mean buoyancy equation $\overline{\mathrm{D}}^L\overline{b}^L + N^2\overline{w}^L = 0$, which is a notable difference compared to the Eulerian-mean buoyancy equation. Indeed, the definition of the residual meridional circulation in (8.49) was motivated in part by the desire to eliminate the disturbance terms in the Eulerian-mean buoyancy equation (8.48). The exact absence of such terms in the Lagrangian-mean buoyancy equation suggests that there is a link between the residual and the Lagrangian-mean circulations. Working to $O(a^2)$ accuracy it is straightforward to show that

$$\overline{v}^* - \overline{v}^L = -\frac{1}{2}(\overline{\eta^2})_{ty} \quad \text{and} \quad \overline{w}^* - \overline{w}^L = -\frac{1}{2}(\overline{\zeta^2})_{tz} - (\overline{\eta\zeta})_{ty}. \tag{11.34}$$

Therefore the residual and Lagrangian-mean circulations in the meridional plane are approximately equal for steady disturbances.

11.2.2 EP flux in GLM theory

Another important point of comparison is that no further manipulations were needed in GLM theory to bring the zonal momentum equation into a form where the time derivative of the zonal pseudomomentum appeared naturally. In Eulerian-mean theory, on the other hand, one either had to use the residual circulations or to restrict attention to the forcing of the mean PV mode in order to achieve the same result.

These issues are related to the definition of the EP flux vector for zonally averaged flows, which by definition is the meridional-plane flux of zonal pseudomomentum. In Eulerian-mean theories this flux vector takes a variety of forms depending on the particular flow model under consideration, and it also depends on the presence of Coriolis forces. For example, in the rotating Boussinesq equations the EP flux was given to $O(a^2)$ accuracy by (8.26), i.e.,

$$\boldsymbol{B} = \left(0, \overline{u'v'}, \overline{u'w'} - \frac{f_0}{N^2}\overline{b'v'}\right) \tag{11.35}$$

On the other hand, in GLM theory the EP flux is always given by the pressure-related meridional flux of zonal momentum, because of the generic fact that these flux components are equal for zonal momentum and zonal pseudomomentum. Thus the corresponding exact expression for the EP flux in GLM theory is always the generic

$$\boldsymbol{B} = \left(0, \overline{p^\xi K_{12}}, \overline{p^\xi K_{13}}\right) = -\left(0, \overline{p^\xi \xi_{j,1} K_{j2}}, \overline{p^\xi \xi_{j,1} K_{j3}}\right), \tag{11.36}$$

which uses the identity (10.73). In other words, the EP flux is *always* the pressure-related flux of zonal momentum across undulating material surfaces.

11.2.3 Rotating vertical slice model in GLM theory

We conclude the material on GLM theory and zonal averaging by reconsidering the rotating vertical slice model of §8.4, which was defined as a two-dimensional Boussinesq model in the xz-plane with Coriolis forces. As discussed, this model is special because it does not have a non-trivial vortical mode that can be forced by zonal forces and it also does not have a meridional circulation. Both aspects cause problems for Eulerian-mean theory.

However, these problems do not arise in GLM theory. Indeed, the zonal momentum equation is given by (11.32) with $m = 3$ only, which implies the usual statement about the EP flux. Also, the meridional momentum equation takes the simple yet exact form

$$\frac{Dv}{Dt} + f_0 u = 0 \quad \Rightarrow \quad \overline{\mathrm{D}}^L \overline{v}^L + f_0 \overline{u}^L = 0. \tag{11.37}$$

The absence of any disturbance terms in this equation is consistent with the existence of the second material invariant in (8.56).

11.3 Langmuir circulations and Craik–Leibovich instability

The GLM equations are particularly well suited for the study of local wave–vortex interactions, a point that will be amplified in part THREE below. For example, such interactions are thought to underpin the celebrated *Langmuir circulations* in the ocean, which is the name given to the ubiquitous arrays of parallel horizontal vortex rolls near the ocean surface. Typically, these vortex rolls are aligned with the prevailing direction of the wind and the surface-wave propagation. The conspicuous advection of surface tracers by the vortex rolls can make them easily visible in aerial photographs, as so-called "wind rows".

The key mechanism for their formation is believed to be the *Craik–Leibovich instability*, which describes a robust nonlinear interaction between surface waves and near-surface vortices. In the simplest model of this instability one assumes that the wave field is unaffected by the weak vortex rolls and hence all wave properties are assumed to be time-independent. It is then simple to investigate the instability using the GLM equations. This simplicity contrasts with the original, more complicated, Eulerian-mean analysis first published by Craik and Leibovich.

In fact, we will investigate the instability from two complementary points of view: first, making full use of zonal averaging along the wave propagation direction, we will derive the instability criterion via the definition of an effective stratification induced by the waves. This is a very nice result, but it is essentially restricted to zonal averaging. In a second approach we will look at the instability from a three-dimensional vortex dynamics perspective, which will make the robust nature of the instability readily apparent. In this perspective the simplicity of the GLM approach stems from the generic fact that $\boldsymbol{\nabla} \times (\overline{\boldsymbol{u}}^L - \mathsf{p})$ is advected by $\overline{\boldsymbol{u}}^L$. Finally, we will note the Eulerian version of the mean-flow equations, which are the ones that are used in computational practice, but whose derivation is still simplest, by far, via the GLM route.

11.3.1 GLM theory for wave-induced stratification

Our fluid model is a Boussinesq system bounded in the vertical by a free surface with rest position at $z = 0$. It is convenient to allow for a uniform background stratification with $N \geq 0$. We align the x-direction with the propagation direction of the surface waves such that in the basic state the steady pseudomomentum is

$$\mathsf{p} = \mathsf{p}_0 = \mathsf{p}_0(z)\,\widehat{\boldsymbol{x}} \tag{11.38}$$

for some suitable zonal profile $\mathsf{p}_0(z) \geq 0$. For example, in the case of irrotational deep-water surface waves with horizontal wavenumber $k > 0$ the relevant profile would be $\mathsf{p}_0 \propto \exp(2kz)$. By assumption, we freeze the pseudomomentum to its basic state, i.e, we assume $\mathsf{p} = \mathsf{p}_0(z)$ throughout. For the mean flow we assume that

$$\overline{\boldsymbol{u}}^L = \overline{\boldsymbol{u}}_0^L + \overline{\boldsymbol{u}}_1^L = \overline{u}_0^L(z)\,\widehat{\boldsymbol{x}} + \overline{\boldsymbol{u}}_1^L \quad \text{and} \quad \overline{b}^L = \overline{b}_1^L. \tag{11.39}$$

Here $\overline{u}_0^L(z)$ is a steady basic zonal flow profile and $\overline{\boldsymbol{u}}_1^L$ is a small deviation therefrom. There is no mean buoyancy disturbance in the basic state. A useful shorthand is

$$R(z) = \overline{u}_0^L(z) - \mathsf{p}_0(z) \quad \text{such that} \quad \boldsymbol{\nabla} \times (\overline{\boldsymbol{u}}_0^L - \mathsf{p}_0) = R'(z)\,\widehat{\boldsymbol{y}}. \tag{11.40}$$

In the special case of strictly irrotational flow (10.100) holds, which implies $\overline{u}_0^L(z) = \mathsf{p}_0(z)$, but we will not restrict to this case. In accordance with the assumption of steady waves (cf. § 10.2.5) we ignore the divergence effect, i.e., we assume that

$$\boldsymbol{\nabla} \cdot \overline{\boldsymbol{u}}^L = \boldsymbol{\nabla} \cdot \overline{\boldsymbol{u}}_1^L = 0. \tag{11.41}$$

Linearizing the GLM equations for small $\overline{\boldsymbol{u}}_1^L$ and $\overline{b}_1^L$ and using $\mathrm{D}_t = \partial_t + \overline{u}_0^L \partial_x$ yields the GLM buoyancy equation

$$\mathrm{D}_t \overline{b}_1^L + N^2 \overline{w}_1^L = 0 \tag{11.42}$$

and the GLM vorticity equation (cf. 11.12)

$$\mathrm{D}_t \boldsymbol{\nabla} \times \overline{\boldsymbol{u}}_1^L + \overline{w}_1^L R''(z)\,\widehat{\boldsymbol{y}} - R'(z)\overline{\boldsymbol{u}}_{1,y}^L - (\overline{v}_{1,x}^L - \overline{u}_{1,y}^L)\overline{u}_{0,z}^L\,\widehat{\boldsymbol{x}} = \boldsymbol{\nabla}\overline{b}_1^L \times \widehat{\boldsymbol{z}}. \tag{11.43}$$

Together, (11.41-11.43) are the governing equations for $(\overline{b}_1^L, \overline{\boldsymbol{u}}_1^L)$ for any choice of averaging. We now restrict to zonal averaging, which implies that the x-derivatives of all mean fields are zero. The z-component of (11.43) then yields

$$\partial_t \overline{u}_{1,y}^L + R'(z)\overline{w}_{1,y}^L = 0, \quad \text{or} \quad \partial_t \overline{u}_1^L + R'(z)\overline{w}_1^L = 0 \tag{11.44}$$

after ignoring integration constants. Of course, this is an approximate version of the exact circulation theorem in the form (10.93), $\overline{\mathrm{D}}^L \overline{u}^L = \overline{\mathrm{D}}^L \mathsf{p}$. Comparison with (11.42) makes obvious that $\overline{u}_1^L$ is simply advected across a mean gradient given by $R'(z)$.

The mean-flow dynamics in the yz-plane follows from combining (11.41) with the x-component of (11.43), which after some cancellations yields

$$\overline{v}_{1,y}^L + \overline{w}_{1,z}^L = 0 \quad \text{and} \quad \partial_t(\overline{w}_{1,y}^L - \overline{v}_{1,z}^L) = \tilde{b}_{1,y} \tag{11.45}$$

where the effective mean buoyancy disturbance $\tilde{b}_1$ is defined by

$$\tilde{b}_1 = \left(\overline{b}_1^L - \mathsf{p}_0'(z)\overline{u}_1^L\right). \tag{11.46}$$

Now, the crucial step is to notice that because $\partial_t \mathsf{p}_0 = 0$ one can multiply (11.44b) by p_0' in order to obtain the evolution law for $\tilde{b}_1$ as

$$\partial_t \tilde{b}_1 + \tilde{N}^2(z)\,\overline{w}_1^L = 0 \quad \text{with} \quad \tilde{N}^2(z) = N^2 - \mathsf{p}_0'(z)R'(z). \tag{11.47}$$

Hence the area-preserving mean flow in the yz-plane described by $(\tilde{b}_1, \overline{v}_1^L, \overline{w}_1^L)$ follows the usual Boussinesq dynamics, but with an effective stratification $\tilde{N}$ in (11.47b) that has been modified by the presence of the waves. This is the sole influence of the frozen wave field on the mean flow.

It is obvious that the stability of the linearized mean flow will depend on the sign of $\tilde{N}^2 = N^2 - \mathsf{p}_0'(z)R'(z)$. In particular, if $N = 0$ then the flow will be unstable to overturning motions whenever $\mathsf{p}_0'(z)R'(z)$ is *positive* in some region. This is the Craik–Leibovich instability criterion.

In the case of evanescent surface waves $\mathsf{p}_0(z) \geq 0$ robustly decreases with depth and therefore the instability criterion reduces to $R'(z) > 0$. The physical interpretation of this criterion follows from (11.40b), namely for instability the basic flow must have positive vorticity in the y-direction. The case of strictly irrotational flow corresponds to $R = 0$ and hence to zero effective stratification and neutral stability. More important is the case where the waves have been subject to prior dissipation and wave breaking, which creates spanwise vorticity in the water by robust processes described in § 13.4 below. The upshot of these considerations is that for wind-generated surface waves propagating in the x-direction the spanwise vorticity created by wave

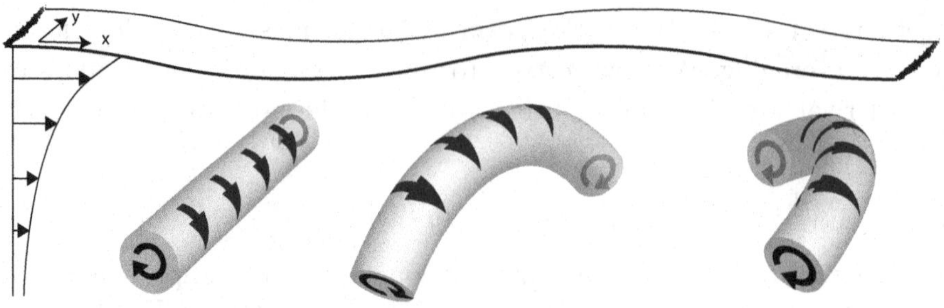

Figure 11.1 Advection of a vortex tube below the waves by the pseudomomentum contribution to the zonal mean flow, which is $\overline{u}^L = \mathsf{p}_0(z)$. Left: unperturbed flat tube with positive y-vorticity, rolling clockwise. Middle: perturbed tube with vertically raised centre. Right: exaggerated view of perturbed tube advected by the zonal flow. The induced velocity by this sheared-over vortex tube reinforces the rising motion in the centre, which is the positive feedback mechanism of the instability. Conversely, this feedback becomes negative if the direction of the basic vorticity is reversed, which explains the instability criterion.

dissipation and breaking will indeed point in the positive y-direction and hence $R'(z)$ will be positive and the mean flow will therefore be robustly unstable to overturning in the yz-plane. This generic state of affairs explain the ubiquity of Langmuir circulations.

11.3.2 Alternative view using vortex dynamics

The notion of a wave-induced effective stratification for the two-dimensional zonally averaged mean flow in the yz-plane is elegant, but it does not generalize to three-dimensional mean flows defined via averages over the rapid time scale of the waves, say. The situation is perhaps akin to the effective sideways stratification induced by Coriolis forces in a two-dimensional xz-model (cf. §8.4), which also did not generalize to three dimensions. Arguably, a more robust understanding of the instability can be obtained by considering the three-dimensional vortex dynamics below the free surface as depicted in figure 11.1. First off, in the absence of stratification and viscous effects the three-dimensional vortex lines are materially advected in the sense that

$$\frac{D}{Dt}(\boldsymbol{\nabla} \times \boldsymbol{u}) = [(\boldsymbol{\nabla} \times \boldsymbol{u}) \cdot \boldsymbol{\nabla}]\boldsymbol{u} \tag{11.48}$$

holds for incompressible flow. The velocity field $\boldsymbol{u}$ is then the sum of a vortical and a wave part. The vortical part is the inversion of $\boldsymbol{\nabla} \times \boldsymbol{u}$ via the standard Biot–Savart law using a no-normal flow boundary condition at $z = 0$, and the wave part is an irrotational flow associated with the motion of the free surface. Now, the analogue of (11.48) for a three-dimensional incompressible Lagrangian-mean flow is (10.99), i.e.,

$$\overline{\mathrm{D}}^L(\boldsymbol{\nabla} \times (\overline{\boldsymbol{u}}^L - \mathbf{p})) = [(\boldsymbol{\nabla} \times (\overline{\boldsymbol{u}}^L - \mathbf{p})) \cdot \boldsymbol{\nabla}]\overline{\boldsymbol{u}}^L. \tag{11.49}$$

As said before, this means that $\boldsymbol{\nabla} \times (\overline{\boldsymbol{u}}^L - \mathbf{p})$ is advected by $\overline{\boldsymbol{u}}^L$. The Lagrangian-mean flow satisfies a no-normal flow boundary condition at $z = 0$ to good approximation, so $\overline{\boldsymbol{u}}^L$ contains only a vortical part, but the dynamics of this vortical part is mixed up with the pseudomomentum of the waves. This is the key for understanding the wave–vortex interactions here.

In general, in order to compute $\overline{\boldsymbol{u}}^L$ from the advected field $\boldsymbol{\nabla} \times (\overline{\boldsymbol{u}}^L - \mathbf{p})$ one inverts the latter using the Biot–Savart law and then adds a contribution equal to the projection of $\mathbf{p}$ onto the space of non-divergent vector fields that satisfy a no-normal flow condition at $z = 0$. In the present case $\mathbf{p} = \mathbf{p}_0$ is horizontal and $\boldsymbol{\nabla} \cdot \mathbf{p}_0 = 0$, which means one simply adds $\mathbf{p}_0$ to the Biot–Savart inversion of $\boldsymbol{\nabla} \times (\overline{\boldsymbol{u}}^L - \mathbf{p})$. The upshot is that vortex tubes formed by $\boldsymbol{\nabla} \times (\overline{\boldsymbol{u}}^L - \mathbf{p})$ are advected by a vertically sheared zonal mean flow equal to $\mathbf{p}_0(z)$ and also by the kind of self-induced velocity field that is familiar from ordinary three-dimensional vortex dynamics.

We can now gain an understanding of the mean-flow instability from the point of view of this kind of vortex tube dynamics. For example, in the basic state we consider a straight spanwise vortex tube pointing in the positive y-direction, as suggested by the instability criterion. This is depicted in the first panel in figure 11.1, and the second panel shows this basic vortex tube in a slightly perturbed state with an initial undulation that raises the tube in the vertical. The differential advection of this raised tube by the strongly sheared pseudomomentum component of the zonal mean flow will cause the elevated parts of the vortex tube to be advected more quickly in x than the lowered parts of it, and therefore the tube will be deformed into a tilted shape as indicated in the third panel.

Crucially, the induced velocity by this tilted and curved vortex tube will point *upwards* in the centre region of the tube, as is obvious from a comparison to the induced velocity of a three-dimensional vortex ring. This reinforces the original vertical undulation of the basic vortex tube and leads to instability. This positive feedback loop is the basic mechanism of the Craik–Leibovich instability from a vortex dynamics perspective. Conversely, if the basic vorticity points in the negative y-direction then the same construction would lead to a *downward* vertical velocity in the centre region and therefore to a negative feedback, and stable vortex tube oscillations. Of course, these conclusions are consistent with those of the previous effective stratification computation, but it is remarkable how different a perspective is offered by these two complementary views of the same physical instability.

11.3.3 Other effects and Eulerian-mean equations

Over long times the initial instability may saturate against viscous and nonlinear effects that have been ignored in our analysis, but which can be accommodated under suitable scaling assumptions. Specifically, if both p_0 and $\overline{u}_0^L$ are $O(a^2)$ in terms of a small wave amplitude $a \ll 1$, then it is possible to augment the equations with additional terms that allow capturing this saturation and subsequent slow evolution of the mean flow. In this case it is clear by inspection of the effective stratification in (11.47) that the instability growth rates will be $O(a^2)$, which suggests a slow time $T = a^2 t$ for the entire mean-flow evolution. It is then consistent to allow for $\overline{\boldsymbol{u}}_1^L = O(a^2)$ as well, in which case the previously neglected nonlinear advection terms such as $(\overline{\boldsymbol{u}}_1^L \cdot \boldsymbol{\nabla})\overline{\boldsymbol{u}}_1^L$ are comparable to the linear terms in the governing equations. Hence these nonlinear terms must be added to the equations, by replacing D_t by $\mathrm{D}_t + (\overline{\boldsymbol{u}}_1^L \cdot \boldsymbol{\nabla})$, among other things. Similar arguments apply to suitable scaled viscous terms as well.

Finally, in practice Eulerian-mean equations are used for the numerical study of Langmuir circulations. These equations are easily obtained in the case of steady surface waves by noting once more from (10.17) that pseudomomentum and Stokes drift are identical at leading order in this special case, and therefore the substitution $\overline{\boldsymbol{u}}^L = \overline{\boldsymbol{u}} + \overline{\boldsymbol{u}}^S = \overline{\boldsymbol{u}} + \mathsf{p}$ readily produces equations for the Eulerian mean flow $\overline{\boldsymbol{u}}$. For example, applying this recipe to (11.49) produces

$$(\partial_t + \overline{\boldsymbol{u}} \cdot \boldsymbol{\nabla})(\boldsymbol{\nabla} \times \overline{\boldsymbol{u}}) - [(\boldsymbol{\nabla} \times \overline{\boldsymbol{u}}) \cdot \boldsymbol{\nabla}]\overline{\boldsymbol{u}} = \boldsymbol{\nabla} \times \boldsymbol{F}, \tag{11.50}$$

where the Eulerian mean flow feels the waves solely via the effective force

$$\boldsymbol{F} = \overline{\boldsymbol{u}}^S \times (\boldsymbol{\nabla} \times \overline{\boldsymbol{u}}). \tag{11.51}$$

This brings a third complementary point of view to the same problem: the Eulerian mean flow feels the effective force (11.51), the Lagrangian mean velocity feels the effective stratification in (11.47), and the Lagrangian mean vorticity feels the additional effective advection by p_0. All points of view offer valuable insights into the nature of the problem, though it is tempting to speculate that for applications other than surface waves the Lagrangian-mean equations might in fact be easier to use, if only because in general $\mathsf{p} \neq \overline{\boldsymbol{u}}^S$ even at leading order. The Lagrangian-mean derivation also makes obvious why in a rotating frame a term $\overline{\boldsymbol{u}}^S \times \boldsymbol{f}$ must be added to the effective force in (11.51).

11.4 Notes on the literature

The application of GLM theory to the Boussinesq equations is described in Bühler and McIntyre (1998), and GLM theory applied to the dissipative pseudomomentum rule and its link to momentum-conserving forces is discussed in Bühler (2000).

The importance of global momentum conservation has long been recognized in atmospheric dynamics, with particular emphasis given to the conservation of angular momentum around the Earth's rotation axis. In particular, the importance of the angular momentum budget for global-scale wave–mean interaction theory involving the residual circulation is described in Haynes et al. (1991) and Andrews et al. (1987), and a more recent account of the importance of this issue for numerical modelling is presented in Shepherd and Shaw (2004).

The Hamiltonian derivation of three-dimensional Eulerian pseudomomentum for small-amplitude waves in the presence of vertical shear is given in the recent Shaw and Shepherd (2008), which can be compared to the results in §11.1.2 here.

The original Craik–Leibovich theory is summarized in §13.2 of Craik (1985) and the contemporary knowledge of Langmuir circulations around that time is reviewed in Leibovich (1983); a more recent review is Thorpe (2004). The relationship between Eulerian and Lagrangian versions of the relevant mean-flow equations that capture the instability was described in Leibovich (1980), where alternative theories for Langmuir circulations were also discussed.

PART THREE

WAVES AND VORTICES

12

A framework for local interactions

We now go beyond classic wave–mean interaction theory, which was based on simple geometry and zonal averaging. Specifically, our aim is to formulate a framework that can be used to investigate the interactions between small-scale waves and large-scale mean flows, where these scale distinctions are based on a local scale separation of the kind familiar from WKB theory and ray tracing. In other words, we want to allow for slowly varying mean flows, but without the classic restriction to zonally symmetric mean flows.

This seemingly harmless extension of the theory leads to profound mathematical and physical complications. For instance, in classic wave–mean interaction theory we made countless use of the key identity

$$(\overline{A})_x = \overline{A_x} = 0 \tag{12.1}$$

for arbitrary zonally periodic flow fields A. Indeed, mean pressure gradients in the zonal direction played no role because $\overline{p}_x = 0$ held by assumption. Mean pressure gradients in the other directions could be nonzero, of course, but their role was suppressed as much as possible by focusing on the zonal mean flow. This focus on the zonal mean flow went hand-in-hand with a focus on the zonal pseudomomentum, which is conserved by assumption in simple geometry. The non-acceleration conditions and other results then followed more or less directly from the consideration of Kelvin's circulation along material contours that traverse the periodic domain.

Yet, when a zonally symmetric mean flow is replaced by a slowly varying mean flow, the picture changes significantly: the zonal mean flow is affected by the mean pressure field and also the zonal pseudomomentum is not necessarily conserved anymore. Indeed, all components of the pseudomomentum vector can now change by refraction, and this will be discussed in depth in §14. However, even in situations in which pseudomomentum is still conserved the mean pressure field alone can indeed lead to significant mean-flow changes, and several examples of this will be given in §12.2 below. What is remarkable in these examples is that the mean flow in a slowly varying version of the problem is different at *leading order* from the mean flow in a zonally symmetric version of the same problem. Basically, as discussed in §12.1, moving away from simple geometry and zonal averaging represents a singular perturbation of the problem, whose physical roots lie in the mean pressure field.

What, then, are the requirements for a theoretical framework designed to capture local wave–mean interactions? First off, it makes sense that we must treat both the velocity field and the pseudomomentum field as vector fields, i.e., we cannot single out a specific component of these fields for primary consideration. Secondly, the special attention given to steady shear flows as prototypical basic flows should be shifted to considering slowly evolving vortex flows as basic states, which occur naturally in three-dimensional and two-dimensional fluid dynamics.

Thirdly, and most importantly, we must disentangle mean pressure effects from other interaction effects that depend intrinsically on the vorticity of the flow, as will be illustrated in §12.3. Indeed, it is only by separating the pressure-like and vortical aspects of the mean-flow response that we can obtain interaction equations that are once again simple enough to be interpreted in terms of a suitable extension of the pseudomomentum rule and so on.

We find it indispensable in this context to focus on the vorticity description of the flow rather than on its momentum description. This is where GLM theory is helpful, because it allows us to keep track of Kelvin's circulation theorem and of vorticity for the mean flow. In particular, we will use GLM theory in §12.4 to formulate an approximate conservation law for the sum of the waves' pseudomomentum and of the mean-flow impulse defined via the Lagrangian-mean potential vorticity of the flow. Arguably, this conservation law is the central statement in the description of interactions between small-scale waves and large-scale vortices.

12.1 A geometric singular perturbation

We begin by pointing out the simple mathematical reason why wave–mean interaction theory for slowly varying mean flows is a singular perturbation to the classic wave–mean interaction theory for zonally symmetric mean flows. For example, under local averaging the equation governing the zonal component of the Eulerian-mean flow for the two-dimensional Boussinesq system as discussed in §6.5 is

$$\overline{u}_t + (\overline{u}\,\overline{u})_x + (\overline{u}\,\overline{w})_z + \overline{p}_x = -(\overline{u'u'})_x - (\overline{u'w'})_z. \tag{12.2}$$

This compares with the earlier

$$\overline{u}_t = -(\overline{u'w'})_z \tag{12.3}$$

that held under zonal averaging. Obviously, these equations differ by the increased dimensionality of the mean flow, because $\overline{w} = 0$ under zonal averaging but not otherwise. Still, the additional term $(\overline{u}\,\overline{w})_z$ would be negligible at leading order in the simplest case of small-amplitude waves relative to a state of rest. Also, the equations differ by the additional Reynolds stress term capturing the zonal flux of zonal momentum. Still, this additional term is a wave property and hence could be computed at leading order from the linear wave solution.

This leaves the mean zonal pressure gradient as the remaining difference. In general, the mean pressure is *not* a wave property, so it cannot be computed at leading order from the linear solution. Also, and this is the heart of the matter, it is not negligible even for slowly varying wavetrains relative to a basic flow at rest. Specifically, if $a \ll 1$ is the wave amplitude and $\epsilon \ll 1$ is a second small parameter that measures the slowness of the wavetrain envelope, then by construction all mean-flow derivatives related to the mean-flow response are $O(\epsilon a^2)$. This includes the mean-flow acceleration and also the mean pressure gradient, i.e.,

$$\overline{u}_t = O(\epsilon a^2) \quad \text{and} \quad \overline{p}_x = O(\epsilon a^2) \tag{12.4}$$

both hold. Now, the key point is that by definition a slowly varying wavetrain has a spatial footprint of size $O(\epsilon^{-1})$ and moves with an $O(1)$ velocity, such as the usual group velocity. This means that a slowly varying wavetrain travels a distance comparable to its own size over a long time $O(\epsilon^{-1})$. Thus, the mean pressure gradient is weak, but it acts over a long time. Indeed, over such a long time the zonal mean-flow acceleration due to the mean pressure gradient can lead to changes in $\overline{u}$ at $O(a^2)$, i.e., at leading order in wave amplitude and *independent* of the scale separation parameter ϵ.

Of course, whether or not the mathematical possibility of such changes are realized in a physical example depends on the geometry of the problem. This is illustrated in the next section.

12.2 Examples of mean pressure effects

We look at some well-known examples of non-local mean pressure effects that arise in wave–mean interaction problems with localized wavetrains. In these examples significant changes in the leading-order mean-flow response arise when a seemingly harmless detail of the set-up is changed. These effects are real, and can be measured in suitable arranged experiments.

12.2.1 Mean-flow response to acoustic wavetrain

We use the polytropic fluid model as described in §1.2.3, which was written such that the polytropic exponent γ and the linear wave speed c_0 could be varied independently. First, we look at one-dimensional flow in the x-direction and consider linear waves. Beginning with a state at rest at $t = 0$, we imagine the waves being generated by a piston that starts oscillating with a certain frequency and small amplitude $O(a)$ around its mean position at $x = 0$, say. This generates non-dispersive acoustic waves that propagate into the region $x > 0$ with linear sound speed c_0. The whole problem belongs to the class of simple wave problems, which involve one-dimensional propagation of acoustic waves into undisturbed regions and which can in fact be solved exactly using Riemann's method of characteristics.

However, we are only interested in the mean-flow response at $O(a^2)$, where the local averaging operation can be defined as a running time-average over a time that is long compared to the oscillation period. It is not critical whether the start-up of the piston is slow or not, but for simplicity we can imagine that the piston oscillation is slowly varying itself, so the entire wave field is a textbook example of a slowly varying wavetrain confined between $x = 0$ and the time-dependent location of the wave front[1] at $x = c_0 t$.

We could solve the $O(a^2)$ mean-flow response problem exactly, but the set-up is simple enough that we can extract information about the salient features within the wavetrain by inspection. To begin with, because the piston oscillation around a fixed location $x = 0$ there can be no mean particle drift within the wavetrain, i.e., $\overline{u}^L = 0$ exactly there. This implies that

$$\overline{u}_2^L = 0 \quad \text{and} \quad \overline{u}_2 = \overline{u}_2^L - \overline{u}_2^S = -\frac{E}{c_0} \tag{12.5}$$

[1] It is part of the nonlinear nature of non-dispersive acoustic waves that simple waves will steepen and form a shock in finite time. In small-amplitude theory the nonlinear steepening manifests itself in a secularly growing mean-flow response travelling with the wave front, but we will neglect this here. Still, it is interesting to note that for sinusoidal waves with amplitude $a \ll 1$ the propagation distance until shock formation is $1/(\pi a(\gamma + 1))$ wavelengths. For example, a shallow-water wave with $\gamma = 2$ and $a = 0.1$ will break after just one wavelength of propagation. The peculiar case $\gamma = -1$ is an exception: there is no wave steepening for this value.

at $O(a^2)$, where

$$E = \overline{u_1' u_1'} = c_0^2 \frac{\overline{\rho_1' \rho_1'}}{\rho_0^2} \tag{12.6}$$

is the mean disturbance energy per unit mass. This uses $\overline{u}_2^S = \mathsf{p}_2 = E/c_0$ (cf. §10.1.2). Notably, (12.5) applies at $x = 0$ as well, so there is a *negative* Eulerian-mean velocity at the rest position of the piston.

How about the Eulerian-mean density change $\overline{\rho}_2$? Here we can use an exact fact from Riemann's theory for simple waves, namely that the nonlinear algebraic relation

$$\frac{\gamma - 1}{2} u + c_0 = c_0 \left(\frac{\rho}{\rho_0}\right)^{\frac{\gamma-1}{2}} \tag{12.7}$$

holds everywhere. Expanding the density term and averaging yields

$$\begin{aligned} \overline{u}_2 = c_0 \frac{\overline{\rho}_2}{\rho_0} + \frac{\gamma - 3}{4} c_0 \frac{\overline{\rho_1' \rho_1'}}{\rho_0^2} &= c_0 \frac{\overline{\rho}_2}{\rho_0} + \frac{\gamma - 3}{4} \frac{E}{c_0} \\ \Rightarrow \quad \frac{\overline{\rho}_2}{\rho_0} &= -\frac{\gamma + 1}{4} \frac{E}{c_0^2} \end{aligned} \tag{12.8}$$

after using (12.5) within the wavetrain. This shows that within the wavetrain the mean density is reduced (if $\gamma > -1$), i.e., the fluid has been dilated by the presence of the waves.[2] Not only is this an effect that would have been hard to guess, but the quantitative details of $\overline{\rho}_2$ depend on γ and therefore on the nonlinear details of the pressure function. This also makes clear that $\overline{\rho}_2$ is not a wave property.

We now consider a two-dimensional version of this wave–mean interaction problem.[3] We work in the xy-plane such that the piston is replaced by a smooth compact wavemaker centred at $x = y = 0$. Specifically, it is easiest (though not essential) to envisage a solid wall at $x = 0$ in which a piston is smoothly embedded at the origin. The y-extent of the piston is proportional to ϵ^{-1} where $\epsilon \ll 1$ is a small parameter. This is a textbook problem for the generation of a slowly varying wavetrain. Specifically, a suitable steady solution for the linear wavetrain is a monochromatic wavetrain with wavenumber $\boldsymbol{k} = (k, 0)$ and pseudomomentum or Stokes drift density

$$\mathsf{p}_2 = (\mathsf{p}_2, 0) \quad \text{with} \quad \mathsf{p}_2 = \frac{E}{c_0} \text{env}(\epsilon y). \tag{12.9}$$

[2] The decreasing fluid volume associated with this mean dilation is compensated by the secularly growing fluid volume travelling with the wave front.
[3] These two-dimensional results carry over without significant changes to three dimensions.

Here $\text{env}(\epsilon y)$ is a smooth envelope for the wavetrain inherited from the piston shape such that $\text{env}(0) = 1$ and $\text{env}(\pm\infty) = 0$.

As described, the linear waves are an obvious slowly varying version of the original one-dimensional problem, but the same is not true for the mean-flow response. To begin with $\overline{\rho}_2$, the previous construction based on Riemann invariants is not valid anymore. However, for irrotational flow with $\boldsymbol{u} = \boldsymbol{\nabla}\phi$ we can employ Bernoulli's theorem, which for a barotropic fluid is the exact statement

$$\boldsymbol{\nabla}\left(\phi_t + \frac{|\boldsymbol{u}|^2}{2} + P(\rho)\right) = 0 \tag{12.10}$$

where P is the enthalpy. Averaging (12.10), expanding in wave amplitude, and assuming a steady mean state yields

$$\frac{E}{2}\text{env}(\epsilon y) + \overline{P}_2 = 0. \tag{12.11}$$

The zero on the right-hand side follows from evaluating the Bernoulli function far away from the wavetrain, as $y \to \infty$, say.[4] In the polytropic case P is given by (1.31) and therefore

$$\overline{P}_2 = c_0^2\left(\frac{\overline{\rho}_2}{\rho_0} + \frac{\gamma-2}{2}\frac{\overline{\rho_1'\rho_1'}}{\rho_0^2}\right) \quad \Rightarrow \quad \frac{\overline{\rho}_2}{\rho_0} = -\frac{\gamma-1}{2}\frac{E}{c_0^2}\text{env}(\epsilon y), \tag{12.12}$$

which is to be compared with the one-dimensional mean density change in (12.8). Unless $\gamma = 3$, these results are clearly different, even in the core of the wavetrain at $y = 0$. Specifically, for $\gamma < 3$ the one-dimensional mean density is smaller than the two-dimensional mean density. This illustrates that the mean densities are different at leading order $O(a^2)$, and not at $O(\epsilon a^2)$; this is the singular perturbation aspect alluded to before.

The two-dimensional result can also be derived via another route, namely by considering the mean pressure at $O(a^2)$. For the envisaged set up it is clear by inspection that in a steady state $\overline{p}_2 = 0$ everywhere, including within the wavetrain. This is because otherwise there would be an imbalance in the y-component of the mean momentum flux across the lateral flanks of the wavetrain; as $v_1' = 0$ this flux is entirely due to $\overline{p}_2$. The polytropic pressure is given by (1.29) and therefore

$$\overline{p}_2 = \rho_0 c_0^2\left(\frac{\overline{\rho}_2}{\rho_0} + \frac{\gamma-1}{2}\frac{\overline{\rho_1'\rho_1'}}{\rho_0^2}\right) = 0, \tag{12.13}$$

[4] This step fails in the one-dimensional problem, where there is no far field permanently at rest.

which by (12.6) is equivalent to (12.12). Notably, the vanishing of the mean pressure response implies a nonzero mean density response, i.e., $\overline{\rho}_2 \neq 0$.

This second derivation also explains the physical origin of the difference between the one-dimensional and the two-dimensional result. In the one-dimensional case there is no lateral confinement of the wavetrain so there is no need for the pressure to match the conditions of the fluid at rest. For instance, in an experiment the one-dimensional case could be realized by using a narrow tube with a piston at one end. In this experiment any mean pressure excess will be borne by the solid sidewalls of the tube, so it does not enter the fluid budget. In the two-dimensional case, on the other hand, this is not possible and the mean pressure inside the wavetrain must match the ambient pressure in the undisturbed fluid, which leads to (12.12).

A similar story of change at leading order unfolds for the steady mean velocity field. As in the one-dimensional problem, the boundary condition $\overline{u}_2^L = 0$ holds at the impermeable wall at $x = 0$, including the piston part of the wall. The structure of $\overline{\boldsymbol{u}}_2^L$ in the interior of the domain can then be computed from

$$\boldsymbol{\nabla} \cdot \overline{\boldsymbol{u}}_2^L = 0 \quad \text{and} \quad \boldsymbol{\nabla} \times \overline{\boldsymbol{u}}_2^L = \boldsymbol{\nabla} \times \mathsf{p}_2. \tag{12.14}$$

The first equation follows from the steady continuity equation and the second is the $O(a^2)$ part of the exact relation $\boldsymbol{\nabla} \times (\overline{\boldsymbol{u}}^L - \mathsf{p}) = 0$ for irrotational flow. Together with the boundary condition, these equations imply that $\overline{\boldsymbol{u}}_2^L$ is the least-squares projection of p_2 onto non-divergent vector fields satisfying a no-normal-flow boundary condition.

It is clear from (12.9) that $\boldsymbol{\nabla} \cdot \mathsf{p}_2 = 0$ in a steady state, so the main issue is the boundary condition at $x = 0$, where $\mathsf{p}_2 \neq 0$ as p_2 points directly into the fluid. It is easy to verify with a concrete computation that the irrotational and non-divergent difference field $\overline{\boldsymbol{u}}_2^L - \mathsf{p}_2$ decays exponentially away from the boundary, with a spatial decay rate proportional to ϵ. Thus, far away from the boundary we have $\epsilon x \gg 1$ and hence $\overline{\boldsymbol{u}}_2^L \approx \mathsf{p}_2$. Near the boundary, on the other hand, $\overline{\boldsymbol{u}}_2^L \neq \mathsf{p}_2$ because of the difference in the respective boundary conditions. The overall picture of the mean-flow circulation is illustrated in figure 12.1.

In summary, for the mean velocity a slowly varying version of the one-dimensional result $\overline{u}_2^L = 0$ is obtained only near the boundary, whilst far away from the boundary the solution is different at leading order, $O(a^2)$ and not $O(\epsilon a^2)$. Moreover, for the mean density a slowly varying version of the one-dimensional result does not hold anywhere. Thus, both mean fields provide examples of the significant changes to do with long-range pressure fields in local wave–mean interactions.

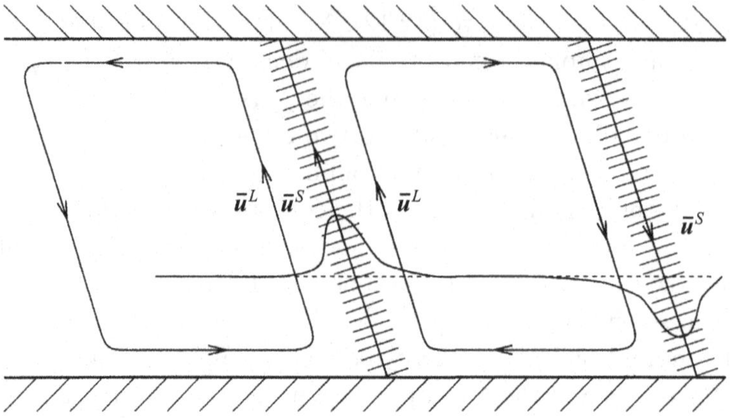

Figure 12.1 Lagrangian-mean circulation induced by acoustic wavetrains. Depicted is a case with several wavetrains in a periodic channel. The waves are generated and absorbed by suitable undulations of the sidewalls (not depicted). The Lagrangian-mean velocity is approximately equal to $\overline{\boldsymbol{u}}_2^S = \mathsf{p}_2$ in the channel centre, but not near the sidewalls. In particular, there is a drift along the walls towards the wave sources and away from the wave absorbers. From Bühler and McIntyre (1998).

12.2.2 Mean force on a wavemaker

In a wave problem with a solid body acting as a wavemaker or wave scatterer, it is natural to consider the mean force exerted on the solid body as part of the mean-flow response. However, generally the force on a solid body involves the second-order pressure field and is therefore not a wave property. This did not play a role in the mountain wave problem studied before, because there the *horizontal* force on the topography was indeed a wave property and equal to the *vertical* flux of *horizontal* pseudomomentum. In GLM theory, this turned out to be an exact result related to the equality in the off-diagonal parts of the flux tensors for mean momentum and pseudomomentum. Such a result, namely that the force on a solid body equals minus the net flux of pseudomomentum across the boundary of the body, can be viewed as another instance of the *pseudomomentum rule*, this time applied to solid bodies. If true, the rule is of obvious value for problems involving active wavemakers or passive wave scattering by solid bodies[5]

However, there is no a priori reason why the pseudomomentum rule should hold in all cases. Indeed, the one-dimensional piston problem studied in the last section demonstrates this point. The leading-order mean flux of x-momentum from piston to fluid is given by $\overline{p}_2^L$ at $x = 0$, or equivalently by its Eulerian counterpart $\rho_0 \overline{u_1' u_1'} + \overline{p}_2$. Either way, per unit area of the piston

[5] The rule is true a priori in problems involving forces on bodies due to electro-magnetic waves propagating through a vacuum. This is because there is then no medium to support a mean pressure field. Alas, fluid dynamics is harder than electro-magnetic dynamics.

the flux is

$$\rho_0\overline{u_1'u_1'} + \overline{p}_2 = \rho_0 E + \overline{p}_2 = \rho_0 E + \rho_0 c_0^2\left(\frac{\overline{\rho}_2}{\rho_0} + \frac{\gamma-1}{2}\frac{E}{c_0^2}\right). \tag{12.15}$$

The first term is equal to the pseudomomentum flux $\rho_o \mathsf{p} c_0 = \rho_0(E/c_0)c_0$, so if $\overline{p}_2 = 0$ then the pseudomomentum rule were true. However, $\overline{p}_2 \neq 0$ in the one-dimensional problem and hence the pseudomomentum rule does not hold. Specifically, by (12.8) the net momentum flux for an infinite tube is

$$\rho_0 E\left(1 - \frac{\gamma+1}{4} + \frac{\gamma-1}{2}\right) = \rho_0 E\left(1 + \frac{\gamma-3}{4}\right). \tag{12.16}$$

Thus, for $\gamma < 3$ the mean force required to hold the piston in place is *less* than predicted by the pseudomomentum rule.[6]

In a situation where $\overline{\rho}_2 = 0$, the net flux (12.15) is

$$\rho_0 E\left(1 + 0 + \frac{\gamma-1}{2}\right) = \rho_0 E\left(1 + \frac{\gamma-1}{2}\right) \tag{12.17}$$

For instance, such a situation could be realized by considering a finite tube capped at the second end by an oscillating piston that acts as a wave absorber. In this case there is no space for a mean density dilation effect and hence $\overline{\rho}_2 = 0$ in a steady mean state. If $\gamma > 1$ then in this situations the mean force on the piston is now *larger* than predicted by the pseudomomentum rule

Generally one can interpret (12.17) as the disturbance-induced excess pressure, i.e., as the part of the mean pressure that exceeds the value of the polytropic pressure function evaluated at the adjusted mean density $\rho_0 + a^2\overline{\rho}_2$. This is the same idea that gave rise to the radiation stress tensor in §10.5 and §10.5.1. For example, after identifying $\tilde{\rho}$ with $\rho_0 + a^2\overline{\rho}_2$ and $\tilde{p}$ with $\rho_0 c_0^2/\gamma$ to leading order it is clear that (12.17) is identical to S_{xx} in (10.191).

We now turn to the two-dimensional case. Here the relevant momentum flux per unit area of the wall is given by (12.15) after multiplication with the envelope function, and the net mean force then follows from integration of that flux over the wall area. However, in this case $\overline{\rho}_2 = 0$ and the net force on the wall is indeed given by the pseudomomentum rule. In other words, the mean density in (12.15) adjusts precisely such that the excess pressure due to the disturbances described by (10.198) is nullified.

6 The non-obvious value of the force on the piston corresponds to the non-obvious manifestation of the momentum in the fluid. Indeed, as $\overline{u}_2^L = 0$ everywhere apart from the wave front, it follows that the entire fluid momentum is confined to the secularly growing mean-flow response at the wave front. This is also consistent with the vanishing of both the net force and of the secular response at the wave front in the peculiar case $\gamma = -1$.

The validity of the pseudomomentum rule in this case goes together with the more obvious way in which the momentum input is realized in the fluid: as the wavetrain expands in the x-direction, it covers more and more fluid area and hence more and more fluid particles are set into motion, i.e., they change their velocity from $\overline{u}_2^L = 0$ to the far-field value $\overline{u}_2^L = \mathsf{p}_2$. The x-momentum for this continual generation of fluid motion is provided by the wavemaker and transported along the wavetrain by the pseudomomentum flux, which in this case equals the mean momentum flux even in its diagonal components. Thus, the wavemaker acts at long range to set the fluid into motion at the wave front. This is analogous to the horizontal mean motion induced by a wave front of mountain lee waves.

The present example provides a useful heuristic guide to the validity of the pseudomomentum rule applied to mean forces on solid bodies: the rule can be expected to hold if the body is sufficiently localized within the multi-dimensional fluid domain such that the diagonal part of the mean momentum flux, which involves the mean pressure field, can adjust to the ambience at rest. This holds true in the two-dimensional version of the present problem, but not in its one-dimensional version. Despite the simplicity of the pseudomomentum rule, and its apparent success in describing realistic mean forces such as the horizontal recoil force on a swimming vessel propelled by a wavemaker at the rear, it seems to be hard to derive its validity from first principles using the full equations.

For our purposes, the main conclusion is a reminder that the wave-induced mean force on a solid body is not in general directly computable at leading order from the linear solution alone, and that in problems involving localized wavetrains the fluid momentum can manifest itself in non-obvious places, such as at the wave front in the one-dimensional problem.

12.2.3 Large-scale return flow beneath surface waves

In this example we consider an incompressible homogeneous fluid under gravity g in a vertical slice described by two-dimensional xz-coordinates where z is the altitude. There is a free surface with rest position $z = 0$ and a flat solid bottom at $z = -H$. The water depth H is deep but finite, i.e., if there is a surface wave with horizontal wavenumber $k > 0$, say, then $kH \gg 1$ but $H < \infty$. This allows using the linear wave structure of deep water waves, e.g., $\omega = \sqrt{gk}$ and all wave fields have the common evanescent form $\exp i(kx - \omega t) \exp kz$, whilst rendering momentum and energy integrals over the whole fluid domain unambiguous and finite.

In the first instance we consider an x-periodic flow with finite period length L, say. We imagine a small-amplitude monochromatic progressive wave is generated from rest by the action of suitable pressure forces applied to the free surface. For instance, one could imagine an undulating solid body being towed along the free surface from left to right with a speed that matches the phase speed of a surface wave with the same undulation wavenumber; this solid body would then act as an efficient wavemaker. We employ zonal averaging and then the mean-flow response is given by the profile of the zonal mean velocity $\overline{u}_2^L(z)$ at steady state. This ignores only the transient Lagrangian-mean velocity $\overline{w}_2^L$ due to the divergence effect described in §10.2.5; the latter raises the mean centre of gravity of fluid particles during the growth phase of the waves in a manner that is consistent with the increase of available potential energy of the fluid as a whole.

The computation of $\overline{u}_2^L$ is completely straightforward because this is a problem in simple geometry and therefore

$$\overline{u}_2^L(z) = \mathsf{p}_2(z) = \frac{k}{\omega}\overline{|\boldsymbol{u}_1'|^2} = \mathsf{p}_2(0)\exp 2kz \tag{12.18}$$

follows from the circulation theorem and the assumption that the flow started from rest. As discussed in §10.1.2, the pseudomomentum equals the Stokes drift here. The last expression makes explicit that the mean flow is sharply evanescent, i.e., it is even more tightly confined to the surface than the linear waves.[7]

Clearly, the pseudomomentum and the mean flow are precisely overlapping in space in this example and it is also clear that the net flow of x-momentum from the wavemaker into the fluid system precisely accounts for both the pseudomomentum and the momentum content of the flow. This already follows from the usual argument about off-diagonal fluxes in GLM theory.

We now consider a slowly varying version of the same problem in which the fluid domain is infinite in the x-direction and the monochromatic wavetrain is localized in x, i.e., it forms a wavepacket whose footprint covers only a compact region with size proportional to ϵ^{-1}. The size of the footprint is large compared to the wavelength, but for simplicity we assume that it is at most comparable to the water depth H; this is illustrated in figure 12.2. As before, for $\epsilon \ll 1$ the linear wave fields are then accurately described by a slowly varying version of the periodic fields, but the same is not true for the mean-flow response. Indeed, in order to compute the mean flow (now

[7] It is worth pausing to note just how rapid the implied decay with depth is. For example, the linear wave fields at a depth of just one wavelength are reduced in amplitude by a factor $\exp(2\pi) \approx 500$ when compared to their size at the free surface; for the pseudomomentum or mean flow the reduction factor is the even more staggering $\exp(4\pi) \approx 250\,000$.

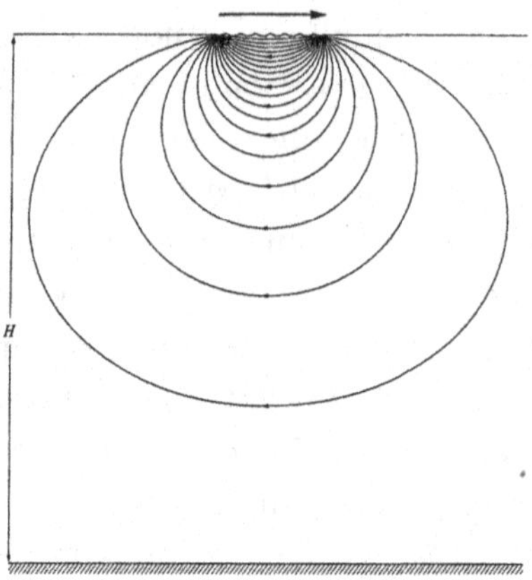

Figure 12.2 Large-scale mean return flow beneath wavepacket with small-scale surface waves. The wavepacket moves from left to right and the return flow at depth moves in the opposite direction, as indicated by the streamlines. The net horizontal mean momentum of the wavetrain together with its return flow is zero. Due to dispersion, any net momentum delivered to the fluid during wave generation has been transported away to large horizontal distances in the form of mean-flow gravity waves. From McIntyre (1981).

defined by a suitable phase average) we need to solve

$$\nabla \cdot \overline{\boldsymbol{u}}_2^L = \frac{1}{2}(\overline{\zeta_1^2})_{,zzt} \quad \text{and} \quad \nabla \times \overline{\boldsymbol{u}}_2^L = \nabla \times \mathbf{p}_2 \tag{12.19}$$

together with the appropriate boundary conditions at $z = 0$ and $z = -H$, which are $\overline{w}_2^L = \frac{1}{2}(\overline{\zeta_1^2})_{,zt}$ and $\overline{w}_2^L = 0$ due to the divergence effect and the solid boundary, respectively. After the action of the wavemaker has ceased the time dependence of the wave fields stems only from the horizontal propagation of the wavepacket with its group velocity. The divergence effect associated with this time dependence serves only to raise and lower the mean centre of gravity of fluid particles currently under the footprint of the wavepacket; direct computation shows this to have a negligible effect on the horizontal motion and so we can ignore it.

Ignoring the divergence effect amounts to setting $\nabla \cdot \overline{\boldsymbol{u}}_2^L = 0$ and using $\overline{w}_2^L = 0$ at $z = 0$ as well as at $z = -H$. It then follows that $\overline{\boldsymbol{u}}_2^L$ is again the least-squares projection of $\mathbf{p}_2$ onto the space of non-divergent vector fields with no-normal-flow boundary conditions. The slowly varying pseudomomentum vector field is purely horizontal and therefore it already satisfies these no-normal-flow boundary conditions.

However, $\nabla \cdot \mathbf{p}_2 = \mathsf{p}_{2,x} \neq 0$ at the ends of the wavepacket. This leads to the large-scale dipolar return flow pattern indicated in figure 12.2, which decays as $1/r^2$ with distance r from the wavepacket. Another way to understand this pattern is to look at the problem of finding $\overline{\boldsymbol{u}}_2^L$ as the task of inverting the known distribution of $\boldsymbol{\nabla} \times \overline{\boldsymbol{u}}_2^L$ by using the appropriate Biot–Savart law for the flow domain. In the present case this vorticity is a thin long tongue of positive y-vorticity equal to $\mathsf{p}_{2,z}$ and therefore confined to the location of the wavepacket; this leads directly to the rolling mean-flow pattern illustrated by the return flow in figure 12.2. It is noteworthy that most of the return flow appears in the region devoid of waves, where Eulerian and Lagrangian mean flows agree with each other and with the actual flow. This highlights that the motion indicated by the streamlines in figure 12.2 is real and physically observable. For example, in a stratified fluid interior this motion would excite internal gravity waves at $O(a^2)$.

The zonal Lagrangian-mean flow below the wavepacket is positive directly at the free surface but negative further underneath it. This raises the question as to what the vertically integrated net zonal momentum is. The following simple fact

$$\frac{\partial}{\partial x}\int_{-H}^{0} \overline{u}_2^L \, dz = -\int_{-H}^{0} \overline{w}_{2,z}^L \, dz = 0 \tag{12.20}$$

yields the surprising conclusion that the vertical integral of zonal mean momentum is independent of x. As the mean flow decays to zero as $|x| \to \infty$, and as we have assumed $H < \infty$ so that there can be no doubt as to the convergence of integrals, we conclude that the vertically integrated zonal momentum must in fact be *zero* everywhere.

This raises the question of where the x-momentum delivered by the wavemaker has manifested itself. Indeed, this slowly varying version of the problem is still covered by the theorem on off-diagonal fluxes of pseudomomentum and mean momentum: positive zonal pseudomomentum has been delivered to the system and there has to be a corresponding amount of positive zonal mean momentum hiding somewhere.

The answer to this puzzle comes from the realization that the leading-order mean flow in this system is capable of non-trival wave motion as well, i.e., there can be mean surface waves. This follows at once from the familiar fact that the mean flow is governed by a forced version of the operator governing the linear dynamics. These mean surface waves were filtered from the dynamics by the application of the rigid-lid boundary condition $\overline{w}_2^L = 0$ at the free surface. This turns out to be accurate after the generation of the wavepacket, but during the generation mean surface waves are in fact excited on the horizontal scale of the wavepacket. The dispersion of surface waves then means that these large-scale mean waves detach from the small-scale wavepacket, because they travel with a different (in fact, a larger) group velocity. The matter is particularly simple if the mean surface waves are long enough to travel with nearly the non-dispersive limiting speed $\sqrt{gH}$. In this case the momentum delivered by the wavemaker is entirely confined to two wave pulses located at $x \approx \pm\sqrt{gH}$.

These examples illustrate a natural difference between the spatial distributions of pseudomomentum and of mean momentum: whilst the pseudomomentum remains attached to the wavetrain by construction, the mean momentum can travel into quite different fluid regions via long-range pressure fields that are not intrinsically linked to the dynamics of the wavetrain. For example, in dispersive systems the linear dynamics of the small-scale waves involves different group velocities than the linear dynamics of the large-scale mean flow. This is a fundamental fact in fluid dynamics, and hence zero-net-momentum examples such as the situation depicted in figure 12.2 are the norm rather than the exception for localized wavetrains. Actually, from this point of view the problems in simple geometry are non-generic, because only in this confined geometry is the natural spreading of mean momentum inhibited and hence pseudomomentum and mean momentum overlap for a long time.

12.3 Vortical mean-flow response

The examples in the previous section made clear that the mean-flow response to localized wavetrains can be quite complicated even for very simple set-ups. As said before, from a physical viewpoint this is mostly due to the long-range action of mean pressure forces, which are sensitive to boundary conditions at large or infinite distances, and which can propagate at large speeds unrelated to the dynamics of the small-scale wavetrain. There is no generic set of equations that describes these wavelike aspects of the mean-flow response.

On the other hand, for the vortical part of the mean-flow response we have generic equations governing the behaviour of mean-flow potential vorticity, including their simplest version $\nabla \times (\overline{\boldsymbol{u}}^L - \mathbf{p}) = 0$ for regions of irrotational flow. The conclusion we draw from this is that general and simple wave–mean interaction results can be expected only for the vortical part of the mean flow, but not for its wavelike part. This is not as bad is it sounds, because the wavelike part of the mean flow is often weak, i.e., it is bounded uniformly in time at $O(a^2)$.[8] This was indeed the case for the examples studied in the previous section.

[8] Of course, it is possible to construct strong wavelike mean-flow responses if resonances can be arranged; in such cases the full mean-flow response must be studied.

Thus we want to focus a priori on the vortical part of the mean flow. We used this idea before, when we focused on strong wave–mean interactions linked to the linear vortical mode of the mean flow. As we shall see, it is possible to build a useful understanding of local wave–mean interactions based on the specific properties of the vortical mean flow. This includes the problems studied in simple geometry as special cases, but it goes significantly beyond these problems in several respects.

We begin investigating these issues by studying a very simple flow set up in the shallow-water system, which is the simplest fluid system in which this is possible.[9] This allows us to observe directly the limited interplay between the wavelike and the vortical parts the mean flow, and it also allows us to find natural counterparts to the interaction effects that we encountered in simple geometry. With these parts of the theory in place we can then include new effects such as three-dimensional refraction, which did not arise in simple geometry.

12.3.1 Local interactions in shallow water

We consider the standard single-layer shallow-water system with a flat lower boundary and some body fore per unit mass $\boldsymbol{F}$, which is governed by

$$\frac{Dh}{Dt} + h\boldsymbol{\nabla}\cdot\boldsymbol{u} = 0 \quad \text{and} \quad \frac{D\boldsymbol{u}}{Dt} + g\boldsymbol{\nabla}h = \boldsymbol{F}. \tag{12.21}$$

We assume small-amplitude disturbances with amplitude $a \ll 1$ relative to a basic state of rest, i.e., $\boldsymbol{U} = 0$ and $H =$const. The Lagrangian-mean equations at $O(a^2)$ written in radiation stress form are (omitting expansion subscripts)

$$\tilde{h}_t + H\boldsymbol{\nabla}\cdot\overline{\boldsymbol{u}}^L = 0 \quad \text{and} \quad \overline{\boldsymbol{u}}^L_t + g\boldsymbol{\nabla}\tilde{h} = -\frac{1}{H}\boldsymbol{\nabla}\cdot\boldsymbol{S} + \overline{\boldsymbol{F}}^L \tag{12.22}$$

where the radiation stress tensor $\boldsymbol{S}$ is defined in §10.5. There was no need to specify the averaging operator so far, but from now on we consider slowly varying waves and work with the corresponding averaging over the rapid wave phase. As discussed in §10.5.1, for slowly varying waves the leading-order pseudomomentum law is governed by

$$\mathsf{p}_t + \frac{1}{2}\boldsymbol{\nabla}E + \frac{1}{H}\boldsymbol{\nabla}\cdot\boldsymbol{S} = \boldsymbol{\mathcal{F}}. \tag{12.23}$$

[9] The only drawback of the shallow-water model is that its waves are non-dispersive, which leads to certain non-generic resonance effects that we do not wish to elaborate on here. One could add Coriolis forces to make the system dispersive.

Here E is the mean disturbance energy per unit mass and we used that $\gamma = 2$ for shallow water. Substituting (12.23) in (12.22b) leads to the mean momentum equation in the convenient form

$$\overline{\boldsymbol{u}}_t^L + g\nabla\tilde{h} = \mathsf{p}_t - \frac{1}{2}\nabla E + \overline{\boldsymbol{F}}^L - \boldsymbol{\mathcal{F}}. \tag{12.24}$$

This makes apparent that the mean flow is forced by a combination of three separate physical effects, namely transience, inhomogeneity, and the action of the body force. This equation also makes apparent that only the body force can change the mean PV, because the curl of (12.24) yields

$$\frac{\partial}{\partial t}\nabla \times (\overline{\boldsymbol{u}}^L - \mathsf{p}) = \nabla \times (\overline{\boldsymbol{F}}^L - \boldsymbol{\mathcal{F}}), \tag{12.25}$$

which is consistent with the exact formula $\overline{q}^L = \nabla \times (\overline{\boldsymbol{u}}^L - \mathsf{p})/\tilde{h}$ in this system. One more aspect of the mean-flow response can be read off (12.24) without solving for a particular problem: in a region with $\boldsymbol{F} = 0$ and steady waves the forcing terms are balanced by the mean-flow pressure term if

$$g\nabla\tilde{h} = -\frac{1}{2}\nabla E, \quad \text{i.e., if} \quad \frac{\tilde{h}}{H} = 1 - \frac{1}{2}\frac{E}{gH}. \tag{12.26}$$

Comparing with (12.12), this density dilation is found to be consistent with the two-dimensional version of the piston problem discussed in §12.2.1.

12.3.2 Bretherton flow

The key component of the typical vortical mean-flow response to a localized wavetrain is a large-scale return of the kind depicted in figure 12.2. However, for stratified fluids the return flow takes place horizontally, i.e., the return flow lies within stratification surfaces. We will call such layerwise two-dimensional return flows *Bretherton flows*. Their mathematical structure is closely related to the mean flows computed in §12.2.1.

In the shallow-water system the Bretherton flow arises clearly in a problem with a compact steady wavetrain. Thus we consider a wavemaker at some location A, say, which sends a monochromatic wavetrain across to a wave absorber at another location B. This situation is depicted in figure 12.3, with loudspeakers depicting the wavemaker and absorber. Both the generation and the absorption of the waves is modelled via suitable irrotational body forces $\boldsymbol{F} = \nabla\phi$ with compact support at A and B, respectively. For example, this would be an appropriate model for the wave generation due to pressure forces applied to the free surface. In order to produce a smooth wavepacket

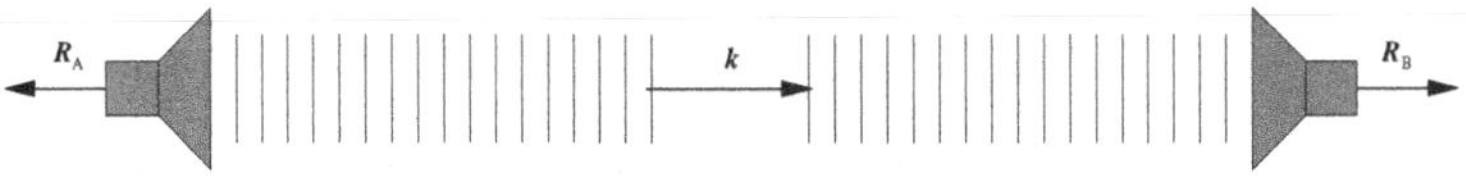

Figure 12.3 Monochromatic steady wavetrain send from wavemaker A on the left towards wave absorber B on the right. Also indicated are the equal-and-opposite recoil forces required to hold the wavemaker and wave absorber in place. From Bühler and McIntyre (2003).

the potential ϕ must be the product of a slowly varying envelope (with gradient scale $O(\epsilon)$ where $\epsilon \ll 1$) times a rapidly varying sinusoidal component (with gradient scale $O(1)$). Clearly, $\boldsymbol{F}$ is not a momentum-conserving force and there is a concomitant net input of both pseudomomentum and momentum into the fluid layer at the generation site A. An equal-and-opposite extraction of both momentum and pseudomomentum occurs at the absorption site B.

In the mean-flow equation we have (using (10.127))

$$\overline{\boldsymbol{F}}^L = \overline{\boldsymbol{\nabla}\phi}^L \quad \text{and} \quad \boldsymbol{\mathcal{F}} = \overline{\boldsymbol{\nabla}\phi}^L - \boldsymbol{\nabla}\overline{\phi}^L, \tag{12.27}$$

and therefore $\overline{\boldsymbol{F}}^L \approx \boldsymbol{\mathcal{F}}$ because their difference is $O(\epsilon a^2)$ rather than $O(a^2)$. Thus the two forcing terms in (12.24) cancel each other. This is consistent with a joint flux of momentum and pseudomomentum into this system: the momentum input from the irrotational force flows simultaneously into the mean momentum and into the pseudomomentum. As we consider a steady mean state with adjusted height field, the pseudomomentum rule for forces on bodies holds, i.e., the recoil forces on the wavemaker and the wave absorber are equal to minus the net flux of pseudomomentum across a control area surrounding A or B, respectively.

The steady mean flow $(\tilde{h}, \overline{\boldsymbol{u}}^L)$ is determined by (12.26), $\boldsymbol{\nabla}\cdot\overline{\boldsymbol{u}}^L = 0$, and the irrotational-flow relation $\boldsymbol{\nabla}\times(\overline{\boldsymbol{u}}^L - \mathsf{p}) = 0$. With suitable decay conditions at infinity, the Lagrangian-mean flow is again the least-squares projection of p onto non-divergent vector fields. The structure of this horizontal Bretherton flow is depicted in figure 12.4. Viewed at large distance, it is a dipole flow centred halfway between A and B, and with strength given by the dipole moment equal to the net pseudomomentum, i.e.,

$$\boldsymbol{\mathcal{P}} = \int \mathsf{p}\, dxdy. \tag{12.28}$$

12.3.3 A wavepacket life cycle

We now consider a time-dependent variant of the previous problem, i.e., we consider a sequence of events beginning with the generation of a wave-

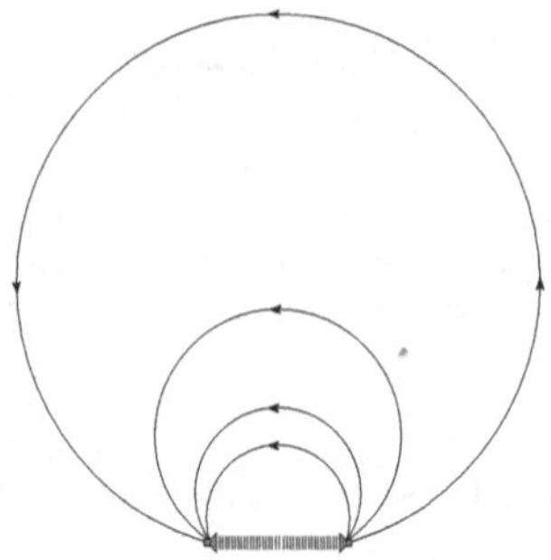

Figure 12.4 Large-scale view of the Bretherton return flow for the steady wavetrain in figure 12.3. Only one half of the symmetric flow pattern is shown. From Bühler and McIntyre (2003).

packet at A, followed by its subsequent propagation from A to B, and the final momentum-conserving dissipation of the wavepacket at B. This allows distinguishing between effects due to wave transience and due to wave dissipation.

As before, we model the generation of the wavepacket at A by a compactly supported irrotational body force $\boldsymbol{F} = \boldsymbol{\nabla}\phi$, which now acts only for a finite time. If the wave generation is rapid enough, then it is consistent to equate $\mathbf{p}_t$ with $\boldsymbol{\mathcal{F}}$ during this process. Moreover it is then consistent to assume $\tilde{h} = H$ and $\overline{\boldsymbol{u}}^L = \mathbf{p}$ as well. One way to see this is to note that $\mathbf{p}_t$ is (12.24) has a monopolar structure, i.e., its integral over A comes to a nonzero vector $\boldsymbol{\mathcal{P}}_t$, whilst the other term $-\frac{1}{2}\boldsymbol{\nabla}E$ has a dipolar structure and integrates to zero. Thus, at the end of the generation phase at time $t = 0$, say, we have a localized wavepacket at A and the mean flow is described by an undisturbed layer depth and a "bullet" of mean momentum such that

$$\tilde{h}(x, y, 0) = H \quad \text{and} \quad \overline{\boldsymbol{u}}^L(x, y, 0) = \mathbf{p}(x, y, 0). \tag{12.29}$$

The subsequent evolution of the linear wavepacket consists of x-propagation with the group velocity plus some diffraction effects to do with the finite wavepacket envelope, which however do not affect our argument here. So, if the initial $\mathbf{p}(x, y, 0) = (E_0/c_0, 0)$ with $c_0 = \sqrt{gH}$, then $E(x, y, t) = E_0(x - c_0t, y)$ and similarly for $\mathbf{p}(x, y, t)$. On the other hand, the subsequent evolution of the $O(a^2)$ mean-flow response is given by a superposition of the solution to the initial-value problem (12.29) together with the response to the forcing terms on the right-hand-side of (12.24). Notably, these forcing terms are all dipolar now, because $\mathbf{p}_t = -c_o\mathbf{p}_x$ during wave propagation.

The time-dependent mean-flow response can be viewed as some kind of "frustrated adjustment", i.e., the frustrated attempt of the Bretherton flow in figure 12.4 to adjust to the moving wavepacket. Why is the adjustment frustrated? Clearly, the adjustment takes place via mean pressure forces, which in the shallow-water system the pressure field can only change with the finite speed c_0.[10] In an incompressible fluid, on the other hand, the mean pressure could react instantaneously to the moving wavepacket and the Bretherton flow can maintain itself as attached to the moving wavepacket. Of course, strict incompressibility is not necessary in order to allow for a less frustrated adjustment process, because dispersion can lead to the same outcome.

For example, with sufficient dispersion we can envisage that the wavepacket moves very slowly compared with the group velocity for the large-scale mean flow. This would be the case, for instance in a three-dimensional version of the problem discussed in §12.2.3, with the vertically integrated radiation stress playing the role of the two-dimensional $\boldsymbol{S}$ here. In this case $\omega^2 = g\kappa$ implies that the small-scale wavepacket moves slowly compared to the large-scale mean flow. This allows the adjustment to be less frustrated, and therefore the establishment of a quasi-steady Bretherton flow attached to the moving wavepacket.

Now, we finally let the wavepacket dissipate at the location B. In contrast to the situation considered in §12.3.2, the dissipation involves a momentum-conserving force $F_i = \Pi_{ij,j}/h$ and hence

$$\overline{F}_i^L = \frac{1}{\overline{h}}\tilde{\Pi}_{ij,j} = O(\epsilon a^2) \quad \Rightarrow \quad |\overline{\boldsymbol{F}}^L| \ll |\boldsymbol{\mathcal{F}}| \tag{12.30}$$

follows from the generic argument given in §11.1.1. This shows that during dissipation, pseudomomentum is extracted from the flow, but not momentum. Indeed, if the dissipation occurs rapidly enough, then it is again consistent to equate the monopolar p_t with the monopolar $\boldsymbol{\mathcal{F}}$, which means that there is no corresponding monopolar forcing of $\overline{\boldsymbol{u}}^L$ in (12.24). This marks a significant difference between the irrotational generation and the momentum-conserving dissipation of the wavepacket.

The dissipation of the wavepacket creates a nonzero PV structure out of nothing, because the leading-order mean PV evolves according to

$$\frac{\partial \overline{q}^L}{\partial t} = \frac{\partial}{\partial t}\left(\frac{\boldsymbol{\nabla}\times(\overline{\boldsymbol{u}}^L - \mathsf{p})}{H}\right) = -\frac{\boldsymbol{\nabla}\times\boldsymbol{\mathcal{F}}}{H}. \tag{12.31}$$

A non-trivial fact that follows from (12.31) is that momentum-conserving

[10] There is also a non-essential resonance effect because the adjustment speed c_0 equals the wavepacket speed c_0. Whilst perfectly real, this resonance effects is peculiar to non-dispersive systems and not of interest here.

dissipative forces *cannot* be irrotational at $O(a^2)$. For example, this fact applies to the dissipation of sounds waves via the usual Navier–Stokes viscous terms. The key observation in such examples is that although the $O(a)$ dissipative force may be irrotational, its correlation with the density disturbance at $O(a^2)$ is not.

A simple form for $\boldsymbol{\mathcal{F}}$ in wave dissipation problems is $\boldsymbol{\mathcal{F}} = -\alpha\mathbf{p}$ with some decay rate $\alpha > 0$, where the sign is determined by the assumed destruction of the wavepacket by the dissipation. Thus (12.31) appears as

$$\frac{\partial \overline{q}^L}{\partial t} = +\frac{\boldsymbol{\nabla} \times (\alpha \mathbf{p})}{H}. \tag{12.32}$$

In other words, the structure of $\overline{q}^L$ changes as if the wavepacket were absent and a body force equal to minus the dissipation rate of pseudomomentum were applied to the flow. This is the dissipative pseudomomentum rule in a form applicable to local wavetrains. Notably, the PV structure implied by (12.32) is necessarily dipolar, as there can be no monopolar PV structure produced by a compact force.

We can note in passing that just as in the zonally symmetric problem, there is no mean-flow acceleration that is intrinsically linked to dissipation. Rather, dissipation destroys pseudomomentum without affecting mean momentum. In other words, dissipation only makes irreversible a mean-flow change that has already occurred due to a combination of wave generation and wave transience effects.

The asymptotic end state of the mean flow is given by the adjustment of the mean flow to the PV structure created in (12.32). In the limit of instantaneous dissipation (i.e., $\alpha = \delta(t - t_d)$ for some dissipation time $t_d > 0$) this end state is equal to the Bretherton flow associated with the wavepacket location and shape at $t = t_d$; otherwise the structure of $\overline{q}^L$ is elongated in the direction of propagation to take into account the propagation of the wavepacket as it dissipates.

Now, some mean-flow gravity waves will radiate away from the dissipation site during the adjustment to the terminal Bretherton flow. Broadly speaking, the strength of this radiation is proportional to the difference between the terminal Bretherton flow and the mean flow just before $t = t_d$. In a system without frustrated adjustment (e.g., in the three-dimensional Boussinesq system), this difference would be zero, and at least in this case it would be clear that a wavepacket does not end with a bang, but a whimper.

12.3.4 Strong interactions and potential vorticity

What have we learned from these examples? In particular, which aspects of the mean-flow response are relevant for strong interactions that can lead to large changes in the mean flow? The natural answer comes from the PV equation: a strong interaction changes the mean PV irreversibly, i.e., it changes the spatial distribution of $\overline{q}^L$. This can occur either via material displacements of pre-exisiting PV structures, or via the generation of new PV gradients due to dissipation. Interactions that do not change the mean PV are viewed as weak and assumed to be bounded uniformly in time at $O(a^2)$.

From this perspective, the generation and propagation parts of the wave-packet life cycle studied in §12.3.3 are both ignorable, as they do not change the PV. Indeed, we have used the irrotational-flow relation $\boldsymbol{\nabla} \times \overline{\boldsymbol{u}}^L = \boldsymbol{\nabla} \times \mathbf{p}$ repeatedly for these parts of the cycle. Thus, the only strong interaction occurs during wave dissipation. If the wavepacket life cycle is repeated many times over, or if the sequence of wavepackets is replaced by a steady wave-train, then the dissipation-induced $\overline{q}^L$ would grow secularly as $O(ta^2)$. Presumably, the long-time evolution of $\overline{q}^L$ would then involve a balance between nonlinear advection of $\overline{q}^L$ and the forcing due to wave dissipation. A practical example of this kind of long-time vortical mean-flow dynamics is given in §13 below.

12.4 Impulse and pseudomomentum conservation

We are looking for a mathematical tool that allows us to focus attention on strong interactions, which change the PV distribution. Ideally, this tool would have broad applicability across fluid systems and it would also work in the presence of refraction by a basic mean flow, a topic that we have not yet considered for local interactions. It appears that in the context of strongly stratified fluid systems, in a sense to be made precise, such a theoretical tool is indeed available in the form of the *Kelvin's impulse* based on the Lagrangian-mean PV. This is reminiscent of the use of the zonal impulse for quasi-geostrophic dynamics in §9, but here the context is significantly broader.

Specifically, we will be able to formulate a conservation law for the sum of this impulse plus the pseudomomentum of the waves, and this conservation law will indeed be valid in the presence of dissipation as well as refraction.

We start by describing the classical impulse concept in incompressible fluid dynamics and then we apply this to strongly stratified flows and GLM theory.

12.4.1 Classical impulse theory

Kelvin's impulse (also called the hydrodynamical impulse) is a classical concept in incompressible homogeneous fluid dynamics. Impulse and momentum are distinct, but closely related quantities. The classical impulse is a vector-valued linear functional of the vorticity defined by

$$\text{impulse} \quad = \frac{1}{n-1}\int \boldsymbol{x}\times(\boldsymbol{\nabla}\times\boldsymbol{u})\,dV \tag{12.33}$$

where $n > 1$ is the number of spatial dimensions and the integral is extended over the whole fluid domain. We are most interested in the two-dimensional case:

$$\text{two-dimensional impulse} \quad = \int (y, -x)\,\boldsymbol{\nabla}\times\boldsymbol{u}\,dxdy. \tag{12.34}$$

The impulse has a number of remarkable kinematic and dynamic properties for incompressible perfect fluid flow. To begin with, the impulse is well defined whenever the vorticity has compact support. If $n = 3$ then this fixes the impulse uniquely, but if $n = 2$ then the value of the impulse depends on the location of the coordinate origin unless the net integral of $\boldsymbol{\nabla}\times\boldsymbol{u}$, which is the total circulation around the fluid domain, is zero. For example, in two dimensions the impulse of a single point vortex with circulation Γ is equal to $\Gamma(Y, -X)$ where (X, Y) is the position of the point vortex. This illustrates the dependence on the coordinate origin. On the other hand, two point vortices with equal-and-opposite circulations $\pm\Gamma$ separated by a distance d yield a coordinate-independent impulse vector with magnitude Γd and direction parallel to the propagation direction of the vortex couple. To fix this image in your mind you can consider the impulse of the trailing vortices behind a tea (or coffee) spoon: the impulse vector is then parallel to the motion of the spoon. A sketch of such a vortex couple and its impulse is given in figure 12.5.

The easily evaluated and unambiguous impulse integral in an unbounded domain with zero net circulation contrasts with the incompressible momentum integral, which in the same situation is not absolutely convergent and therefore is ambiguous. For instance, in the case of a two-dimensional vortex couple[11] the velocity field at large distances has a dipolar structure and

[11] The vortex couple can be smoothed, i.e., the point vortices can be replaced by smooth

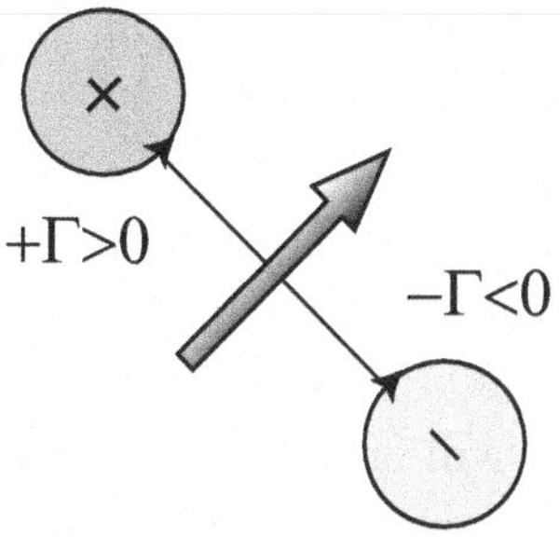

Figure 12.5 Two-dimensional vortex couple consisting of two smooth vortices separated by a distance d and with equal-and-opposite circulations $\pm\Gamma$. The large arrow indicates the net impulse vector, which has magnitude Γd and runs parallel to the propagation direction of the couple. Notably, the impulse decreases as the distance d decreases, even though the propagation velocity of the vortex couple, which is $\Gamma/(2\pi d)$, increases.

decays as $1/r^2$ with distance r from the couple. This is not fast enough to make the momentum integral

$$\int \boldsymbol{u}\, dxdy \tag{12.35}$$

absolutely convergent, i.e., the value of (12.35) depends on the limiting shape of the integration domain as the size of the integration domain goes to infinity. Thus a vortex couple in an unbounded domain has a unique impulse, but no unique momentum. As an example, the momentum contained in a large circle with radius R centred at the vortex couple converges to 1/2 of the impulse as $R \to \infty$. The same argument applies to a vortex ring in three dimensions, where the momentum contained in a large sphere centred at the ring converges to 2/3 of the impulse.

Consistent with the ambiguity of the total momentum volume integral there is a corresponding ambiguity of the surface integral for the pressure-induced momentum flux across the boundary of a large domain. The instantaneous pressure field is computed from a Poission equation with $\boldsymbol{\nabla} \cdot \boldsymbol{F}$ as a source term, which induces non-compact pressure fields even for compact force fields. In particular, the net surface flux due to the dipolar part of the pressure field, whose strength is proportional to the second integral in (12.36) and which decays as $1/r^{n-1}$ with distance r from the forcing region, is responsible for the ambiguity. Thus, only the sum of the volume-integrated momentum plus the surface-integrated momentum flux due to pressure is independent of the domain shape, and equal to the net momentum delivered by the body force. In other words, in incompressible unbounded flow it is not

finite-core vortices such that the velocity field is bounded, without changing the conclusions about the far field.

possible to distinguish between momentum at large distances and pressure waves.

Now, as far as dynamics is concerned, it can be shown that the unforced incompressible Euler equations in an unbounded domain conserve the impulse. The proof involves time-differentiating (12.33) and using integration by parts together with a mild estimate of the decay rate of $\boldsymbol{u}$ based on a compact vorticity field. We will encounter an example of this calculation in two dimensions in (12.40) below.

The conservation of impulse for unbounded unforced flow is clearly useful. Even more useful is the fact that the time rate of change of impulse exactly equals the net momentum input by a compact body force per unit mass $\boldsymbol{F}$. This follows at once from the time derivative of the impulse together with the following integration-by-parts identity that holds for arbitrary vector fields with compact support:

$$\int \boldsymbol{F}\, dV = -\int \boldsymbol{x}\, \boldsymbol{\nabla}\cdot\boldsymbol{F}\, dV = \frac{1}{n-1}\int \boldsymbol{x}\times(\boldsymbol{\nabla}\times\boldsymbol{F})\, dV. \qquad (12.36)$$

The integrals are extended over the support of $\boldsymbol{F}$ and the attractive second integral is included for completeness; it illustrates that $\boldsymbol{\nabla}\cdot\boldsymbol{F}$ and $\boldsymbol{\nabla}\times\boldsymbol{F}$ are not independent for compact vector fields. Note that (12.36) does not apply to the velocity $\boldsymbol{u}$ because $\boldsymbol{u}$ does not have compact support. The simple relation between $\boldsymbol{F}$ and the rate of change of impulse holds for arbitrary time-dependence of $\boldsymbol{F}$, even in the limit as $\boldsymbol{F}$ becomes "impulsive", i.e., in the limit in which $\boldsymbol{F} = \delta(t)\boldsymbol{F}_0(\boldsymbol{x})$ for some spatial shape $\boldsymbol{F}_0$, say. Historically, this application gave the impulse its evocative name.[12]

Notably, in the tea spoon scenario the impulse of the trailing vortex couple can be related to the net force exerted by the spoon. This illustrates how classical impulse concepts can be useful for fluid–body interaction problems. For example, similar impulse concepts have been used to study the locomotion of fish and of water-walking insects.

In a bounded domain the impulse and momentum behave very differently. For instance, in an enclosed domain the incompressible momentum integral is convergent and in fact zero, because the centre of mass of an enclosed body of homogeneous fluid cannot move. This implies that any momentum input due to a body force is simply transmitted into the enclosing boundary via pressure forces. The impulse, on the other hand, is nonzero in general and also not constant in time anymore. This is obvious by considering the example of a vortex couple propagating towards a wall, which increases the

[12] Momentum and impulse are truly false friends in languages such as Russian or German, where the word *Impuls* means momentum, leading to no end of confusion.

separation d of the vortices and thereby increases their impulse. However, the instantaneous rate of change of the impulse due to a compact body force $\boldsymbol{F}$ is still given by the net integral of $\boldsymbol{F}$. Intermediate cases such as a zonal channel geometry are also possible, in which the flow domain is periodic or unbounded in x, but is bounded by two parallel straight walls in y. In this case the x-component of impulse is still exactly conserved under unforced flow, but not the y-component.

Finally, mutatis mutandis, the impulse concept can also be applied to flows in spherical geometry. For example, consider the two-dimensional incompressible flow of a homogeneous fluid confined to the thin gap between two rigid concentric spheres with radius a and $a + h$ where $h \ll a$. This is a crude model of an atmospheric flow along stratification surfaces. Due to the rigidity of the confining spheres the flow dynamics is given by the two-dimensional area-preserving flow that leaves the absolute vorticity

$$q(\lambda, \theta, t) = f(\theta) + \frac{1}{a \cos\theta}\left(v_\lambda - (u \cos\theta)_\theta\right) \tag{12.37}$$

materially invariant. Here λ and θ are longitude and latitude, respectively, (u, v) are the corresponding two-dimensional velocity components, and the Coriolis parameter $f = 2\Omega \sin\theta$ in the presence of frame rotation Ω. Area preservation takes the form $u_\lambda + (v \cos\theta)_\theta = 0$. It is then straightforward to show that the impulse defined by the surface integral

$$\int (a \sin\theta) q \, dA \quad \text{with} \quad dA = a^2 \cos\theta \, d\lambda d\theta \tag{12.38}$$

is conserved for unforced fluid motion, and indeed equal to the total angular momentum around the rotation axis divided by the layer depth h. Here the factor $a \sin\theta$, which is the elevation above the equatorial plane, replaces the meridional coordinate y that is familiar from the Cartesian zonal impulse.

12.4.2 Impulse and pseudomomentum in GLM theory

We now consider the question of whether the impulse concept can be applied to wave–vortex interactions. The idea is to define a suitable mean-flow impulse that evolves in a useful way under such interactions. This raises two issues. First, the classical impulse concept is restricted to incompressible flow, i.e, if compressible flow effects are allowed then most of the useful properties of the impulse are lost. Still, the vortical mean-flow dynamics, especially in the geophysically relevant regime of slow layer-wise two-dimensional flow, is often characterized by weak two-dimensional compressibility; a case in point is standard quasi-geostrophic dynamics, in which the horizontal flow divergence is negligible at leading order. This suggests that two-dimensional impulse theory may still be useful for strongly stratified flows. The second issue is the question as to which vorticity field to use to form the impulse as in (12.34). For instance, once could base the GLM impulse on the curl of $\overline{\boldsymbol{u}}^L$, but it turns out to be more convenient to base it on the curl of $\overline{\boldsymbol{u}}^L - \mathsf{p}$ instead, because this curl appears in the definition of the mean PV $\overline{q}^L$. Indeed, $\nabla \times \overline{\boldsymbol{u}}^L$ is changed by transient interaction effects whereas $\nabla \times (\overline{\boldsymbol{u}}^L - \mathsf{p}) \propto \overline{q}^L$ is not.

We thus define the GLM impulse in the shallow water system as

$$\mathcal{I} = \int (y, -x)\, \overline{q}^L \tilde{h}\, dxdy = \int (y, -x)\, \nabla \times (\overline{\boldsymbol{u}}^L - \mathsf{p})\, dxdy. \qquad (12.39)$$

Clearly, $\mathcal{I}$ is well defined if $\overline{q}^L$ has compact support. Also, $\mathcal{I}$ is obviously zero in the case of irrotational flow. This suggest that $\mathcal{I}$ is targeted on the vortical part of the flow, which is what we want, but the important question is how $\mathcal{I}$ evolves in time. The easiest way to find the time derivative of $\mathcal{I}$ in the case of compact $\overline{q}^L$ is by interpreting the integral in (12.39) as an integral over a material area that contains the support of $\overline{q}^L$. The time derivative of such a material integral can then be evaluated by applying $\overline{\mathrm{D}}^L$ to the entire integrand, including $dxdy$. At first we ignore the body force, which means that both $\overline{q}^L$ and $\tilde{h}dxdy$ are mean material invariants and hence the only nonzero term comes from $\overline{\mathrm{D}}^L(y, -x) = (\overline{v}^L, -\overline{u}^L)$. After some integration by parts this yields

$$\frac{d\mathcal{I}}{dt} = \int (\overline{\boldsymbol{u}}^L - \mathsf{p})\, \nabla \cdot \overline{\boldsymbol{u}}^L\, dxdy + \int (\nabla \overline{\boldsymbol{u}}^L) \cdot \mathsf{p}\, dxdy + \text{remainder}. \qquad (12.40)$$

Here the p contracts with $\overline{\boldsymbol{u}}^L$ and not with ∇. Explicitly,

$$(\nabla \overline{\boldsymbol{u}}^L) \cdot \mathsf{p} = (\overline{u}^L_x \mathsf{p}_1 + \overline{v}^L_x \mathsf{p}_2, \overline{u}^L_y \mathsf{p}_1 + \overline{v}^L_y \mathsf{p}_2) \qquad (12.41)$$

if $\mathsf{p} = (\mathsf{p}_1, \mathsf{p}_2)$. The remainder in (12.40) consists of integrals over derivatives such as $\overline{v}^L_x \overline{v}^L = \frac{1}{2}\partial_x (\overline{v}^L)^2$ or $(\overline{v}^L \mathsf{p}_2)_x$, which yield vanishing contributions in an unbounded domain if $\overline{\boldsymbol{u}}^L$ and p decay fast enough with distance r. For example, a decay $\overline{\boldsymbol{u}}^L = O(1/r)$ or $\overline{\boldsymbol{u}}^L = O(1/r^2)$ is sufficient, respectively, depending on whether p is compact or not. We assume that p is compact in our examples and hence we can safely ignore this remainder.

So far we have used the exact GLM equations, but now we assume that the term involving $\nabla \cdot \overline{\boldsymbol{u}}^L$ in (12.40) can be neglected, which is equivalent to neglecting changes in the effective mean layer depth $\tilde{h}$. This is the key assumption of strong stratification that has to be made in order for the GLM impulse budget to work. In the present shallow-water case this will be justified if the flow has a small enough Froude number $\sqrt{|\boldsymbol{u}|^2/(gH)}$, i.e., if the stratification is strong. In practice, one might expect the requirement to be slightly less stringent, because a moderate amount of mean-flow gravity waves would simply lead to reversible oscillations of the $\overline{q}^L$ structure, and therefore to fluctuations of $\mathcal{I}$ around a useful mean value.

Either way, from here on we approximate the impulse evolution by

$$\frac{d\mathcal{I}}{dt} = + \int (\nabla \overline{\boldsymbol{u}}^L) \cdot \mathsf{p} \, dx dy. \tag{12.42}$$

Up to a minus sign, on the right-hand side we recognize the generic pseudomomentum source term due to mean-flow refraction, as written down (10.126). Indeed, neglecting $\nabla \cdot \overline{\boldsymbol{u}}^L$ and changes in $\tilde{h}$, it is a direct consequence of (10.126) and the assumption of a compact disturbance field that

$$\mathcal{P} = \int \mathsf{p} \, dx dy \quad \Rightarrow \quad \frac{d\mathcal{P}}{dt} = - \int (\nabla \overline{\boldsymbol{u}}^L) \cdot \mathsf{p} \, dx dy, \tag{12.43}$$

which is the equal-and-opposite of (12.42). This means the sum of impulse and total pseudomomentum is conserved, i.e.,

$$\frac{d}{dt}(\mathcal{I} + \mathcal{P}) = 0. \tag{12.44}$$

This conservation law is the relevant tool that we can use to discern strong, irreversible interaction effects. The present derivation makes clear that it is only the neglect of $\nabla \cdot \overline{\boldsymbol{u}}^L$ and $\nabla \tilde{h}$ that limits the validity of (12.44). In particular, there is no essential restriction to small wave amplitudes.

Before discussing the application of this conservation law, we include a compact body force $\boldsymbol{F}$. The impulse law (12.42) and the pseudomomentum law (12.43) are then augmented by domain integrals over $\overline{\boldsymbol{F}}^L - \mathcal{F}$ and $\mathcal{F}$, respectively. This uses the mean PV law in the incompressible form

$$\overline{\mathrm{D}}^L(\nabla \times (\overline{\boldsymbol{u}}^L - \mathsf{p})) = \nabla \times (\overline{\boldsymbol{F}}^L - \mathcal{F}) \tag{12.45}$$

as well as the identity (12.36). Hence, in the presence of forcing we obtain

$$\frac{d\mathcal{I}}{dt} = \int (\overline{\boldsymbol{F}}^L - \mathcal{F}) \, dx dy, \quad \frac{d\mathcal{P}}{dt} = \int \mathcal{F} \, dx dy,$$
$$\text{and} \quad \frac{d}{dt}(\mathcal{I} + \mathcal{P}) = \int \overline{\boldsymbol{F}}^L \, dx dy. \tag{12.46}$$

In particular, for momentum-conserving dissipative forces the right-hand side of (12.46) is negligible and hence $\mathcal{I} + \mathcal{P}$ are constant during wave dissipation. Overall, the conservation law (12.44,12.46) is remarkable both in its simplicity and its scope.

Without forcing, or with forcing that is momentum-conserving, the conservation law (12.44) makes clear that there is zero-net-sum game of exchange going on between the impulse $\mathcal{I}$ and the pseudomomentum $\mathcal{P}$. Conversely, only in the presence of an external force that adds or subtracts momentum from the fluid does $\mathcal{I}+\mathcal{P}$ change. This simple state of affairs allows a concise description of wave–mean interactions involving wavetrains or wavepackets.

For example, at the beginning of the wavepacket life cycle described in §12.3.3, both $\mathcal{I}$ and $\mathcal{P}$ were zero. During wave generation by the irrotational force, $\mathcal{P}$ changed whilst $\mathcal{I}$ remained zero because $\overline{q}^L$remained equal to zero. During the subsequent propagation of the wavepacket both $\mathcal{I}$ and $\mathcal{P}$ remained constant. Notably, in the presence of mean-flow refraction both $\mathcal{I}$ and $\mathcal{P}$ would have changed at this stage, but in a manner such that their sum $\mathcal{I} + \mathcal{P}$ had remained constant. Finally, during wave dissipation $\mathcal{P}$ was reduced to zero and $\mathcal{I}$ changed irreversibly such that $\mathcal{I} + \mathcal{P}$ remained a constant, and equal to the original momentum input by the wave-making force (divided by the still layer depth H).

This example suggests considering changes in $\mathcal{I}$ as the hallmark of strong wave–mean interactions, and this is the view we will adopt. In the present case, $\mathcal{I}$ changes only at the end of the cycle, when the wavepacket dissipates. However, as already noted, in principle $\mathcal{I}$ can also change due to mean-flow refraction, so dissipation is sufficient but not necessary for strong interactions involving wavepackets and slowly varying mean flows.

12.4.3 GLM theory for barotropic three-dimensional flow

We note in passing a version of the GLM impulse and pseudomomentum conservation law that is useful in cases such as the homentropic flow of an ideal gas, or the incompressible flow of a density-homogeneous fluid. In these cases the flow is barotropic (i.e., the density is a function of the pressure) and therefore the perfect vorticity dynamics is described simply by the three-dimensional material advection and stretching of the vortex lines. In this case the three-dimensional definitions

$$\mathcal{I}_3 = \frac{1}{2}\int \boldsymbol{x} \times (\boldsymbol{\nabla} \times (\overline{\boldsymbol{u}}^L - \mathbf{p}))\, dV \quad \text{and} \quad \mathcal{P}_3 = \int \mathbf{p}\, dV \tag{12.47}$$

combined with the assumptions that $\boldsymbol{\nabla} \cdot \overline{\boldsymbol{u}}^L$ and $\tilde{\rho} - 1$ are both negligible then yield the conservation law

$$\frac{d}{dt}(\mathcal{I}_3 + \mathcal{P}_3) = \int \overline{\boldsymbol{F}}^L\, dV. \tag{12.48}$$

This is useful to understand the generation of mean-flow vortex rings by dissipating packets of sound waves, for instance.

12.5 Notes on the literature

The pitfalls associated with mean pressure effects and the significant changes in the wave–mean interactions once simple geometry is dropped are described in McIntyre (1981) and Bühler and McIntyre (1998). The typical dipolar PV structure generated by dissipating wavepackets is discussed in McIntyre and Norton (1990) and Bühler (2000) and the dipolar Bretherton flow on stratification surfaces was introduced in Bretherton (1969). Classical impulse theory is described in the textbooks Batchelor (1967) and Saffman (1993); an application of impulse theory to a bio-locomotion problem is offered in Bühler (2007). The conservation law for impulse plus pseudomomentum and its use in wave–mean interaction theory was pointed out in Bühler and McIntyre (2005).

13

Wave-driven vortex dynamics on beaches

The similarities and differences between zonal-mean theory and local averaging are illustrated nicely by the problem of wave-driven circulations in the nearshore regions on beaches. This problem is both significant in coastal oceanography as well as directly observable in everyday life, which is an attractive feature.

We first describe the classic theory of wave-driven longshore currents, which is based on zonal averaging and simple geometry, and then we consider the changes in the problem once localized wavetrains are allowed. This will lead to a discussion of vorticity generated by breaking waves and also to a consideration of vortex dynamics in a sloping domain, which are interesting fluid-dynamical topics in their own right.

We conclude with a consideration of how the long-term mean-flow behaviour may differ significantly from the predictions of classic theory in the presence of non-trivial topography features such as barred beaches.

13.1 Wave-driven longshore currents

The basic situation is as envisaged in the left panel of figure 13.1: looking down on the xy-plane ocean waves are obliquely incident from the left on a beach with a straight shoreline located at $x = 0$, say. The waves are refracted and turned towards the shoreline by the decreasing water depth as the shoreline is approached. To fix terminology, the x-direction is called the cross-shore direction and the y-direction is called the longshore direction. It is implicit in this picture that the flow is periodic in the longshore direction and that the wavetrain is homogeneous in y as well. Now, the picture suggests, and analysis readily confirms, that there is a wave-induced flux of momentum into the nearshore region. Specifically, there is a wave-induced flux of longshore momentum in the cross-shore direction such that the off-

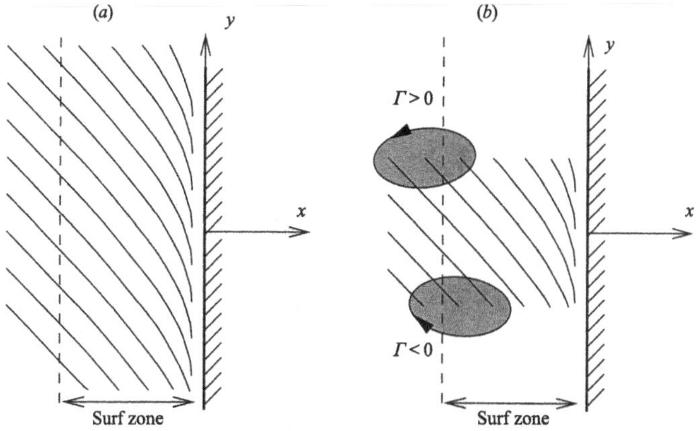

Figure 13.1 Left: crests of homogeneous wavetrain obliquely incident on beach with shoreline on the right. The waves break in the surf zone and drive a longshore current in the positive y-direction there. There are no vortices. Right: crests of inhomogeneous wavetrain. The breaking location is flanked by a vortex couple generated by wave breaking; the indicated vorticity signatures are the vertical outcroppings of the three-dimensional vorticity banana described in §13.4. Due to the oblique wave incidence, the vortex couple is slightly tilted relative to the shoreline and therefore it has a positive impulse in the y-direction. From Bühler and Jacobson (2001).

diagonal component of the radiation stress $S_{xy} > 0$, for instance. As we know, S_{xy} equals the cross-shore flux of the longshore component of both pseudomomentum and Lagrangian-mean momentum.

Now, the decreasing water depth towards the shoreline leads to growth in wave amplitude and eventually to nonlinear wave breaking and three-dimensional turbulence in the so-called surf zone. The details of this turbulent wave breaking cannot be known, but it is reasonable to assume that the destruction of the wave reduces the wave-induced momentum flux to zero. Furthermore, observation shows that wave reflection is weak on beaches, i.e., there is little scope for the incoming momentum flux being carried back out the sea via reflected waves, and hence the conclusion must be that there is a net convergence of longshore momentum in the surf zone. This flux convergence drives a *longshore current* in the surf zone, which grows in amplitude until it saturates because of the combined effects of horizontal mixing by turbulence and, most importantly, of bottom friction in a viscous boundary layer at the sandy and pebbly beach floor.

Longshore currents are of appreciable magnitude, with typical speeds of one metre per second and a typical width of a hundred metres or so. These currents can interact with and co-produce rip currents, i.e., narrow and fast currents driving out to sea, they can contribute to beach erosion and evolution, and they can impact on engineering structures in the nearshore region. The swift alongshore advection by these currents also makes them relevant for boat landing manoeuvres, so there has been a long-standing naval interest in the relationship between wave conditions further offshore and the strength and the structure of wave-driven longshore currents in the nearshore region.

The classical theory for longshore currents was formulated by Longuet–Higgins. It is a model example of wave–mean interaction theory using zonal averaging and we will summarize its results in their simplest form here.

13.2 Classic theory based on simple geometry

We assume a slowly varying wavetrain and employ phase-averaging for the definition of the mean fields. In the present case of a homogeneous wavetrain this implies that all mean fields, including the wave properties, are y-independent. It is possible to compute the horizontal radiation stress tensor for water waves of arbitrary depth (using integration over the vertical direction), but a three-dimensional mean flow is much more complicated to compute than a two-dimensional mean flow and therefore we will confine ourselves to the theory applied to the shallow-water equations, which broadly speaking is valid in the surf zone and out to water depths of a few metres.[1] For ease of reference, the governing equations are

$$\frac{Dh}{Dt} + h\nabla\cdot\boldsymbol{u} = 0 \quad \text{and} \quad \frac{D\boldsymbol{u}}{Dt} + g\nabla(h - H) = \boldsymbol{F} + \boldsymbol{B}. \tag{13.1}$$

where $H(x, y)$ is the still water depth (cf. §1.6). The body force on the right-hand side has been split into a part $\boldsymbol{F}$ related to wave dissipation and a part $\boldsymbol{B}$ related to bottom friction. This is usually taken to be a quadratic drag term modelling a turbulent viscous boundary layer at the bottom of the water column:

$$\boldsymbol{B} = -\frac{c_f}{h}\boldsymbol{u}|\boldsymbol{u}|. \tag{13.2}$$

Here c_f is a non-dimensional coefficient related to the roughness of the bottom; a typical value is $c_f = 0.01$ for beaches.

In the simplest form of the theory the topography varies only in the cross-shore direction, i.e., we allow $H(x)$ only. This is in fact required to obtain a homogeneous wavetrain in the longshore direction. A direct consequence of this is that the net y-momentum of the fluid is conserved by frictionless motion, because there can be no pressure-related form stress in the

[1] The main drawback of shallow-water theory is that it is very inaccurate in the description of dispersive water waves further offshore, so in practice one often couples an improved wave model to a mean-flow model based on shallow-water theory.

y-direction exerted on the topography in this case. Also, for linear waves relative to a basic state at rest, i.e., $\boldsymbol{u} = 0$ and $h = H(x)$, there is now conservation of the net y-pseudomomentum because the basic state is y-symmetric. The saturated mean flow is typically small compared to the speed of the waves, so Doppler-shifting effects are weak and can be neglected for simplicity.

13.2.1 Wave structure

The local wavetrain structure is that of shallow-water plane waves based on the local value of $H(x)$, i.e., if the wavenumber vector is defined as $\boldsymbol{k} = (k, l)$ then the waves are non-dispersive longitudinal waves obeying equipartition of energy in the form $\bar{E} = \overline{gh'^2} = H\overline{|\boldsymbol{u}'|^2}$ and also the local dispersion relation

$$\omega = \kappa\sqrt{gH(x)} \quad \text{such that} \quad \boldsymbol{c}_g = \frac{\boldsymbol{k}}{\kappa}\sqrt{gH} \quad \text{where} \quad \kappa = \sqrt{k^2 + l^2}. \tag{13.3}$$

The quadratic bottom friction can be ignored for linear waves and therefore the small-amplitude pseudomomentum law for the evolution of $H\mathsf{p} = \boldsymbol{k}\bar{E}/\hat{\omega}$ in a slowly varying wavetrain can be read off from the general (10.126), say, as

$$\frac{\partial \mathsf{p}_i}{\partial t} + \frac{1}{H}B_{ij,j} = -\frac{\bar{E}}{2H^2}H_{,i} + \mathcal{F}_i. \tag{13.4}$$

Note that the refractive source term stems from $\tilde{\rho} = H$ in (10.126); this is because $\boldsymbol{\nabla}H$ can be treated as a constant for a slowly varying wavetrain and therefore the additional term $\overline{\xi_{j,i}(H_{,j})^{\xi}} = 0$. The pseudomomentum flux tensor B_{ij} can also be written as

$$B_{ij} = H\overline{u_i' u_j'} = \bar{E}\,\frac{k_i k_j}{\kappa^2} = H\mathsf{p}_i c_{gj}, \tag{13.5}$$

which are useful expressions in ray tracing. Now, the structure of the steady wave field is computed in two stages. The first stage consists of inviscid propagation without breaking towards the shoreline. This takes place in the region $x_0 < x < x_b$, where x_0 is a reference location offshore and x_b is the breaker line, i.e., the location of the first occurrence of wave breaking. The symmetries of the basic state in y and t imply the conservation of l and ω along rays and in the present homogeneous and steady configuration this means that l and ω have the same value everywhere. Using a subscript zero to denote offshore values we thus have

$$\kappa(x) = \kappa_0 \sqrt{\frac{H_0}{H(x)}}, \quad k(x) = \kappa_0 \sqrt{\frac{H_0}{H(x)} - \frac{l_0^2}{\kappa_0^2}} \quad \text{and} \quad l = l_0. \tag{13.6}$$

The increase of k with decreasing water depth H quantifies the refraction towards the shoreline that is apparent in figure 13.1: ocean waves always come in parallel to the shore.

The y-component of the steady pseudomomentum law without dissipation is

$$B_{yx,x} + B_{yy,y} = \frac{\partial}{\partial x}(H\overline{v'u'}) + \frac{\partial}{\partial y}(H\overline{v'v'}) = 0. \tag{13.7}$$

The second term is zero because of the homogeneity of the wavetrain. Alternatively, we could also invoke the convenient *small-angle approximation*, which is based on $k \approx \kappa$ and $l \ll \kappa$. In this approximation only terms of first order in the small parameter l/κ are retained, e.g., $k = \kappa$ and the second term in (13.7) is then negligible even for two-dimensional wavetrains. The small-angle approximation obviously applies sufficiently close to the shoreline, but it may be inaccurate further offshore; its use is convenient but not essential for what follows.

Using this approximation (13.7) reduces to the constancy of

$$B_{yx} = B_{xy} = H\overline{u'v'} = \frac{l_0}{\kappa_0}\sqrt{\frac{H}{H_0}}\,\bar{E} \quad \Rightarrow \quad \bar{E}(x) \propto H(x)^{-1/2}. \tag{13.8}$$

The implied increase of $\bar{E}$ with decreasing H leads to an inexorable amplitude growth as the shoreline is approached and therefore to a breakdown of linear theory. Specifically, the non-dimensional amplitude A of plane shallow-water waves is the ratio of the depth disturbance to the still water depth, i.e., if $h' = \hat{h}\cos\theta$ then

$$A = \frac{\hat{h}}{H} \quad \Rightarrow \quad A(x) = \sqrt{\frac{2\bar{E}}{gH^2}} \propto H(x)^{-5/4}. \tag{13.9}$$

This sharp amplitude increase makes clear that wave breaking is inevitable and we assume that it occurs first at the breaker line $x = x_b$. This brings in the second stage, which concerns the structure of the wave field in the surf zone $x_b < x < 0$. This problem cannot be solved from first principles because the wave breaking involves three-dimensional turbulence and a breakdown of the assumption underlying the shallow-water equations. Various empirical schemes have been formulated to parametrize the decay of the breaking wave field in the surf zone; basically, these amount to a model for $\boldsymbol{\mathcal{F}} = (\mathcal{F}, \mathcal{G})$ to be used in the steady pseudomomentum law (13.4), i.e.,

$$\frac{1}{H}\nabla\cdot\left(\frac{\boldsymbol{k}\boldsymbol{k}}{\kappa^2}\bar{E}\right) + \frac{\bar{E}}{2H^2}\nabla H = \boldsymbol{\mathcal{F}}. \tag{13.10}$$

In the small-angle approximation this yields the convenient expressions

$$\mathcal{F} = \frac{1}{H^{3/2}}\frac{\partial}{\partial x}\left(H^{1/2}\bar{E}\right) \quad \text{and} \quad \mathcal{G} = \frac{l_0}{\kappa_0\sqrt{H_0}H}\frac{\partial}{\partial x}\left(H^{1/2}\bar{E}\right), \tag{13.11}$$

which are consistent with $\boldsymbol{\mathcal{F}} = 0$ outside the surf zone. The ratio between the force components is $\mathcal{G}/\mathcal{F} = l/k$, which is very small and together with $H_x < 0$ confirms that $-\boldsymbol{\mathcal{F}}$ is parallel to p, as expected.

The simplest parametrization scheme is that of wave saturation, in which the plane wave structure is maintained but the non-dimensional amplitude is pinned at a certain value, the saturation value $A = A_s$, say. This corresponds to wave breakers in which the surface elevation remains at a constant fraction of the local water depth; this is not inconsistent with casual observations of wave breaking on beaches. Notably, saturation reverses the usual logic in ray tracing: rather than using conservation of a wave activity to deduce the wave amplitude, in a saturation scheme we *fix* the wave amplitude and deduce the dissipation of wave activity.

Now, if the waves do not stop breaking in the surf zone, which will be justified if $H(x)$ decreases monotonically, then saturation implies

$$\bar{E}(x) = \frac{A^2}{2}gH^2 = \frac{A_s^2}{2}gH^2 \propto H(x)^2. \tag{13.12}$$

This decaying structure can be compared to the growing structure in (13.8). Substitution in (13.11) yields

$$\mathcal{F} = \frac{5}{4}A_s^2 g\, H_x \quad \text{and} \quad \mathcal{G} = \frac{5}{4}A_s^2 g\frac{l_0}{\kappa_0}\sqrt{\frac{H}{H_0}}\, H_x \tag{13.13}$$

in the surf zone. It would be hypercredulous to believe the exact numerical values in (13.13). A more defensible use of (13.13) is to extract plausible scalings of the dissipative forces, which can then be tested against observations. For example, on a planar beach with constant H_x this theory predicts that the cross-shore force $\mathcal{F}$ is constant and that the longshore force $\mathcal{G}$ is proportional to the square root of the distance from the shoreline. This is a useful and testable prediction.

13.2.2 Mean-flow response

The leading-order mean-flow equations for variable H are closely similar to (12.22):

$$\tilde{h}_t + \nabla\cdot(H\overline{\boldsymbol{u}}^L) = 0 \quad \text{and} \quad \overline{\boldsymbol{u}}_t^L + g\nabla(\tilde{h} - H) = -\frac{1}{H}\nabla\cdot\boldsymbol{S} + \overline{\boldsymbol{B}}^L. \tag{13.14}$$

In the first equation $\tilde{h} = \overline{h}$ holds at $O(a^2)$ for slowly varying wavetrains. The second equation uses that ∇H is treated as a constant for a slowly varying wavetrain and that $\overline{\boldsymbol{F}}^L \approx 0$ for momentum-conserving dissipative forces. Here S_{ij} is the radiation-stress tensor, which as discussed in §10.5.1 is related to the pseudomomentum flux tensor B_{ij} via

$$S_{ij} = B_{ij} + \frac{\gamma - 1}{2}\bar{E}\delta_{ij} = B_{ij} + \frac{1}{2}\bar{E}\delta_{ij}. \tag{13.15}$$

The relevant components in the small-angle approximation are

$$S_{xx} = H\overline{u'^2} + \frac{\bar{E}}{2} = \frac{3\bar{E}}{2} \quad \text{and} \quad S_{yx} = S_{xy} = H\overline{u'v'} = \frac{l_0}{\kappa_0}\sqrt{\frac{H}{H_0}}\,\bar{E}. \tag{13.16}$$

In addition, combining (13.4) and (13.15) yields the physical decomposition of the radiation-stress convergence (12.23):

$$-\frac{1}{H}\nabla\cdot\boldsymbol{S} = \mathsf{p}_t - \nabla\left(\frac{\bar{E}}{2H}\right) - \boldsymbol{\mathcal{F}}. \tag{13.17}$$

Both (13.16) and (13.17) are useful for different aspects of the mean-flow problem. In particular, for a steady wavetrain (13.17) allows the easy deduction that the radiation stress due to non-dissipating waves can be balanced by mean pressure forces via a suitable adjustment of the height field $\tilde{h}$. This non-acceleration result is not an obvious fact in the presence of refraction, because the radiation-stress is nonzero in that case. Conversely, for dissipating waves it is the *vortical part* of $\boldsymbol{\mathcal{F}}$ that cannot be balanced by mean pressure forces. Indeed, a consideration of $\nabla\times\boldsymbol{\mathcal{F}}$ will be the central topic in the local wavetrain theory in §13.3.

For a homogeneous wavetrain (13.14a) implies that $H\overline{u}^L$ is constant and therefore $\overline{u}^L = 0$ as there is no mass flux across the shoreline. This reduction to a one-dimensional mean shear flow is reminiscent of other two-dimensional mean-flow problems in simple geometry. If the x-component of the bottom friction force is ignored[2] then the x-component of (13.14b) reduces to

$$g\frac{\partial}{\partial x}(\tilde{h} - H) = -\frac{1}{H}S_{xx,x} = -\frac{3}{2H}\frac{\partial\bar{E}}{\partial x}. \tag{13.18}$$

[2] See the discussion of (13.21) below.

This equation includes the effect due to nonzero $\mathcal{F}$. It can be used to compute the 'set-up', i.e., the changes in the mean surface elevation $\tilde{h} - H$ due to the waves, both inside and outside the surf zone. It turns out that this 'set-up' has an interesting structure centred at the breaker line $x = x_b$: seawards from the breaker line $\bar{E}$ increases towards the shoreline and therefore the mean surface slopes down towards the beach. Specifically, (13.17) makes clear that $g(\tilde{h} - H) + \bar{E}/(2H)$ is constant in this region.

On the other hand, shorewards from the breaker line the situation is reversed and the mean surface slopes up towards the beach. In principle, this should lead to a mean surface minimum at the breaker line, which is associated with a pressure minimum and which may have implications for the existence of an undertow flow towards the breaker line. This is a beautiful result, but it is not clear whether this one-dimensional result about the mean pressure is robust against two-dimensional perturbations, for the reasons that were discussed in §12.2.1.

The y-component of (13.14b) involves the longshore current:

$$\overline{v}^L_t = -\frac{1}{H} S_{yx,x} + \overline{B}^L_y = -\mathcal{G} + \overline{B}^L_y. \tag{13.19}$$

Here $\overline{B}^L_y$ is the y-component of $\overline{\boldsymbol{B}}^L$. As expected, there is no wave-induced force outside the surf zone and inside the surf zone the effective mean force is given by the dissipative pseudomomentum rule. For a steady longshore current

$$\mathcal{G} = \overline{B}^L_y, \tag{13.20}$$

so we need to balance $\mathcal{G} = O(a^2)$ with $\overline{B}^L_y$ at this order. This indicates that it is the functional form of the friction term that determines the scaling of the steady longshore current. For example, for linear friction (which would be appropriate for a laminar boundary layer at the beach floor) $\overline{B}^L_y = O(a^2)$ if $\overline{v}^L = O(a^2)$, i.e., in this case the steady current is $O(a^2)$. Thus, for linear friction doubling the wave amplitude would quadruple the steady current strength, albeit from a low base compared to the $O(a)$ wave velocities. Also, if the wave amplitude were changing in time then the current would adjust on the same time scale to this change, i.e., the mean current is in approximate equilibrium with the wave forcing at all times.

However, for the more realistic quadratic friction law the same argument yields the much larger $\overline{v}^L = O(a)$ for the saturated current. Doubling the wave amplitude now yields a doubling of the steady current strength. Of course, it takes a long time $t = O(a^{-1})$ to grow such a steady current with an $O(a^2)$ force $-\mathcal{G}$, and therefore the mean current can be expected to lag in time a change in wave forcing. This is the hallmark of strong interactions, even though in this case the mean flow does not grow to $O(1)$.

Overall, the saturation of the mean flow at $O(a)$ makes this problem qualitatively different from the QBO problem described in §7.3.3. For example, even a current of strength $O(a)$ is negligible for the linear wave dynamics, so the approximation of a basic rest state for the incoming waves is self-consistent even for long times.

It is easy to compute $\overline{v}^L$ using the wave saturation model, although the evaluation of

$$\overline{B}_y^L = -\frac{c_f}{H}\overline{v|\boldsymbol{u}|} \tag{13.21}$$

at $O(a^2)$ is somewhat awkward. Substituting the linear wave solution in (13.21) gives zero because the force averages to zero by symmetry. However, there is a nonzero correlation between the $O(a)$ mean flow and the $O(a)$ wave field. In practice, one often assumes that u' and $\overline{v}^L$ are both $O(a)$ but that u' is significantly bigger than $\overline{v}^L$. In this case (13.21) can be evaluated as

$$\overline{B}_y^L = -\frac{c_f}{H}\frac{2}{\pi}\sqrt{\frac{2\bar{E}}{H}}\,\overline{v}^L. \tag{13.22}$$

Combining this with (13.12) in (13.20) yields

$$\begin{aligned} &\frac{5}{4}A_s^2 g\frac{l_0}{\kappa_0}\sqrt{\frac{H}{H_0}}H_x = -\frac{c_f}{H}\frac{2}{\pi}\sqrt{\frac{2\bar{E}}{H}}\,\overline{v}^L \qquad (13.23)\\ \Leftrightarrow\quad &-\frac{5\pi}{8c_f}A_s\sqrt{gH_0}\frac{l_0}{\kappa_0}\frac{H}{H_0}H_x = \overline{v}^L \end{aligned}$$

inside the surf zone. The same cautionary comment as below (13.13) applies here as well. For a planar beach with constant H_x this predicts a linear profile of $\overline{v}^L$ with distance from the shoreline, and that this profile terminates with a jump to zero at the breaker line. In practice, (13.23) tends to predict excessive current velocities at the breaker line, and it also does not predict the typically observed current maximum inside the surf zone. Naturally, adding horizontal mixing to the longshore mean-flow equation reduces the current maximum and it also moves the maximum of the current profile inside the surf zone. Of course, the strength of the horizontal mixing is then another parameter that needs to be tuned.

In summary, the classic theory for longshore currents in simple geometry yields detailed and interesting results, which have been the backbone of longshore current modelling for several decades. For instance, it is easy to build more elaborate models for wave saturation, or models that go beyond monochromatic waves, or which add reasonable extra processes such as horizontal mixing. Of course, for any such model the quantitative fit with observations depends on tuning the adjustable parameters. If one understand this basic model, then these more complicated models in simple geometry are easily understood as well.

In the next section we will investigate a model for wave-driven currents that allows for localized wavetrains. As we shall see, this changes the nature of the mean-flow response profoundly, especially on beaches with non-trivial topography such as barred beaches, which feature a local minimum of $H(x)$ at some location offshore.

13.3 Theory for inhomogeneous wavetrains

We now look at the inhomogeneous wavetrain in the right panel of figure 13.1. We retain phase-averaging and the y-periodicity of the flow, but the $O(a^2)$ mean fields are not y-independent anymore. Physically, an inhomogeneous wavetrain, which of course implies inhomogeneous wave breaking, can be due to either inhomogeneous wave sources or inhomogeneous topography. The latter possibility is the most common in practice, where underwater topography features can easily lead to 'point breaks' and to an inhomogeneous wavetrain structure in the longshore direction. However, the simplest situation for the theory retains one-dimensional $H(x)$ and allows the wave source to have a slowly varying envelope in the y-direction; this is the case we look at here. For ease of comparison with the homogeneous case, one can imagine repeating this inhomogeneous wavetrain pattern in the y-direction and comparing the resultant mean flow with that due to a homogeneous wavetrain with the same average momentum flux per unit longshore distance.

The equations derived in the previous section remain valid with obvious modifications. In particular, the initial mean-flow response is still governed by (13.14) and (13.17). For example, without wave breaking the radiation-stress convergence for a steady wavetrain can still be balanced by a suitable adjustment of $\tilde{h}$. This works even though in the present case the radiation-stress convergence is a vector with two nonzero components, but (13.17) makes clear that this vector is irrotational.

In the presence of dissipation we have

$$\frac{\partial}{\partial t}\boldsymbol{\nabla} \times (\overline{\boldsymbol{u}}^L - \mathsf{p}) = -\boldsymbol{\nabla} \times \boldsymbol{\mathcal{F}} = \mathcal{F}_y - \mathcal{G}_x, \tag{13.24}$$

which produces the characteristic dipolar structure depicted in figure 13.1: wave dissipation goes hand-in-hand with the generation of a vortex couple. Such a vortex couple would also be generated if the waves were normally incident, the influence of oblique incidence is that the couple is slightly tilted relative to the shoreline. Notably, although $\mathcal{F}$ played no role for the mean current in the homogeneous case it, $\mathcal{F}$ is now significant for the generation of the vortex couple.

We know from §12.3.3 that the occurrence of such horizontal vortex couples with vertical vorticity is a natural phenomenon in shallow-water theory with wave dissipation. However, shallow-water theory in only a model for the realistic three-dimensional dynamics of breaking ocean waves. It is perhaps not immediately obvious how the generation of vertical vorticity by wave dissipation in the shallow-water model is connected to the generation of three-dimensional vorticity by breaking waves in the real ocean. Moreover, in a numerical simulation of the nonlinear shallow-water equations wave breaking would manifest itself in the form of *shock formation*, which involves discontinuous flow fields and therefore may raise doubts about the validity of the wave–mean interaction theory, which assumed smooth fields.

Layerwise two-dimensional fluid models such as the shallow-water system are of central importance in geophysical fluid dynamics, and therefore we now consider in more detail the generation of vorticity and especially potential vorticity due to three-dimensional wave breaking and shock formation. This allows us to judge the scope and limitations, both theoretically and numerically, of modelling these phenomena with the shallow-water equations.

13.4 Vorticity generation by wave breaking and shock formation

It is a basic fluid-dynamical fact that wave breaking can generate vorticity and that this can occur even if there has been no vorticity prior to the breaking.[3] Here wave breaking is understood in a generalized sense, meaning the irreversible overturning of material lines or surfaces that would otherwise undulate reversibly under the dynamics of small-amplitude, linear waves. Examples of such material entities includes stratification surfaces of constant

[3] As an aside we note that by their mathematical construction both vorticity and potential vorticity always satisfy conservation laws in the integral sense, i.e. their evolution laws can always be written in flux divergence form. For instance, the Rossby–Ertel PV is defined by $q = \nabla \times \boldsymbol{u} \cdot \nabla\theta/\rho$, where θ is potential temperature for atmospheric applications. This can be rewritten as $\rho q = \nabla \cdot (\theta \nabla \times \boldsymbol{u})$, which makes obvious that $(\rho q)_t$ is always the divergence of some flux, even in the presence of forcing or heating. Thus, the generation of vorticity or PV by localized processes is always constrained to have zero net integral: there can be no monopolar source of vorticity or PV due to forcing processes with compact support.

entropy for internal waves, the air–water interface for surface waves, and the contours of potential vorticity on isentropic surfaces for Rossby waves.

Non-uniform shock waves in compressible flow models can also generate vorticity out of nothing, as is well known in gas dynamics. The shallow-water model is a two-dimensional compressible model for the three-dimensional incompressible flow of a thin fluid layer, and as such it inherits the capacity for shock formation and for the concomitant vorticity generation. We will look at both breaking and shock formation in turn in order to understand clearly the connection with the dissipative forces that appear in equations such as (13.24).

13.4.1 Vortex tubes generated by breaking waves

An idealized wave breaking scenario is depicted in figure 13.2, which on the left shows a stratification surface that is irreversibly deformed and rolled-up by a breaking wave travelling from left to right, i.e., in the positive x-direction. For example, this surface could be the air–water interface in the case of a violent plunging surface wave breaker. Alternatively, it could be a stratification surface of constant entropy that is being rolled up by a breaking internal gravity wave. We will focus on the surface wave interpretation.

Eventually the overturning wave crest crashes onto the fluid just before the crest. The flow can remain essentially two-dimensional and inviscid up to this moment, in particular up to this point Kelvin's circulation theorem holds for material contours lying within the stratification surface. After the crash, there are violent viscous boundary layer effects at the overturned water–water interface and a rapid transition to three-dimensional turbulence occurs, which is clearly visible, and audible, by the foam and bubbles in the breaking region. The presence of the three-dimensional turbulence alone indicates a significant creation of vorticity, but to us the more or less disorganized vorticity of the turbulence is not of primary interest. Rather, there is also an organized, large-scale component of vorticity in the y-direction, which results from the conversion of the overturning irrotational water mass into a rolling water mass with an appreciable amount of solid-body rotation in the y-direction, i.e., spinning clockwise in the xz-plane. In other words, a vortex tube aligned with the y-direction has been created as sketched in the middle of figure 13.2! Notably, such spanwise vortex tubes played a crucial role in the theory of Langmuir circulations described in §11.3.2.

Now, vorticity lines cannot end inside a fluid body and this has important consequences if wave breaking is possible only in some finite lateral region $y \in [0, D]$, with some lateral breaking width D. This would be appropriate

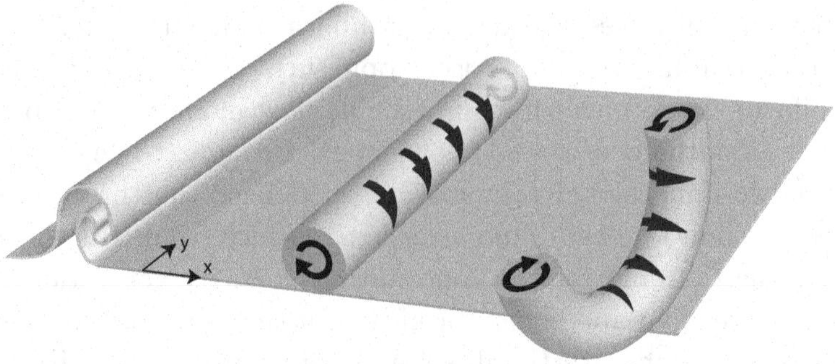

Figure 13.2 The generation of a vortex tube by a breaking wave. Left: a breaking wave moves in the x-direction and rolls up part of a stratification surface. Middle: the rolled-up fluid becomes a vortex tube, i.e., an organized bundle of spanwise vorticity pointing in the y-direction, spinning clockwise as indicated (this motivated the basic vorticity profile assumed in §11.3.2). Right: if the wave breaking region is finite in the spanwise direction then the vortex tube is likewise finite, and its three-dimensional vortex lines must then inevitably pierce the stratification surfaces. For surface waves this leads to the indicated banana-shaped vortex tube, and to the characteristic dipolar pattern of vertical vorticity that was depicted in figure 13.1.

for the localized wave breaking on a beach with a wavetrain of size D in the y-direction, for instance. In this case the vorticity must be confined to this lateral interval and because vorticity lines cannot end in the fluid that means the organized vortex tube in the core of the breaker must burrow out and connect to the boundaries of the fluid. If the water is deep enough it stands to reason that this will occur at the air–water boundary. This leads to the image of a banana-shaped tube of organized vorticity, with the core of the banana lying in $y \in [0, D]$ and the upward-bending ends of the banana[4] connecting to the free surface as depicted on the right of figure 13.2.

To understand the sense of circulation around the ends of the tube it is useful to envision a rolling banana at the moment when its curved ends are pointing upwards: this makes clear that viewed from above (i.e., looking down onto the xy-plane) the vertical vorticity is such that there is positive vorticity to the left of the propagation direction of the breaking wave and negative vorticity to the right, as in figure 12.5. Consequently, the impulse associated with this vertical vorticity points in the same direction as the horizontal phase velocity of the wave. This is consistent with a breaking-induced conversion of horizontal pseudomomentum (which is parallel to the intrinsic phase speed) into mean-flow impulse.

As noted before, fundamentally the same considerations can be applied to the overturning and breaking of internal waves in a continuously stratified fluid, with stratification surfaces of constant entropy replacing the water–air

[4] If you feel lucky, you can replace the vorticity banana with a horseshoe vortex.

interface and with Rossby–Ertel PV replacing the vertical vorticity. Again, the lateral confinement of the breaking region sets the width of the resultant vortex couple that is being produced on the stratification surface.

Hence in the context of wave breaking the vertical vorticity in the shallow-water system can be interpreted as the intersection of three-dimensional vortex tubes with the relevant stratification surface, which is the free surface in the shallow-water model. Of course, the original scaling assumptions underlying the validity of the shallow-water model are defeated by the presence of the strong organized horizontal vorticity due to wave breaking. For example, with wave breaking it is clearly not true that the horizontal flow velocities are constant throughout the water column. What rescues the utility of the shallow-water system in the presence of wave breaking is the interpretation of the shallow-water flow fields as vertical integrals over the water column. In this interpretation the shallow-water model has a consistent mass and horizontal momentum budget and the principal error in its formulation is then associated with the hydrostatic approximation for the pressure field throughout the water column. This interpretation is the basis for the modelling of three-dimensional wave breaking by two-dimensional shock formation.

13.4.2 Shocks as models for wave breaking

As said before, waves in the nonlinear shallow-water equation naturally form shocks, and the speed and the structure of these shocks is determined by the usual jump conditions that follow from the conservation laws for mass and horizontal momentum across a shock.

On the other hand, real ocean wave breaking regions are turbulent bores, i.e., turbulent transition regions between water masses with significantly different depth and vertically integrated horizontal velocity. The simplest model for the jump conditions across such bores is derived by assuming that the boundary pressure at the free surface, including a smooth cut through the turbulent region, is the constant air pressure whilst the interior pressure at some distance away from the bore is the standard hydrostatic pressure distribution. For example, this is the usual model for the familiar tidal bores, which can propagate over large distances along canals. Remarkably, the jump conditions at bores derived from this model are equivalent to the jump conditions at shallow-water shocks. This means that the shallow-water equations can be used to predict such bores.

Now, it is a basic fact from gas dynamics that curved shocks, or shocks with variable strength along the shock front, can create vorticity. This is a consequence of the jump conditions at the shock for the velocity component normal to the shock front: basically, the jump in this velocity component is a function of the shock strength and hence a shock with variable strength along the front creates vorticity due to the variation of this normal velocity along the front. Thus, the vertical vorticity generation by a shock front in shallow water will agree with that produced by a turbulent bore to the extent that the respective jump conditions agree.

Finally, there is no need to admit truly discontinuous flow fields into the equations of wave–mean interaction theory. The addition of momentum-conserving viscous terms, such as those occurring in the usual Navier–Stokes equations for compressible flows, allows for smooth flow fields across jumps, and by making the viscosity coefficient small enough one can confine the importance of viscosity to the jump locations. This standard argument from gas dynamics implies that wave–mean interaction theory, and its conclusions such as the pseudomomentum rule, remain valid for any nonzero value of the viscosity coefficient. Crucially, it is the combination of local mass and momentum conservation that yields the correct vorticity generation; this echoes the importance of momentum-conserving dissipative forces that was pointed in connection with the dissipative pseudomomentum rule in §11.1.1.

This motivates the use of numerical schemes that conserve mass and momentum on the grid scale, because the shock size will always shrink to the grid scale. In practice, this means that finite-volume schemes, in which such conservation laws are hard-wired into the definition of the discrete variables as cell averages, are the best choice for flow simulations with shocks. Other schemes are possible as well, of course, but they require diligent care to ensure local mass and momentum conservation on the grid scale.

In summary, the one-layer shallow-water model does much better than could have been expected as a model for breaking ocean waves and for the concomitant vortical mean-flow response. The ultimate reason for this is that the shallow-water equations can be viewed as a consistent, if approximate, mass and momentum budget for the vertically integrated flow, and that the jump conditions in shallow water agree with the standard model for the jump conditions at a turbulent bore.

The situation is harder for internal wave breaking in three-dimensional stratified flow. There the shallow-water equations are indeed relevant if the pressure is approximated as hydrostatic and if the flow is formulated using the stratification variables as the vertical coordinate. However, there is no standard formulation for a bore jump condition in three-dimensional stratified flow: for example, the conservation laws for mass, momentum and energy are not sufficient to determine the jump conditions in multi-layer flows. Therefore it is hard to judge the verisimilitude of a shallow-water version of internal wave breaking.

13.5 Vortex dynamics on sloping beaches

The $O(a)$ saturated longshore current in the homogeneous problem was a steady solution of the governing equations so, stability issues aside, there was no need to include the nonlinear terms $(\overline{\boldsymbol{u}}^L \cdot \boldsymbol{\nabla})\overline{\boldsymbol{u}}^L$ in the mean momentum equations even though these terms are $O(a^2)$ for the saturated current and therefore of the same order as the forcing and bottom friction terms. The situation is clearly different in the inhomogeneous version of the problem, because a saturated $O(a)$ vortical mean flow will not be a steady solution, and hence will evolve according to nonlinear vortex dynamics. Vortex dynamics on sloping terrain such as a beach has a number of interesting features and therefore we will look at this dynamical system more closely here.

The vortical mean flow is a slow low-Froude number flow and it is convenient to study its dynamics using simpler shallow-water equations based on replacing the free surface by a rigid lid. The unforced rigid-lid shallow-water equations are

$$\boldsymbol{\nabla} \cdot (H\boldsymbol{u}) = 0 \quad \text{and} \quad \frac{D\boldsymbol{u}}{Dt} + \boldsymbol{\nabla} P = 0. \tag{13.25}$$

Here P is a normalized pressure at the rigid lid, which as in incompressible flow is determined non-locally from the constraint imposed by the continuity equation. Indeed, for constant H we just retrieve the equations for two-dimensional incompressible homogeneous flow. This indicates that the three fields (u, v, P) are coupled together such that there is only a single degree of freedom, which is the vortical mode in the linear problem.

These equations obviously satisfy the conditions for Kelvin's circulation theorem and we have the standard PV law

$$q = \frac{\boldsymbol{\nabla} \times \boldsymbol{u}}{H} \quad \text{such that} \quad \frac{Dq}{Dt} = 0. \tag{13.26}$$

Non-constant H leads to vortex stretching effects as fluid parcels move through regions of variable depth. The easiest way to gain insight into the full effect of nonzero $\boldsymbol{\nabla} H$ is via the vorticity stream function formulation, which in this case is based on the mass stream function ψ such that $\psi_x = Hv$ and $\psi_y = -Hu$ and therefore the dynamics can be written as

$$\boldsymbol{\nabla} \cdot \left(\frac{\boldsymbol{\nabla} \psi}{H} \right) = Hq \quad \text{and} \quad q_t + \frac{1}{H} \partial(\psi, q) = 0, \tag{13.27}$$

where the Jacobian $\partial(\psi, q) = \psi_x q_y - \psi_y q_x$. The diagnostic relation between

ψ and q in (13.27a) involves a linear self-adjoint operator, as is easy to show in the case of a bounded domain with $\psi = 0$ on the boundary, say. As in §9.1.2, this implies the conservation of the vortical energy in the generic form

$$-\frac{1}{2}\int \psi q H\, dxdy = \frac{1}{2}\int |\boldsymbol{u}|^2 H\, dxdy, \tag{13.28}$$

which is just the total kinetic energy. There is potential energy, of course, but due to the rigid lid none of it is available for conversion into kinetic energy.

13.5.1 Impulse for one-dimensional topography

For general two-dimensional $H(x,y)$ neither the momentum nor the impulse is conserved, because of the pressure-related momentum exchange between the fluid and the undulating topography. However, for one-dimensional $H(x)$ the y-component of momentum is conserved, and we note here that there is a corresponding definition of longshore impulse that is also conserved in this special case. For simplicity, we consider a y-periodic channel geometry with solid walls at two locations in x, say at the shoreline and at some offshore location, such that $u = 0$ there.

If we inspect the derivation of how the y-impulse with density $-xq dxdy$ was conserved in the case of constant H, then we find that the key steps were that $-Dx/Dt = u$ and $D(dxdy)/Dt = 0$. For variable $H(x)$ we need to adapt both of these, and playing with the equations then leads to replacing $dxdy$ by $H dxdy$ and $-x$ by a function $L(x)$ such that

$$\frac{dL}{dx} = -H(x) \quad \Rightarrow \quad \frac{DL}{Dt} = uL_x = -uH. \tag{13.29}$$

The y-component of the impulse is then

$$I_y = \int L(x)H(x)q\, dxdy \quad \Rightarrow \quad \frac{dI_y}{dt} = 0. \tag{13.30}$$

The proof uses $DL/Dt = -Hu$, integration by parts, periodicity in y, and that $u = 0$ at the channel side walls. In the constant-depth case $H = 1$ we have $L(x) = -x$ and therefore (13.30) recovers the classical impulse. On a planar constant-slope beach with $H = x$, say, we obtain $L = -x^2/2$ and so on. In general, I_y equals the net y-momentum plus some constant terms related to the conserved circulation along the channel walls. If forcing is added then it is straightforward to show that I_y changes according to the net input of longshore momentum.

Analogously, for wave–vortex interactions we can define a shallow-water mean-flow impulse in the y-direction at $O(a^2)$ by

$$\mathcal{I}_y = \int L(x)H(x)\overline{q}^L \, dxdy \quad \Rightarrow \quad \frac{d}{dt}(\mathcal{I}_y + \mathcal{P}_y) = 0 \tag{13.31}$$

under unforced evolution, or momentum-conserving dissipation. Here the definition of the net longshore pseudomomentum $\mathcal{P}_y$ includes a factor of $H(x)$ in the integrand. This conservation law also holds for wave refraction by the mean flow, the only difference is that the production term $-(\overline{\boldsymbol{u}}_y^L) \cdot \mathbf{p}$ then also acquires a factor of $H(x)$.

To illustrate (13.30), we again consider I_y due to a vortex couple with circulations $\pm\Gamma$ and separation d. We assume the couple makes an angle θ with the shoreline such that $\theta = 0$ corresponds to the vortex separation being parallel to the shoreline, i.e., θ is the angle of wave incidence in the right panel of figure 13.1. Thus $\Delta x = d \sin\theta$ is the cross-shore separation of the vortices. Now, if the left vortex has positive circulation then in the case of constant $H = 1$ this produces $I_y = \Gamma\Delta x$, so a constant longshore impulse implied a constant cross-shore separation. For variable H we obtain $I_y = \Gamma\Delta L$ instead, where ΔL is the difference of $L(x)$ between the two vortex locations.

For example, on a planar beach with $H = -x$ and $L = x^2/2$ the conclusion would be that $\bar{x}\Delta x$ is constant as the couple moves across the topography. Here $\bar{x}$ is the average x-position of the two vortices. This has the peculiar consequence that the cross-shore separation Δx of a vortex couple climbing a planar beach towards the shoreline (i.e., propagating towards $x = 0$ if $H = -x$) will *increase* as the distance to the shoreline decreases. In other words, whilst wave crests always come in parallel to the shoreline, vortex couples always come in crookedly.

13.5.2 Self-advection of vortices

The unit of understanding in vortex dynamics is the Green's function for the diagnostic relation (13.27a), i.e., the stream function belonging to a point vortex with circulation Γ centred at some (x_0, y_0), say. For simplicity, we put the vortex centre at the origin of the coordinate system, thus we want the solution to

$$\nabla^2\psi - \frac{\boldsymbol{\nabla}\psi \cdot \boldsymbol{\nabla}H}{H} = H^2 q = H\Gamma\delta(x)\delta(y). \tag{13.32}$$

This equation can be solved analytically in a few simple cases, such as that of exponential topography where $H \propto \exp(x)$, but in general it must be solved numerically. However, for the self-advection of a vortex only the stream function in the vicinity of the vortex location is relevant, and we can extract some asymptotic information about this from (13.32). This is based on the fact that the highest-order term in (13.32) is the Laplacian and therefore the local structure of the stream function near the point vortex is dominated by a logarithmic term, just as in the case of flat topography. Let us consider weak topography such that $\nabla H = O(\epsilon)$ and make a formal expansion $\psi = \psi_0 + \epsilon\psi_1 + \ldots$ in the small parameter $\epsilon \ll 1$. We assume that $H_y = 0$ at the vortex location, which we can do without loss of generality by a coordinate rotation if necessary. It then follows that

$$\nabla^2\psi_0 = H\Gamma\delta(x)\delta(y) \quad \text{and} \quad \nabla^2\psi_1 = \psi_{0x}\frac{H_x}{H}, \tag{13.33}$$

from which we can formally derive that

$$\psi_0 = \frac{\Gamma H(0,0)}{2\pi}\ln r \quad \text{and} \quad \nabla^2\psi_1 = \frac{\Gamma}{2\pi}\frac{x}{r^2}H_x(0,0) \tag{13.34}$$

in the neighbourhood of the vortex. Thus on a planar beach with $H_x < 0$ the effective vorticity for computing ψ_1 has the dipolar structure of a vortex couple that implies propagation in the positive y-direction for positive Γ. In other words, the topography slope leads to self-advection of the vortex along a line of constant water depth.

This formal conclusion is qualitatively robust and correct, but it turns out that a point vortex model is ill-posed on sloping terrain. Specifically, if the point vortex is replaced by a finite-core vortex with radius b and $b \to 0$ then the self-advection velocity diverges as $\ln b$. Both the direction of the self-advection and the divergence as $b \to 0$ can be understood heuristically by two enjoyable arguments. First, we can consider the special case of a wedge-shaped topography such that $H = x$, say. This corresponds to a strictly planar beach. In this case we can make use of a virtual three-dimensional extension of a finite-core vortex into a closed circular tube of vorticity whose axis coincides with the shoreline. By construction, the three-dimensional velocity field belonging to this vortex ring satisfies the no-normal-flow condition at both $z = 0$ and $z = -H(x)$ and therefore the restriction of this field to the wedge presumably provides a good model for the rigid-lid flow on a planar beach. As is well known, the self-advection of a thin vortex ring with radius R and core radius $b \ll R$ is approximately

$$\frac{\Gamma}{4\pi R}\left(\ln\left(\frac{8R}{b}\right) - \frac{1}{4}\right), \tag{13.35}$$

which illustrates the logarithmic divergence with core radius. Here the appropriate ring radius is the distance to the shoreline $R = |H_x|/H$. The self-advection of the vortex ring moves it along its axis, i.e., it moves the finite-core vortex parallel to the shoreline, following a line of constant depth, and with a sense of orientation such that a vortex with positive circulation moves to the right when viewed from the shoreline.

The second argument uses arbitrary one-dimensional topography $H(x)$ in an unbounded domain under the assumption that H goes to finite limits as $x \to \pm\infty$. The rate of change of the longshore position of the vorticity centroid is then given by

$$\frac{d}{dt}\int yHq\,dxdy = \int vHq\,dxdy \qquad (13.36)$$

and in the case of a small vortex this is approximately equal to Γ times the self-induced y-velocity of the vortex. On the other hand, integration by parts, using $H\nabla\cdot\boldsymbol{u} = -uH_x$, and disregarding all boundary terms leads to the exact relation

$$\frac{d}{dt}\int yHq\,dxdy = \int vHq\,dxdy = -\int u^2\frac{H_x}{H}\,dxdy. \qquad (13.37)$$

For monotonic H this makes clear that the vortex will indeed slide along in the y-direction. Moreover, as the core radius $b \to 0$ the integral will be dominated by the $1/r$ structure of u near the origin, which yields an asymptotic $\ln b$ term for this integral, as expected. This argument also makes clear that it is the sign of H_x at the vortex location that determines the sign of the self advection.

Now, the essentially local nature of the self-induced flow suggests a useful approximate formula for the self-advection velocity for general $H(x, y)$, namely

$$\frac{\Gamma}{4\pi}\frac{\nabla H\times\hat{\boldsymbol{z}}}{H}\left(\ln\left(\frac{8H}{b|\nabla H|}\right) - \frac{1}{4}\right). \qquad (13.38)$$

This is based on the vortex ring formula fitted with a radius R based on the local value of $|\nabla H|/H$. This formula should be accurate as long as b is small compared to the scale of the topography. By construction, the self-advection moves the vortex along a line of constant depth.

Thus the general conclusion is that a small finite-core vortex on sloping terrain is subject to self-advection along lines of constant water depth, and that the speed of this advection is proportional to $-\Gamma\ln b\nabla H\times\hat{\boldsymbol{z}}/H$. Importantly, the sense of self-advection depends on the sign of Γ and for a vortex couple as depicted in figure 13.1 this means that the vortex couple will tend to separate due to the effect of self-advection.

13.5.3 Mutual interaction of vortices and rip currents

The mutual interaction of vortices involves the far-field of the stream function for a single vortex, which depends globally on the structure $H(x, y)$ and therefore does not have a generic structure. However, one generic piece of information is available: we know that the circulation around a single vortex is independent of the distance r from the vortex and this implies that the velocity cannot decay faster than $1/r$, for example. Thus we can qualitatively think of the mutual interaction between two vortices as being similar to the familiar interaction between two-dimensional vortices in a domain of constant H.

Thus a vortex couple as depicted in figure 13.1 will tend to move forwards, normal to its separation vector, at a speed proportional to Γ/d where d is the distance between the vortices. Moreover, the no-normal-flow condition at the straight shoreline can be qualitatively taken into account by adding equal-and-opposite image vortices in the region $x > 0$, which is the standard procedure in the case of constant H. The impact of the image vortices is to slow down the vortex couple and to increase their separation distance d as they approach the shoreline. The upshot of both the shoreline effect and the self-advection effect is that a vortex couple that tries to climb a planar beach gets separated and therefore slows down.

Conversely, a vortex couple moving in the other direction, i.e., moving into deeper water, would have its separation d decreased by the same mechanisms. This is clear at once, because this process is simply the time-reverse of the other. Thus, a vortex couple driving out to sea *speeds up* as it propagates into deeper water. In this case the region of fast-moving water between the vortex couple can be viewed as a rip current, and the speed-up of the vortex couple in deeper water illustrates the danger of such currents for swimmers. This also suggests why swimming parallel to the shoreline is the right method to defeat a rip current: the longshore width of the vortex couple decreases in deeper water.

A link between water depth and vortex separation is also apparent from the mass budget: a vortex with radius b moving into deeper water conserves its volume and therefore b^2H remains constant. This implies that b must decrease, and therefore there is a horizontal convergence of area flow into the vortex. This is true for both members of a vortex couple, and therefore their separation d decreases because the horizontal flow is contracting.

The picture that emerges from these considerations is one in which regions of deep water are more attractive to vortex couples than regions of shallow water. Simply put, once a vortex couple is in a deep region it will find it difficult to climb out again, and the converse is true for couples in shallow regions. Arguably, vortices prefer deep water.

13.5.4 A statistical argument for vortex locations

Freely-evolving flows that consist of a large number of well-separated vortices allow the application of methods from the theory of statistical mechanics for discrete systems. This requires knowing the energy (and any other conserved integral quantity such as the longshore impulse in the case of one-dimensional topography $H(x)$) as a function of the vortex locations and strengths. For example, for constant H the classic results due to Kirchhoff show that the interaction energy between two point vortices with circulations $\Gamma_{1,2}$ is proportional to $-\Gamma_1\Gamma_2 \ln r_{12}$, where r_{12} is the distance between the vortices. Thus for a vortex couple with $\Gamma = \Gamma_1 = -\Gamma_2$ and $r_{12} = d$ we have $\Gamma^2 \ln d$, i.e., the couple is more energetic if the distance increases. Similarly, the self-energy of a vortex due to the presence of a straight wall is proportional to $\Gamma^2 \ln r_0$, say, where r_0 is the distance to the wall. The self energy increases with distance from the wall, which is obvious from the implied increased separation from the oppositely-signed image vortex in this case.

Now, for variable H the situation is less simple, but we can point out some simple heuristic facts. Most importantly, the approximation (13.38) for the self-advection velocity implies that the self energy is an increasing function of depth H. In particular, disregarding changes in b as the vortex moves around, the self energy is proportional to $\Gamma^2 \ln H$. One way to derive this result is to consider the action of a compact force in the y-direction applied to a small vortex with $\Gamma > 0$ on a planar beach. If the support of the force matches the support of q then the curl of this force will move the centroid of q offshore, i.e., into deeper water. At the same time, the work done by the force can be computed by integrating the dot product of the force and the fluid velocity over the vortex, which is basically the self-advection velocity. This product is positive, so deeper water corresponds to higher energy. Thus on a planar beach the self energy again increases with distance from the shoreline.

That the self energy is proportional to $\Gamma^2 \ln H$ is in fact consistent with the vortex couple energy being proportional to $\Gamma^2 \ln d$: a vortex couple moving into deeper water is pushed together, which decreases its interaction energy, whilst the self energy of either vortex increases to compensate.

Now, the methods of statistical mechanics were first applied to ensembles of point vortices in a bounded domain by Onsager, who made the key observation that depending on the overall energy level two distinct statistical regimes are possible. Not surprisingly, in the regime of low total energy the vortices tend to be found in low-energy configurations, i.e., close to walls, in loose groups of same-signed vortices, and in tight groups (typically couples) of oppositely-signed vortices. In the regime of high total energy the reverse is true. The surprising twist in Onsager's theory is that these trends are strongly amplified for vortices with larger absolute circulations, i.e., in a mixture of strong and weak vortices as measured by values of Γ, the strong vortices will behave more orderly and less randomly than the weak vortices. In the regime of high total energy this predicts the clustering of strong vortices in tight groups of same-signed vortices and Onsager interpreted this behaviour as a possible statistical underpinning of the well-observed tendencies of two-dimensional turbulent flows to form strong large-scale vortices.[5]

Viewed from this statistical perspective, regions of deep water would appear to be preferred by strong vortices, at least if the total energy level is high enough. Conversely, if the total energy is low, strong vortices would prefer shallow regions. In both regimes, the weaker vortices would roam the domain more or less at random, with little preference for either deep or shallow regions.

It is possible to firm up the details of this heuristic statistical argument at least in the case of one-dimensional topography $H(x)$, where a variable transformation $H dx = dX$ and $dy = dY$ allows writing the flow as an incompressible flow in XY-space, i.e., using the transformed stream function derivatives as $\psi_X = v = V$ and $-\psi_Y = Hu = U$, we have $U_X + V_Y = 0$. The vorticity and stream function formulation (13.27) is then replaced by

$$\psi_{XX} + \frac{1}{H^2}\psi_{YY} = q \quad \text{and} \quad q_t + Uq_X + Vq_Y = 0. \tag{13.39}$$

Note that finding the function $H(X)$ from $H(x)$ is part of this transformation[6] and that (13.39a) is the appropriate self-adjoint form because a delta function in vorticity transforms as $Hq = \Gamma\delta(x)\delta(y) = H\Gamma\delta(X)\delta(Y)$.

This variable transformation neatly takes care of the neglected impact of changes in the vortex size, and it provides canonical coordinates for the statistical mechanics vortex theory. Basically, in addition to the heuristic energy effects described above, this more detailed theory contains a purely kinematic effect, namely that a region of deep water in xy-space covers "more"

[5] There are obvious difficulties in applying such a point-vortex theory to the continuous vorticity fields of real two-dimensional turbulence, but Onsager's pioneering work has certainly provided motivation and inspiration to several generations of turbulence theorists.

[6] For example, on a planar beach with $x \geq 0$ and $H = x$ we find $H(X) = \sqrt{2X}$, whilst on an exponential beach with $H = \exp x$ we find $H(X) = X$.

XY-space than a region of shallow water. This is due to the weighting with H in the map between the spaces, which adds a kinematic, energy-independent tendency for vortices to favour deep water: deep regions correspond to more phase space than shallow regions.

Of course, these arguments neglect the impact of bottom friction and we will see in §13.6.2 that for realistic values of c_f bottom friction strongly affects the flow on a beach. However, these friction-free considerations may be of use in other fluid systems, which is why we have included them here.

13.6 Barred beaches and current dislocation

As mentioned before, the classic theory based on homogeneous wavetrains together with its modifications and extensions forms the backbone of present-day numerical modelling of longshore currents. To be sure, the quantitative fit with data from laboratory or field experiments depends on the tuning of the adjustable parameters, but the qualitative agreement with nature is impressively good, at least on beaches where $H(x)$ decreases monotonically towards the shoreline.

However, there have been systematic qualitative differences between predictions and observations in the case of non-monotonic beaches such as barred beaches, on which $H(x)$ has a local minimum at some offshore bar location $x = x_b$, say. Ray tracing then predicts *two* locations of wave breaking, one at bar and one at the final approach to the shoreline. There is a hiatus of wave breaking in the bar trough shorewards of x_b, because the deeper water in the trough makes the wave amplitude drop to non-breaking levels. Consistent with this picture of wave breaking on a barred beach, which is easily confirmed as realistic from observations, the classical theory predicts two currents, one over the bar and a second, weaker one near the shoreline. This is illustrated in figure 13.3 by a direct numerical simulation of both waves and currents using a shallow-water model.

However, observation on real barred beaches frequently (though not always) show a *single* current situated in the bar trough, i.e., the current is situated away from either wave breaking region. In other words, the current has been dislocated shorewards, and away from the region of wave breaking over the bar.

This observed current dislocation poses a conundrum for the classical theory, which by (13.19,13.20) is based fundamentally on a local balance between wave breaking and bottom friction, i.e., there can be no current without wave breaking in this theory. Small changes such as adjusting horizontal diffusion make little difference in the case of barred beaches. This

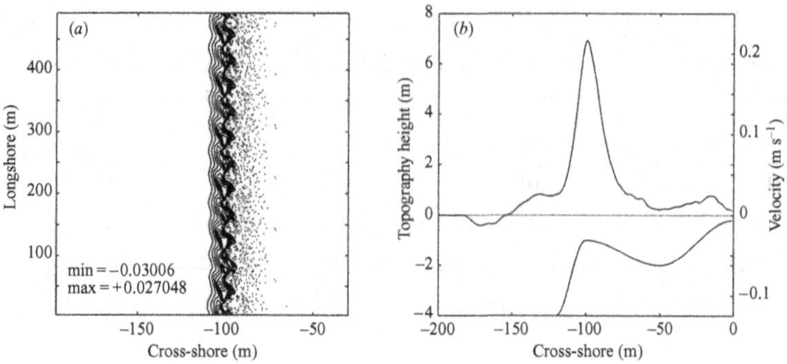

Figure 13.3 Shallow-water simulations with a homogeneous wavetrain after 1200 wave periods on a barred beach. Left: PV contours. Solid or dotted contours correspond to positive or negative PV, respectively. Right: mean longshore velocity and topography shape. From Bühler and Jacobson (2001).

qualitative disagreement between data and the predictions of classical theory is good news scientifically, because such a strong, qualitative disagreement suggests the presence of hitherto neglected physical processes.

For example, a well-received suggestion for such a physical process in coastal oceanography are so-called 'wave rollers', which are meant to describe the rolling water mass gliding on top of a breaking wave crest. It is fair to say that there is no fundamental theory for such rollers, but it is assumed that these rollers can store momentum and then release it into the mean current with some delay as the breaking wave crest travels along. Not surprisingly, the delayed release of the roller momentum allows a current dislocation effect to take place within classical theory. By tuning the various parameters a very good fit between the roller models and observations can usually be obtained.

Of course, in any parametrization scheme with tunable parameters there is a real scientific question about the interpretation of such a fit. For example, changing the topography structure $H(x)$ typically necessitates retuning many of the roller parameters, as there is no fundamental fluid-dynamical theory for wave rollers that could be used to predict how the parameters ought to change. Also, apparently sometimes the best model fit is obtained by setting parameters to values outside their physical bounds, which questions the fundamental physical interpretation of these parameters. Perhaps the most fundamental scientific question is this: if wave rollers are essential on barred beaches, why are they not essential on planar beaches, where the classic theory applies well without them?

Of course, other physical processes may also contribute significantly to the current structure on barred beaches. For example, here we briefly sketch an argument based on inhomogeneous wavetrains and simple vortex dynamics.

The attraction of this argument is that it does not require new fluid dynamics, and that it seems to explain why current dislocation is significant on barred beaches and not so on planar beaches. On the downside, the argument relies on significant amounts of vertical mean-flow vorticity being generated, and this assumption may or may not be valid on real beaches.

13.6.1 Current dislocation by vortex dynamics

The basic idea is disarmingly simple and readily apparent from the right panel of figure 13.1: a mean-flow vortex couple formed by inhomogeneous wave breaking will grow in strength until it starts moving under the joint effects of self-advection and mutual advection. On a planar beach, the vortices will then separate in the longshore direction due to the oppositely-signed self-advection of the two vortices, and hence the shorewards motion of the couple due to mutual advection will weaken until, eventually, the vortices will decay due to bottom friction. If we identify the fast moving body of water between the vortices with the main current, then this argument shows that nonlinear vortex dynamics does not yield much change to the current location on a planar beach, i.e., current dislocation by vortices is weak on a planar beach.

On the other hand, on a barred beach the shoreward motion of the vortex couple moves it into deeper water, which narrows the vortex couple and accelerates its shorewards motion. Clearly, as the couple descends into the bar trough the topography slope is such that the self-advection pushes the couple closer together, thus enhancing the mutual advection. The upshot is that the vortex couple quickly makes it into the bar trough. Once there, the couple will try to climb the topography towards the shoreline, but now it will be subject to the same dynamics as the original couple on a planar beach. Thus the couple will not make much headway in its approach up to the shoreline, and bottom friction will once again terminate its progress.

Overall, this scenario predicts that on a barred beach the current should be dislocated into the bar trough, and that the most likely current location is just shorewards of this trough. This is illustrated in figure 13.4, which is based on shallow-water simulations with an inhomogeneous wavetrain.

A by-product of these vortex-based dynamical ideas is a natural mechanism to generate rip currents even on a planar beach. This follows from a periodic repetition in the longshore direction of the situation depicted in figure 13.1: if the friction is low enough then the upper positive vortex, which propagates in the positive y-direction, should meet with a lower negative vortex from the next wavetrain above, which travels in the negative

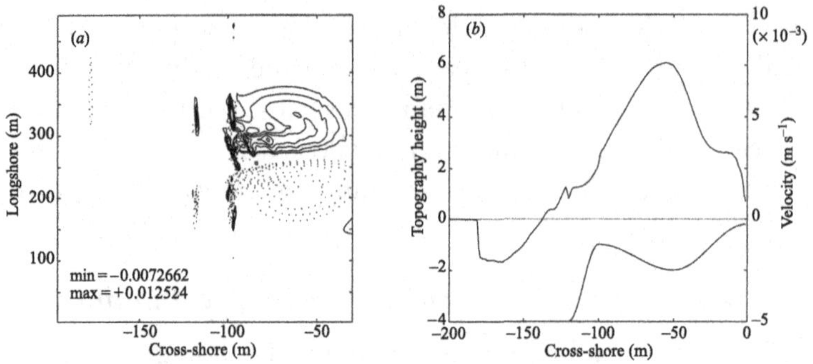

Figure 13.4 Shallow-water simulations with an inhomogeneous wavetrain after 1200 wave periods on a barred beach. Left: PV contours. Solid or dotted contours correspond to positive or negative PV, respectively. Right: mean longshore velocity and topography shape. From Bühler and Jacobson (2001).

y-direction, and then these vortices can mingle to form a new vortex couple, but this time a vortex couple driving outwards to sea as a rip current.

Finally, the vortex dynamics that was described here is generic and does not depend on the details of how the vortices have been generated by the breaking waves, or by other dynamical processes. For instance, it is well known that a homogeneous longshore current may break up into vortices due to shear flow instability, and these vortices are quite unrelated to the details of the original wave breaking that drove the current. Still, it stands to reason that the further development of these vortices will again lead to the kind of vortex-induced current dislocation described here.

13.6.2 Bottom fricton and turbulence

It is intriguing to speculate whether the concepts of two-dimensional turbulence, such as vortex merging and a concomitant up-scale cascade of energy, could apply to wave-driven vortex dynamics on beaches. However, it turns out that for realistic values of bottom friction the range of shallow-water wavenumbers that can be actively turbulent is very small indeed. This suggests that two-dimensional vortex dynamics on beaches is essentially laminar and tightly controlled by bottom friction.

The argument for this is an application of results from studies of two-dimensional turbulence in a constant-depth domain. In these studies the governing equations are

$$\nabla \cdot \boldsymbol{u} = 0 \quad \text{and} \quad \frac{D\boldsymbol{u}}{Dt} + \nabla P = -r\boldsymbol{u}|\boldsymbol{u}| + \boldsymbol{F} \tag{13.40}$$

for quadratic friction, where the constant r has units of one over length and can be identified with c_f/H in our beach notation. In the simplest set-up the flow domain is doubly periodic and the velocity field is forced by $\boldsymbol{F}$ within some finite range of high wavenumbers, corresponding to a forcing scale that is small compared to the domain size. A two-dimensional turbulent energy cascade to lower wavenumber then ensues, and in a sufficiently large domain this cascade is eventually arrested at a wavenumber comparable to a frictional arrest wavenumber κ_a, say, which by dimensional analysis[7] must be proportional to r. Numerical experimentation shows that the constant of proportionality is about 50, i.e., vigorous two-dimensional turbulence is possible only for wavenumbers

$$\kappa \geq \kappa_a \approx 50r. \tag{13.41}$$

If we assume that the results for constant H are relevant for beach vortex dynamics by using the local value of the variable H, then we have *two* inequality constraints on shallow-water wavenumbers κ lying in the turbulent range:

$$1 \geq \kappa H \geq \kappa_a H \approx 50c_f. \tag{13.42}$$

The first inequality expresses the basic scaling assumption for the aspect ratio of shallow water dynamics. Now, for the realistic value of $c_f = 0.01$ we find that $1 \geq \kappa H \geq 0.5$, which makes coherent turbulent dynamics virtually impossible. The appropriate value of c_f for a given beach is subject to some uncertainty, of course, but it seems clear that unless $c_f \ll 0.01$, two-dimensional shallow-water flows on real beaches are essentially laminar.

13.7 Notes on the literature

The classic theory for homogeneous longshore currents was formulated by Longuet–Higgins in the ground-breaking papers Longuet-Higgins (1970a,b). The potential importance of vorticity and vortex dynamics for nearshore circulations was pointed out by Peregrine in Peregrine (1998, 1999) and the theory for wave-driven vortex dynamics in shallow water as described here is given in Bühler and Jacobson (2001). An experimental study of vortex dynamics in a wedge geometry is presented in Centurioni (2002). Recent idealized studies of rigid-lid vortex dynamics on non-uniform topography are Johnson et al. (2005); Hinds et al. (2007). The phenomenon of current dislocation on barred beaches is described in Church and Thornton (1993); an example of recent modelling work for this situation is Ruessink et al.

[7] This assumes that the forcing scale and the total energy level are both irrelevant for the frictional arrest scale, which is confirmed by numerical simulations.

(2001). The study on frictional arrest scales used in §13.6.2 can be found in Grianik et al. (2004). Statistical theories for a broad range of geophysical flows are described in Majda and Wang (2006).

14
Wave refraction by vortices

We now consider wave refraction due to velocity strain and shear associated with vortical mean flows. Such refraction changes the waves' pseudomomentum field and, arguably, the central topic of wave–mean interactions outside simple geometry is how such pseudomomentum changes are related to the leading-order mean-flow response. The same question was satisfactorily answered in simple geometry by the pseudomomentum rule. However, refractive changes in the pseudomomentum do not rely in any essential way on wave dissipation or external forces, and yet they can irreversibly change the total amount of pseudomomentum in the wave field. This makes clear that the usual pseudomomentum rule of simple geometry, which equates such changes to an effective force exerted on the mean flow, must be modified.

As we shall see, the conservation law for the sum of pseudomomentum and GLM impulse is the key for understanding the wave–mean interactions in the presence of refraction. We will illustrate this by a number of examples consisting of wavepackets and confined wavetrains. The most important result is the following: if the concept of an effective mean force makes sense at all, then this force is *not* exerted at the location of the wavepacket, but at the location of the vortices that induce the straining field. This gives the wave–mean interactions a non-local character that was clearly absent in simple geometry, where the effective mean force was always exerted at the location of the wavepacket. Essentially, this is because of the slowly varying mean pressure field, which, as discussed in §12, can mediate mean-flow changes at long range in this context.

The outline of the chapter is as follows. We will first recall in §14.1 the generic refraction equations for the wavenumber vector $\boldsymbol{k}$ and the partial analogy between wave phase and a passive tracer. After illustrating these basic kinematic features with the example of a bath-tub vortex in §14.1.1, we then turn in §14.2 to the refraction of a wavetrain in shallow water by a single isolated vortex, which leads to the phenomenon of *remote recoil.* This is followed by an extension of the pseudomomentum plus impulse conservation law to three-dimensional Boussinesq flow in §14.3.1 and the study of *wave capture* for internal waves and the concomitant mean-flow response in §14.3.2-14.3.3. We conclude by discussing the peculiar and intimate relationship between wavepackets and vortex couples in §14.4, with a special emphasis on the transmutation of the former to the latter via dissipation.

14.1 Anatomy of wave refraction

We want to focus on refraction effects due to the basic velocity $\boldsymbol{U}(\boldsymbol{x}, t)$, so we neglect for simplicity any explicit dependence on $\boldsymbol{x}$ or t of the intrinsic dispersion function $\hat{\Omega}(\boldsymbol{k})$. The standard ray tracing equation in §4.4.2 for $\boldsymbol{k} = \boldsymbol{\nabla}\theta$ with wave phase θ then reduces to

$$\frac{d\boldsymbol{k}}{dt} = -(\boldsymbol{\nabla}\boldsymbol{U}) \cdot \boldsymbol{k} \quad \Leftrightarrow \quad \frac{dk_i}{dt} = -U_{j,i}k_j. \tag{14.1}$$

As described in §4.5.5, the refraction of $\boldsymbol{k}$ is inherited by the pseudomomentum vector $\mathbf{p} = \boldsymbol{k}E/\hat{\omega}$, say. In fact, the refraction term for the pseudomomentum vector of ray tracing has an exact counterpart in (10.126) of GLM theory, with $\overline{\boldsymbol{u}}^L$ identified with $\boldsymbol{U}$, so (14.1) and its implications for $\mathbf{p}$ are very robust predictions from ray tracing.

This robustness is also illustrated by the aforementioned partial analogy between the wave phase θ and a passive tracer ϕ, say, which evolves according to

$$\mathrm{D}_t\phi = 0 \quad \Rightarrow \quad \mathrm{D}_t(\boldsymbol{\nabla}\phi) = -(\boldsymbol{\nabla}\boldsymbol{U}) \cdot \boldsymbol{\nabla}\phi. \tag{14.2}$$

As noted before, this is a partial analogy because (14.2) involves the $O(1)$ material derivative D_t whereas (14.1) involves the derivative along group velocity rays. These two operators differ by the advection with the intrinsic group velocity, i.e.,

$$\frac{d}{dt} - \mathrm{D}_t = \widehat{\boldsymbol{c}}_g \cdot \boldsymbol{\nabla}. \tag{14.3}$$

Nevertheless, the analogy is useful because it explains the generic behaviour of $\boldsymbol{k}$ in the presence of multi-dimensional strain fields.

For example, the standard straining behaviour in two horizontal dimensions, which is relevant for stratified flow, is illustrated in figure 14.1. Here the flow $\boldsymbol{U}(x, y)$ is assumed to be incompressible (i.e., $U_x + V_y = 0$) and its gradient to be steady. As is well known, the ultimate fate of the advected

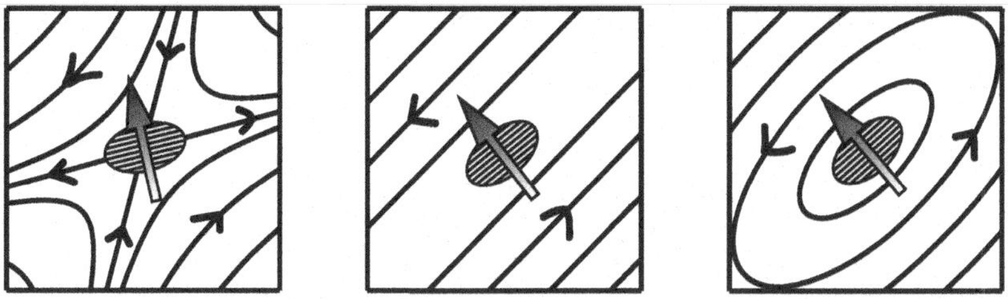

Hyperbolic D>0 **Parabolic D=0** **Elliptic D<0**

Figure 14.1 Two-dimensional straining pictures depending on the sign of D in (14.4). The streamlines are drawn in a frame moving with the local velocity and also shown is a patch of an advected tracer ϕ together with its gradient $\nabla\phi$ indicated by an arrow. Left: $D > 0$, hyperbolic case, open streamlines. The tracer contours align with the axis of extension, the tracer gradient turns normal to this axis and grows exponentially in time. Middle: $D = 0$, parabolic case with shear flow. Tracer contours align with shear direction and $\nabla\phi$ grows linearly in time. Right: $D < 0$, elliptic, vorticity-dominated flow with closed streamlines. Tracer contours rotate in time and $\nabla\phi$ oscillates in direction and magnitude. From Bühler and McIntyre (2005).

tracer gradient is then governed by the sign of

$$D = U_x^2 + \left(\frac{V_x + U_y}{2}\right)^2 - \left(\frac{V_x - U_y}{2}\right)^2, \tag{14.4}$$

which demarcates whether the streamlines are open or closed. In particular, if $D > 0$ the streamlines are open and the gradient follows exponential growth in time in the long run. If intrinsic wave propagation is negligible, then the same is true for $\boldsymbol{k}$ in the analogous ray tracing scenario.

Most important for the wave dynamics is that in the long run $\boldsymbol{k}$ would be asymptotically oriented in a direction that is completely determined by the eigenvectors of $\nabla \boldsymbol{U}$, and which does not depend on the initial conditions of $\boldsymbol{k}$ on the ray. In other words, the wavenumber gradient "forgets" its initial state, and,consequently, a strained wavepacket can be expected to forget its initial pseudomomentum as well. This is obviously very different from refraction by zonally symmetric mean flows, in which the zonal wavenumber and pseudomomentum component are conserved in time.

Of course, this simple scenario ignores the time-dependence of $\nabla \boldsymbol{U}$ along group-velocity rays. Still, in the case of a passive tracer it is known that such mean-flow time dependence generally slows down, but does not prevent, the ultimate exponential growth in time of the gradient. This essentially comes down to a comparison between the strength of $\nabla \boldsymbol{U}$ and the inverse of its auto-correlation time along group-velocity rays.

Even for a steady flow $\boldsymbol{U}$ it is possible that $\nabla \boldsymbol{U}$ is time-dependent along group-velocity rays. The familiar example of a bath-tub vortex illustrates this.

14.1.1 Refraction by a bath-tub vortex

Wave refraction by vortical mean flows can be easily observed in a bath tub: surface waves propagating towards a bath-tub vortex over an open plug-hole are refracted, and their tilting wave crests give a vivid, if often misleading, impression of the spinning flow around the vortex. The impression is often misleading because the wave crest pattern spins mostly clockwise if the vortex spins, in fact, anti-clockwise!

To illustrate this we look at a simple two-dimensional model of shallow-water waves refracted by the flow outside a vortex core, where the basic flow is irrotational, steady, and *weak* in the sense that at the wavepacket location the Froude or Mach number $|\boldsymbol{U}|/c = O(\epsilon)$ with $c = \sqrt{gH}$ and $\epsilon \ll 1$. This blatantly ignores the fact that bath-tub waves are almost never shallow-water waves, but allowing for finite-water depth would not change the basic anatomy of the refraction.

For weak flows in shallow water the layer depth variations in the basic state are $O(\epsilon^2)$ and these small variations can then be ignored compared to $\boldsymbol{U} = O(\epsilon)$. Furthermore, we can make use of the classic result described in §4.4.3, namely that non-dispersive wave rays through an irrotational background flow are straight lines to $O(\epsilon)$.

This allows solving the ray tracing equations for $(\boldsymbol{x}, \boldsymbol{k})$ analytically to $O(\epsilon)$:

$$\boldsymbol{x}(s) = \boldsymbol{x}_0 + \left(\boldsymbol{U}_0 + c\frac{\boldsymbol{k}_0}{\kappa_0}\right) s \quad \text{and} \quad \boldsymbol{k}(s) = \boldsymbol{k}_0 - \frac{\kappa_0}{c}(\boldsymbol{U} - \boldsymbol{U}_0), \tag{14.5}$$

where $\boldsymbol{U}$ is evaluated along the ray, the subscript zero refers to the initial conditions at the start of the ray, and $s \geq 0$ is the distance along the straight ray. It is easy to check that $\boldsymbol{k}(s)$ satisfies (14.1) to $O(\epsilon)$ provided that $U_{i,j} = U_{j,i}$.

We apply (14.5) to the case of a single circular vortex with counterclockwise circulation $\Gamma > 0$ centred at the origin of the coordinate system. The basic flow outside the vortex is then

$$\boldsymbol{U} = (U, V) = \frac{\Gamma}{2\pi} \frac{(-y, +x)}{x^2 + y^2} \tag{14.6}$$

and we consider the fate of a wavepacket that starts at $\boldsymbol{x}_0 = (-L, -D)$ with $\boldsymbol{k}_0 = (\kappa_0, 0)$. Here D and L are two constants such that $L \gg D$. The initial wavenumber vector together with $U_0 \approx 0$ makes clear that the wave propagates along a line of constant $y = -D$, and passes below the vortex at minimum distance D at the point $\boldsymbol{x} = (0, -D)$. By (14.5), the wavenumber vector along the ray is

$$\boldsymbol{k} = (\kappa_0, 0) - \frac{\kappa_0 \Gamma}{2\pi c} \frac{(D, x)}{x^2 + D^2} \quad \text{such that} \quad \frac{l}{k} = -\frac{\Gamma}{2\pi c} \frac{x}{x^2 + D^2} \tag{14.7}$$

to $O(\epsilon)$. If we denote by β the angle that $\boldsymbol{k}$ makes with the x-axis and by $\alpha \in [0, \pi]$ the bearing angle of the vortex relative to the wavepacket such that $\tan \alpha = -D/x$, then (14.7b) can be rewritten as

$$\frac{l}{k} = \tan \beta = +\frac{\Gamma}{2\pi c D} \sin \alpha \cos \alpha. \tag{14.8}$$

Thus, the initial rotation of $\boldsymbol{k}$ (and hence of the wave crest pattern) is *counter-clockwise* such that the wavepacket seems to glance towards the vortex. As α reaches 45 degrees, this glancing reaches its maximum, and thereafter the crests are now turning *clockwise*, reaching $\beta = 0$ at the point of minimal distance $\alpha = \pi/2$, and then passing through increasingly negative values $\beta < 0$ until α reaches 135 degrees. Thereafter, the crests are straightened out again, and $\beta \to 0$ as the vortex is left behind and $\alpha \to \pi$.

Now, in the bathtub setting the most prominent part of the wave refraction occurs near the vortex, where $\alpha \in [45, 135]$ degrees and the crests are rotating clockwise if the vortex is spinning counter-clockwise. The easily observed retrograde rotation of the wave crests near the vortex hence gives a misleading impression of the vortex rotation sense, as was mentioned before.

14.2 Remote recoil

We now turn to the wave–vortex interactions that go together with refraction by a single vortex. We can make use of the shallow-water set-up of the previous section and turn it into a steady wavetrain problem by placing a wave source and a wave sink at $\boldsymbol{x} = (-L, -D)$ and $\boldsymbol{x} = (+L, -D)$, respectively. This scenario is illustrated by the somewhat busy figure 14.2, which shows the irrotational loudspeakers and the vortex at a distance D from the steady wavetrain. Here we have assumed that $L < D$ such that the bearing angle $\alpha \in [45, 135]$ degrees and hence the wavetrain undergoes clockwise tilting.

The circumferential basic velocity with magnitude $\tilde{U}$ is indicated by the dashed line. The loudspeakers are slightly angled such that the y-component of the intrinsic group velocity at the wavemaker A on the left cancels the y-component of the basic velocity there, and vice versa at the wave absorber

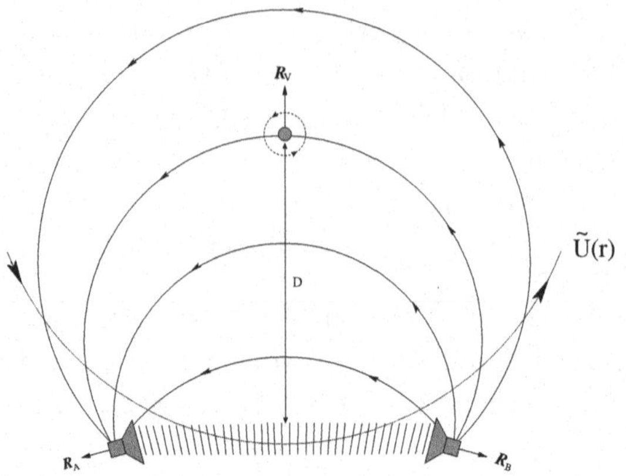

Figure 14.2 A steady wavetrain travels from left to right from the wavemaker A to the wave absorber B. The vortex flow refracts the wavetrain and there is a net recoil force $\boldsymbol{R}_A + \boldsymbol{R}_B$ in the negative y-direction on the loudspeakers. Concomitant is an effective remote recoil force $\boldsymbol{R}_V = -(\boldsymbol{R}_A + \boldsymbol{R}_B)$ "felt" by the vortex. The recoil manifests itself by a leftward material displacement of the vortex at $O(a^2)$ due to the Bretherton return flow. From Bühler and McIntyre (2003).

B on the right. This allows the wavetrain to follow the straight line $y = -D$, as before.

Now, following the discussion in §12.2.2, in steady state the net force on the localized loudspeakers can be equated to minus the net pseudomomentum flux away from the loudspeakers. For irrotational forcing we have $\overline{\boldsymbol{F}}^L = \boldsymbol{\mathcal{F}}$ and therefore

$$\boldsymbol{R}_{A,B} = -\int_{A,B} \overline{\boldsymbol{F}}^L \, dxdy = -\int_{A,B} \boldsymbol{\mathcal{F}} \, dxdy \tag{14.9}$$

are the respective recoil forces exerted on the wavemaker and on the wave absorber (we use $H = 1$ for simplicity). By construction, there is an equal-and-opposite push in the x-direction because the net fluxes of x-pseudomomentum are equal at the two loudspeakers and therefore cancel. On the other hand, the net fluxes of y-pseudomomentum do not cancel, i.e., there is a net recoil in the negative y-direction due to the refraction:

$$\boldsymbol{R}_A + \boldsymbol{R}_B + \boldsymbol{R}_V = 0 \quad \text{where} \quad \boldsymbol{R}_V = (0, R_V) \quad \text{with} \quad R_V > 0, \tag{14.10}$$

say. So $-\boldsymbol{R}_V$ is the net pseudomomentum generation per unit time and $\boldsymbol{R}_V$ is the corresponding net holding force in the y-direction that is necessary to keep the loudspeakers fixed in their positions. This leads to the following question: where does this external input of y-momentum manifest itself?

The answer comes from the conservation law for pseudomomentum plus impulse in (12.46), which for a steady wavetrain yields

$$\frac{d(\mathcal{I}+\mathcal{P})}{dt} = \frac{d\mathcal{I}}{dt} = \int \overline{\boldsymbol{F}}^L \, dxdy = -(\boldsymbol{R}_A + \boldsymbol{R}_B) = \boldsymbol{R}_V. \tag{14.11}$$

This shows that the mean-flow impulse should change in order to compensate for the net recoil in the y-direction exerted on the loudspeakers. The total impulse due to a single vortex with nonzero net circulation depends on the origin of the coordinate system, but the *changes* in the impulse due to movement of the vortex are origin-independent. In particular, the mean-flow impulse is

$$\mathcal{I} = \int (y, -x)\overline{q}^L \, dxdy = (Y, -X)\Gamma \tag{14.12}$$

if (X, Y) are the coordinates of the vortex centroid. Therefore (14.11) implies

$$\frac{dY}{dt} = 0 \quad \text{and} \quad \frac{dX}{dt} = -\frac{R_V}{\Gamma}. \tag{14.13}$$

At first sight this is a surprising result because it means that the vortex must move to the left in figure 14.2. However, this is consistent with the Bretherton return flow induced by the finite wavetrain. The structure of this $O(a^2)$ flow is indicated by the thin solid lines and it clearly pushes the vortex to the left. It can be checked by solving for $\overline{\boldsymbol{u}}_2^L$ from

$$\boldsymbol{\nabla} \times \overline{\boldsymbol{u}}_2^L = \boldsymbol{\nabla} \times \mathsf{p}_2 \quad \text{and} \quad \boldsymbol{\nabla} \cdot \overline{\boldsymbol{u}}_2^L = 0 \tag{14.14}$$

that the $O(a^2)$ Bretherton flow has precisely the right magnitude at the vortex core to be consistent with (14.13).

We call this action-at-a-distance of the wavetrain on the vortex the *remote recoil* in order to stress the non-local nature of this wave–vortex interaction. After all, the waves and the vortex do not overlap in physical space. The term 'recoil' is also apt because the movement of the vortex is consistent with the action of a compact body force on the vortex with net integral equal to $\boldsymbol{R}_V$. Such a force would be relevant in a parametrization problem in which the small-scale wavetrain is not explicitly resolved, but modelled via an effective mean force acting on the PV field. Such an effective force pattern would produce positive vorticity to the left of the vortex and negative vorticity to its right, which would lead to the movement of the vortex centroid to the left, as required.

This quasi-advective impact of the non-local effective mean force contrasts sharply with the impact of a local mean force that is exerted at the location of the wavetrain, say. Such a force could also balance the momentum and impulse budgets, but it would do so by creating a new dipolar PV structure at the wavetrain, which is completely spurious. For example, if the wavetrain is switched off after some finite time then in the original problem the vortex would stop moving and the flow would settle down to rest. With the local mean force, on the other hand, the spurious vortex couple would continue to move and the flow does not settle down to a steady state. This example makes clear that generally we must allow for non-local effective mean forces in problems involving wave refraction by vortices.

It can be shown that the remote recoil idea remains valid if the loudspeakers recede to infinity, in which case the net pseudomomentum generation is due to $O(\epsilon^2)$ terms that we have neglected here. One can show that these terms lead to the scattering of the waves into the lee of the vortex and the concomitant recoil is again consistent with (14.11). So whilst the set-up in figure 14.2 is certainly very special, it is not artificial: the remote recoil is real.

Overall, the recoil scenario makes clear that generally there is no component of pseudomomentum that is automatically conserved in non-dissipative flow. For instance, the conservation of angular pseudomomentum on a rotating Earth is lost once zonally asymmetric mean flows are allowed to act on the waves. This has obvious consequences for parametrization schemes that seek to model to vertical transport of angular momentum due to internal waves, because such schemes are typically based on simple geometry and hence on the conservation of angular pseudomomentum in the absence of dissipation. The remote recoil example here and the wave capture example in the next section make clear that this limited framework misses refraction effects that can modify the momentum budget of the mean flow in a systematic and irreversible manner.

14.3 Wave capture of internal gravity waves

The analogy between tracer advection and wave refraction described in §14.1 suggests an interesting possibility for wave–vortex interactions: the unbounded exponential growth in time of $\boldsymbol{k}$ along group-velocity trajectories. By the conservation of wave action, such growth would imply unbounded growth of $\boldsymbol{\mathcal{P}}$, which is obviously important for wave–mean interactions. However, unbounded growth of $\boldsymbol{k}$ is not possible in the shallow-water wave system, at least not in the simplest case of a steady and sub-critical basic flow $\boldsymbol{U}(\boldsymbol{x})$. This is because then the ray invariance of ω implies an a priori bound on κ:

$$\omega = \boldsymbol{U}\cdot\boldsymbol{k} + c\kappa = \omega_0 \quad \Rightarrow \quad \kappa \leq \frac{\omega_0/c}{1-||\boldsymbol{U}/c||_\infty}. \tag{14.15}$$

This is finite if the maximal Froude/Mach number $||\boldsymbol{U}/c||_\infty$ is less than unity, so for sub-critical flows there can be no unbounded wavenumber growth.[1] This is based on the growth of $|\hat{\omega}|$ with κ in shallow water, so we can expect that there will be no such bound if $|\hat{\omega}|$ can remain finite as $\kappa \to \infty$.

This is possible for internal gravity waves, in which $\hat{\omega}$ is zeroth-degree homogeneous in the components of $\boldsymbol{k}$. Specifically, we recall from §8.2.2 that the relevant dispersion relation is

$$\omega = \boldsymbol{U} \cdot \boldsymbol{k} + \hat{\omega} = \boldsymbol{U} \cdot \boldsymbol{k} + \frac{1}{\kappa}\sqrt{N^2(k^2+l^2) + f^2 m^2}. \tag{14.16}$$

Comparing with (14.15), there is no a priori bound for κ in this case.

Now, $\boldsymbol{U}$ could have three nonzero components but for atmosphere–ocean applications a useful restriction is to consider $\boldsymbol{U} = (U, V, 0)$ with $U_x + V_y = 0$, which models the quasi-horizontal layerwise flow familiar from quasi-geostrophic dynamics. This restriction implies that the refraction problems for the horizontal wavenumbers $\boldsymbol{k}_H = (k, l, 0)$ and that for the vertical wavenumber m decouple, i.e., we find that

$$\frac{d}{dt}\begin{pmatrix} k \\ l \end{pmatrix} = -\begin{pmatrix} U_x & V_x \\ U_y & V_y \end{pmatrix}\begin{pmatrix} k \\ l \end{pmatrix} \quad \text{and} \quad \frac{dm}{dt} = -kU_z - lV_z \tag{14.17}$$

By incompressibility, (14.17a) also holds with V_y replaced by $-U_x$. Thus $\boldsymbol{k}_H$ evolves as suggested by the arrow in the two-dimensional §14.1. In particular, ignoring time-dependence of $\boldsymbol{\nabla}_H \boldsymbol{U}$ (here $\boldsymbol{\nabla}_H = (\partial_x, \partial_y, 0)$) along group-velocity rays, for almost all initial conditions $\boldsymbol{k}_H$ will eventually align itself with the growing eigenvector of (14.17a), and exponential growth of $\boldsymbol{k}_H$ will then ensue. For example, assuming $V_x \neq 0$, we can expect that

$$\boldsymbol{k}_H(t) \propto (-V_x, U_x + \sqrt{D}, 0)\exp(\sqrt{D}t) \tag{14.18}$$

asymptotically, where $\sqrt{D}$ is the positive root of (14.4). Moreover, inspection of (14.17b) shows that exponential growth of m will then ensue as well, because

$$m(t) \approx -\frac{U_z k(t) + V_z l(t)}{\sqrt{D}} \tag{14.19}$$

once $\boldsymbol{k}_H$ is growing exponentially. Notably, if the vertical and horizontal gradient scales for $\boldsymbol{U}$ are H and L, respectively, then (14.18) and (14.19) imply that $|\boldsymbol{k}_H/m| \approx H/L$ asymptotically. This means that the aspect ratio of the internal wave approaches the aspect ratio of the basic flow.

The analogy between wave phase and passive tracer advection rests on

[1] The same conclusion also holds for rotating shallow water.

the assumption that the intrinsic wave propagation is negligible, i.e., that $|\widehat{\boldsymbol{c}}_g|$ is small compared to $|\boldsymbol{U}|$. This was manifestly untrue in the sub-critical shallow-water system, but it can be true for internal waves. Moreover, for internal waves $|\widehat{\boldsymbol{c}}_g|$ *decreases* as κ increases, so there is a positive feedback cycle in which wavenumber straining leads to an improved analogy with passive advection, and therefore to more straining and so on. Explicitly, we recall from §8.2.2 that

$$|\widehat{\boldsymbol{c}}_g| = \frac{1}{\kappa}\sqrt{\frac{(N^2 - \hat{\omega}^2)(\hat{\omega}^2 - f^2)}{\hat{\omega}^2}}. \tag{14.20}$$

Hence exponential growth of κ goes together with exponential decay of $|\widehat{\boldsymbol{c}}_g|$. This occurs regardless of the behaviour of $\hat{\omega}$, although if $|\boldsymbol{k}_H/m| \to H/L$ then $\hat{\omega}$ would approach a value corresponding to the aspect ratio of the basic flow. For example, if $H/L = f/N$ then $\hat{\omega} \to \sqrt{2}f$ in the limit as $\kappa \to \infty$. Notably, such near-inertial waves are ubiquitous in the atmosphere and the ocean, and hence a refraction process of this kind would be inconspicuous in spectral data.

Thus, the wavepacket gets "glued" into the basic flow because its group velocity converges to the basic flow velocity. By definition, this strengthens the analogy between passive advection and wave refraction, which then leads to more stretching of $\boldsymbol{k}$ and to even more reduced $|\widehat{\boldsymbol{c}}_g|$, reinforcing the cycle. This process together with the attendant wave–vortex interactions will be called *wave capture* here.

The key question now is the following: how does the mean flow react to the exponentially growing amount of pseudomomentum $\boldsymbol{\mathcal{P}}$ that is contained in a wavepacket? After all, the experience from simple geometry was that the mean flow reacts locally (i.e., at the wavepacket location) to the pseudomomentum budget, which would suggest a surge in the local mean flow in response to the surge in the pseudomomentum. Again, the actual answer is quite different and involves a non-local interaction between the wavepacket and the vortices that cause the refraction. This follows once we have written down the impulse plus pseudomomentum conservation law for three-dimensional stratified flow.

14.3.1 Impulse and pseudomomentum for stratified flow

Here we adapt the derivation of the horizontal impulse plus pseudomomentum conservation law from shallow-water theory to three-dimensional stratified flow in the Boussinesq system. We first consider the case without rotation, i.e., $f = 0$. The key assumption in shallow-water theory was to neglect the compressibility of the Lagrangian-mean flow. The corresponding assumption for the Boussinesq model is to neglect the vertical Lagrangian-mean motion, which also implies that the mean stratification surfaces remain flat. Specifically, with $\overline{\boldsymbol{u}}_H^L = (\overline{u}^L, \overline{v}^L, 0)$ we assume that[2]

$$\nabla_H \cdot \overline{\boldsymbol{u}}_H^L = 0 \quad \text{and} \quad \overline{w}^L = 0, \tag{14.21}$$

and also that the mean stratification surfaces $\overline{\Theta}^L = \text{const}$ are flat horizontal planes. The exact GLM PV then takes the simple form

$$\overline{q}^L = \hat{\boldsymbol{z}} \cdot \nabla \times (\overline{\boldsymbol{u}}^L - \mathsf{p}) \equiv \nabla_H \times (\overline{\boldsymbol{u}}_H^L - \mathsf{p}_H) \quad \Rightarrow \quad \overline{\mathrm{D}}^L \overline{q}^L = 0. \tag{14.22}$$

Here p_H is the horizontal pseudomomentum vector and $\overline{q}^L$ is by construction a materially invariant of the layerwise two-dimensional Lagrangian-mean flow. We now define the net horizontal mean-flow impulse and pseudomomentum by

$$\mathcal{I}_H = \int (y, -x, 0)\overline{q}^L \, dxdydz \quad \text{and} \quad \mathcal{P}_H = \int \mathsf{p}_H \, dxdydz. \tag{14.23}$$

By essentially the same manipulations as in the shallow-water case (including the assumptions that the flow is unbounded horizontally and that the pseudomomentum and PV are compact) it is straightforward to show that

$$\frac{d\mathcal{I}_H}{dt} = \int \left((\nabla_H \overline{\boldsymbol{u}}_H^L) \cdot \mathsf{p}_H + \overline{\boldsymbol{F}}_H^L - \mathcal{F}_H \right) dxdydz \quad \text{and} \tag{14.24}$$

$$\frac{d\mathcal{P}_H}{dt} = \int \left(-(\nabla_H \overline{\boldsymbol{u}}_H^L) \cdot \mathsf{p}_H + \mathcal{F}_H \right) dxdydz \tag{14.25}$$

both hold. We therefore have the conservation law

$$\frac{d(\mathcal{I}_H + \mathcal{P}_H)}{dt} = \int \overline{\boldsymbol{F}}_H^L \, dxdydz. \tag{14.26}$$

As before, both $\mathcal{I}_H$ and $\mathcal{P}_H$ vary individually due to refraction, but their sum remains constant unless the flow is forced externally. This makes obvious that during wave capture any exponential growth of $\mathcal{P}_H$ must be compensated by an exponential decay of $\mathcal{I}_H$. Because the value of $\overline{q}^L$ on mean trajectories cannot change, this must be achieved via material displacements of the basic-state PV structure, just as in the remote recoil situation in shallow water.

In other words, when we look at the horizontal straining flow in figure 14.1 we must investigate how this straining flow has been induced by distant vortices, say. Only then can we understand how (14.26) regulates the exchange

[2] We use the GLM notation for the Lagrangian-mean flow; in the context of small-amplitude theory $\overline{\boldsymbol{u}}^L = \boldsymbol{U} + a^2\overline{\boldsymbol{u}}_2^L$ and so on.

between mean-flow impulse and pseudomomentum. A simple example of this is considered in §14.3.2 below. However, before looking at this example we complete the derivation of the central conservation law (14.26) by adding nonzero rotation $f \neq 0$.

Coriolis forces now cause significant vertical displacements of mean stratification surfaces, but for small Rossby number flows such mean vertical displacements are still compatible with $\overline{w}^L = 0$ at leading order. Of course, this amounts to applying the usual quasi-geostrophic theory to the Lagrangian-mean flow. Thus we first augment the PV definition by

$$\overline{q}^L = \boldsymbol{\nabla}_H \times (\overline{\boldsymbol{u}}_H^L - \mathbf{p}_H) + \frac{\partial}{\partial z}\left(\frac{f\overline{b}^L}{N^2}\right) \tag{14.27}$$

where $\overline{b}^L$ is the mean buoyancy and $N(z)$ may depend on z as usual. The assumptions about the mean velocity field are then extended to include the thermal wind-shear relations for the mean flow:

$$\boldsymbol{\nabla}_H \cdot \overline{\boldsymbol{u}}_H^L = 0, \quad \overline{w}^L = 0, \quad f\left(\frac{\partial \overline{u}^L}{\partial z}, \frac{\partial \overline{v}^L}{\partial z}\right) = \left(-\frac{\partial \overline{b}^L}{\partial y}, +\frac{\partial \overline{b}^L}{\partial x}\right). \tag{14.28}$$

The definition of $\boldsymbol{\mathcal{I}}_H$ is unchanged, but there is now a strong coupling between the layerwise flows at different altitudes, which was absent in the non-rotating case. This can be highlighted by considering the layerwise density of horizontal impulse $\boldsymbol{\mathcal{J}}_H(z,t)$, say, defined as

$$\boldsymbol{\mathcal{J}}_H(z,t) = \int (y, -x, 0)\overline{q}^L \, dxdy \tag{14.29}$$

such that $\boldsymbol{\mathcal{I}}_H$ is the z-integral of $\boldsymbol{\mathcal{J}}_H$. It then follows that

$$\frac{\partial \boldsymbol{\mathcal{J}}_H}{\partial t} + \frac{\partial}{\partial z}\left(\frac{f}{N^2}\int (-\overline{v}^L, +\overline{u}^L, 0)\overline{b}^L \, dxdy\right) = \tag{14.30}$$

$$\int \left((\boldsymbol{\nabla}_H \overline{\boldsymbol{u}}_H^L)\cdot \mathbf{p}_H + \overline{\boldsymbol{F}}_H^L - \boldsymbol{\mathcal{F}}_H\right) dxdy. \tag{14.31}$$

This shows a vertical flux of horizontal impulse, which is analogous to the Eliassen–Palm flux of quasi-geostrophic theory. The fact that horizontal impulse can be transported in the vertical is linked to the fact that the horizontal distribution of $\overline{q}^L$ at some altitude z generally affects the mean flow at all other altitudes. Of course, this follows from the usual PV inversion problem for finding the quasi-geostrophic stream function ψ such that

$$(\overline{u}^L, \overline{v}^L, \overline{b}^L) = (-\psi_y, \psi_x, f\psi_z) \tag{14.32}$$

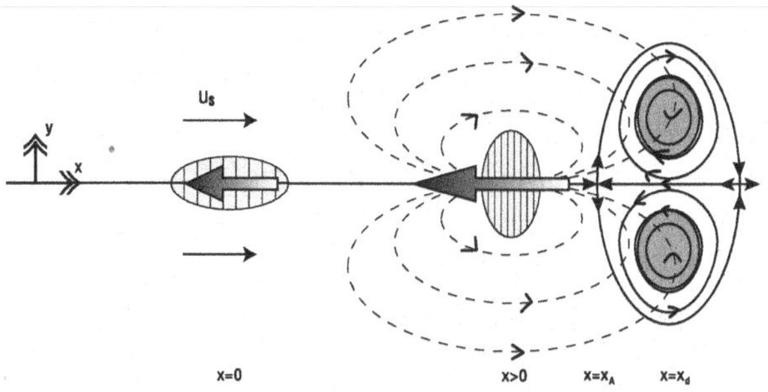

Figure 14.3 A wavepacket indicated by the wave crests and arrow for the net pseudomomentum is squeezed by the straining flow due to a vortex couple on the right. As the wavepacket slides towards the stagnation point in front of the couple, its x-extent decreases, its y-extent increases, and so does its total negative zonal pseudomomentum. This pseudomomentum increase is precisely compensated by a decrease in the vortex couple impulse caused by the Bretherton flow of the wavepacket, which is indicated by the dashed lines. From Bühler and McIntyre (2005).

and therefore (14.27) implies

$$\psi_{xx} + \psi_{yy} + \frac{\partial}{\partial z}\left(\frac{f^2}{N^2}\psi_z\right) = \overline{q}^L + \nabla_H \times \mathbf{p}_H. \tag{14.33}$$

With $f = 0$ this equation can be inverted separately at each altitude z and therefore the mean flow at some altitude z only depends on $\overline{q}^L$ and $\mathbf{p}_H$ at that altitude. With $f \neq 0$ that is no longer true: Coriolis forces lead to quasi-geostrophic mean flows that are intrinsically coupled in the vertical. Finally, as far as the net horizontal impulse

$$\mathcal{I}_H = \int \mathcal{J}_H \, dz \tag{14.34}$$

is concerned, we again obtain (14.24) and (14.26) provided the mean-flow EP flux in (14.30) is zero at the domain boundaries in z.

14.3.2 Wavepacket and vortex dipole example

We can now give a simple example of how the straining of a wavepacket and the concomitant changes in its pseudomomentum are compensated by impulse changes due to non-local wave–mean interactions. The specific example we look at is the refraction of a wavepacket by a vortex couple as in figure 14.3, which shows a horizontal cross-section of the flow on a stratification surface defined by some fixed altitude z. Here the basic flow consists of a constant zonal flow $U_s > 0$ plus a counter-propagating vortex dipole that holds itself steady against U_s, say. The impulse of the couple is zonal, negative, and proportional to the separation of the two vortices. If we assume that $f = 0$ for simplicity, then this layerwise two-dimensional flow can be studied without consideration of the flow conditions at other altitudes.

We now consider a wavepacket intersecting this horizontal layer as indicated in the figure. For example, we could view this wavepacket as the intersection of a tall three-dimensional wavetrain with the stratification surface under consideration. Moreover, we may assume that initially the zonal group velocity of the packet is zero. It is easy to deduce by inspection that the area-preserving strain flow due to the vortex couple squeezes the wavepacket in the x-direction such that its negative zonal pseudomomentum increases in magnitude. The concomitant decrease in zonal intrinsic group velocity then implies that the wavepacket is sliding backwards, and towards the basic-flow stagnation point at the front of the couple. This is shown by the second snapshot of the wavepacket in the figure.

Now, the conservation law (14.26) requires that the growth of negative zonal pseudomomentum *must* be compensated by a reduction in negative zonal pseudomomentum of the vortex couple. This reduction is mediated by the layerwise Bretherton return flow associated with the wavepacket. Indeed, the instantaneous structure of the $O(a^2)$ Bretherton flow follows immediately from (14.33) with $f = 0$ and $\overline{q}_2^L = 0$, i.e.,

$$\psi_{xx} + \psi_{yy} = \nabla_H \times \mathbf{p}_H. \tag{14.35}$$

As before, this means that $\overline{\boldsymbol{u}}_H^L$ is the layerwise least-squares projection of $\mathbf{p}_H$ onto non-divergent vector fields, which yields the usual return flow as indicated by the dashed lines in figure 14.3. Crucially, this flow pushes the vortex couple closer together, thereby reducing its negative zonal impulse by precisely the correct amount to compensate for the pseudomomentum growth of the wavepacket. This clearly illustrates the fundamentally non-local nature of the wave–mean interactions that take place here.

With $f \neq 0$ the flow fields at different altitudes are coupled together. On the one hand, this allows a wavepacket at $z = z_1$ to be refracted by a straining flow induced by a vortex couple at $z = z_2$, say. On the other hand, the Bretherton flow due to the wavepacket at $z = z_1$, which is found by inverting (14.33) in three dimensions, is then present at $z = z_2$ as well. Again, this leads to a non-local compensation between pseudomomentum and impulse changes, where "non-local" in this case include differences in horizontal location as well as in altitude.

14.3.3 Mean-flow response at the wavepacket

The previous section made clear that the exponential surge in the net pseudomomentum is compensated by the exponential loss of impulse of the vortex couple, and how this non-local exchange is mediated by the Bretherton flow of the wavepacket. This involves the far-field structure of ψ far away from the wavepacket. Still, there might be a lingering concern about the local structure of ψ at the location of wavepacket itself, because the right-hand side of (14.35) contains the exponentially growing p_H. If there is an exponentially growing local mean-flow response $\overline{\boldsymbol{u}}^L$ at the wavepacket then this would be important because a large $\overline{\boldsymbol{u}}^L$ might induce wave breaking or other nonlinear effects.

We can study this problem in a simple two-dimensional set-up where we look at a wavepacket centred at the origin of an (x, y) coordinate system such that at $t = 0$ the pseudomomentum is $\mathsf{p} = \boldsymbol{k}A = (1, 0)A$ for some compact envelope function $A(x, y)$ that equals the wave action density $E/\hat{\omega}$. The local $O(a^2)$ Lagrangian-mean flow induced by the wavepacket is the solution of

$$\overline{u}^L_x + \overline{v}^L_y = 0 \quad \text{and} \quad \overline{v}^L_x - \overline{u}^L_y = \boldsymbol{\nabla} \times \mathsf{p} = -A_y(x, y). \tag{14.36}$$

We now imagine that the wavepacket is exposed to a pure straining basic flow $\boldsymbol{U} = (-x, +y)$, which squeezes the wavepacket in x and stretches it in y. Specifically, for $t > 0$ we have the flow map $(x, y) \to (x/\alpha, \alpha y)$ where $\alpha(t) = \exp(t) \geq 1$. We ignore intrinsic wave propagation relative to $\boldsymbol{U}$, which implies that the wave action density A is advected by $\boldsymbol{U}$, i.e., $\mathrm{D}_t A = 0$. The wavenumber vector changes according to $\mathrm{D}_t \boldsymbol{k} = -(\boldsymbol{\nabla U}) \cdot \boldsymbol{k}$ and we obtain the advected pseudomomentum as

$$\mathsf{p} = (\alpha, 0)A(\alpha x, y/\alpha) \quad \text{and} \quad \boldsymbol{\nabla} \times \mathsf{p} = -A_y(\alpha x, y/\alpha). \tag{14.37}$$

This shows that p_1 grows exponentially whilst $\boldsymbol{\nabla} \times \mathsf{p}$ does not; in fact, it is easy to check that $\boldsymbol{\nabla} \times \mathsf{p}$ is simply advected by $\boldsymbol{U}$. This is a consequence of the stretching in the transverse y-direction, which diminishes the curl because it makes the x-pseudomomentum vary more slowly in y. Thus whilst there is an exponential surge in p_1 there is none in $\boldsymbol{\nabla} \times \mathsf{p}$.

We now compute $\overline{u}^L$ at the core of the wavepacket by using the Fourier transforms pair

$$\mathrm{FT}\{f\}(k, l) = \int e^{-i[kx+ly]} f(x, y)\, dxdy \tag{14.38}$$

and

$$f(x,y) = \frac{1}{4\pi^2}\int e^{+i[kx+ly]}\mathrm{FT}\{f\}(k,l)\,dkdl. \tag{14.39}$$

It follows from (14.35) that the transforms of $\overline{u}^L$ and of p_1 are related by

$$\mathrm{FT}\{\overline{u}^L\}(k,l) = \frac{l^2}{k^2+l^2}\mathrm{FT}\{\mathsf{p}_1\}(k,l). \tag{14.40}$$

In order to discern the effect of changing $\alpha(t)$ we denote the initial p_1 for $\alpha = 1$ by p_1^1 and the later pseudomomentum for other values of α by

$$\mathsf{p}_1^\alpha(x,y) = \alpha\mathsf{p}_1^1(\alpha x, y/\alpha) \quad \Rightarrow \quad \mathrm{FT}\{\mathsf{p}_1^\alpha\}(k,l) = \alpha\mathrm{FT}\{\mathsf{p}_1^1\}(k/\alpha, \alpha l). \tag{14.41}$$

The value of $\overline{u}^L$ at the wavepacket core $x = y = 0$ is the total integral of (14.40) over the spectral plane, which using (14.41) can be written as

$$\begin{aligned}\overline{u}^L(0,0) &= \frac{1}{4\pi^2}\int \frac{l^2}{k^2+l^2}\mathrm{FT}\{\mathsf{p}_1^\alpha\}(k,l)\,dkdl \\ &= \frac{\alpha}{4\pi^2}\int \frac{l^2}{\alpha^4k^2+l^2}\mathrm{FT}\{\mathsf{p}_1^1\}(k,l)\,dkdl\end{aligned} \tag{14.42}$$

after renaming the dummy integration variables. This is as far as we can go without making further assumptions about the shape of the initial wavepacket.

Now, if the wavepacket is circularly symmetric initially then p_1^1 depends only on the spatial radius $r = \sqrt{x^2+y^2}$ and $\mathrm{FT}\{\mathsf{p}_1^1\}$ depends only on the spectral radius $\kappa = \sqrt{k^2+l^2}$. In this case (14.42) is

$$\overline{u}^L(0,0) = \frac{\alpha}{\alpha^2+1}\,\mathsf{p}_1^1(0,0) = \frac{1}{\alpha^2+1}\,\mathsf{p}_1^\alpha(0,0). \tag{14.43}$$

The first equality relates $\overline{u}^L$ to the initial value of p_1 whilst the second equality uses the instantaneous (i.e., larger) value of p_1. At the initial time $\alpha = 1$ and we see that

$$\alpha = 1 \quad \Rightarrow \quad \overline{u}^L(0,0) = \frac{1}{2}\mathsf{p}_1^1(0,0) \tag{14.44}$$

holds at the core of the circular wavepacket, i.e., the induced mean flow is *half* of the local pseudomomentum magnitude. As time progresses and α grows, this ratio decreases monotonically to zero, which demonstrates that there is no growth at all in the local mean flow, even though the local pseudomomentum grows exponentially at the wavepacket core. Hence there is no surge in $\overline{u}^L$ at all, not even at the core of the wavepacket.

Finally, we can now understand how the relationship between $\overline{u}^L$ and p_1

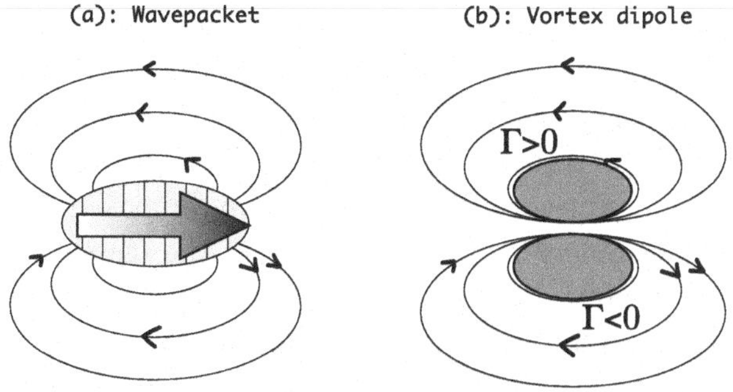

Figure 14.4 Wave–vortex duality. Left: wavepacket together with streamlines indicating the Bretherton flow; the arrow indicates the net pseudomomentum. Right: a vortex couple with the same return flow; the shaded areas indicate nonzero PV values with opposite signs. Dissipation transmutes a wavepacket into a vortex couple. From Bühler and McIntyre (2005).

converges to the classical $\overline{u}^L = \mathsf{p}_1$ familiar from interactions in simple geometry. To this end we must consider a very long, cigar-shaped wavepacket, i.e., a wavepacket that is long in x and narrow in y. This corresponds to values of α close to zero, which can be reached in the present example by reversing the direction of $\boldsymbol{U}$, for example. In this case the second equality in (14.43) shows that $\overline{u}^L \to \mathsf{p}_1^\alpha$, i.e., the velocity field of the mean-flow response at the wavepacket core converges to the instantaneous value of the pseudomomentum field as the wavepacket cigar gets longer and longer. This is the link to the familiar result from simple geometry.

14.4 Wave–vortex duality and dissipation

Let us take another look at the similarity between a wavepacket and a vortex couple in the two-dimensional cross-section depicted in figure 14.4. This is a snapshot in a stratified flow described by the Boussinesq system and we assume $f = 0$ for simplicity. In particular, we consider the similarity of the horizontal Lagrangian-mean flow induced by the wavepacket and the vortex couple. By inspection of (14.35), it is clear that the instantaneous mean flow due to the wavepacket is identical to that of a vortex couple with vertical vorticity equal to $\boldsymbol{\nabla}_H \times \mathbf{p}_H$. Of course, the wavepacket moves with the group velocity, which differs from the material velocity by the intrinsic group velocity. During wave capture, however, the intrinsic velocity goes to zero and the wavepacket behaves more and more like a vortex couple. For instance, we concluded before that during wave capture the refraction by the horizontal mean flow results in the advection of $\nabla_H \times \mathbf{p}_H$ by the mean material flow, which is of course how the PV of a vortex couple evolves.

Indeed, if the wavepacket in figure 14.3 were replaced by a vortex couple, then we would recognize that figure 14.3 displays the early stage of the classical leap-frogging of vortex rings, with two-dimensional vortex couples replacing the three-dimensional vortex rings of the classical example. This suggests a 'wave–vortex duality' during wave capture, because the wavepacket acts and interacts with the remaining flow precisely *as if* it were a vortex couple.

Now, if we allow a wavepacket to dissipate instantaneously then the wavepacket on the left in figure 14.4 would simply be transmuted into the dual vortex couple on the right. The Lagrangian-mean flow $\overline{\boldsymbol{u}}^L$ would be unchanged during the dissipation, but $\overline{q}^L = \nabla_H \times (\overline{\boldsymbol{u}}^L_H - \mathbf{p}_H)$ would adjust instantaneously as $\nabla_H \times \mathbf{p}_H$ is reduced to zero. This leads to an intriguing conclusion: once a wavepacket has been captured by the mean flow (i.e., once its intrinsic group velocity has become negligible) then whether or not the wavepacket actually dissipates has no further effect on the Lagrangian-mean flow. In other words, wave capture is a peculiar form of dissipation: the loss of intrinsic group velocity is equivalent, as far as wave–vortex interactions are concerned, to the loss of the wavepacket altogether.

All of these considerations point to the profound role of intrinsic wave propagation: it is the life blood of both linear wave dynamics and also of nonlinear wave–vortex interactions. Once intrinsic wave propagation ceases, wavepackets and vortex couples are both frozen into the fluid, and, like the proverbial old married couple, they become indistinguishable in their interactions with the outside world.

14.5 Notes on the literature

The nature of passive advection by the quasi-two-dimensional layerwise flows that are typical for atmospheric dynamics is described in Haynes and Anglade (1997) and its references. The use of ray tracing to study wave–vortex interactions outside simple geometry has been advocated in the case of sound waves by Nazarenko et al. (1995). The remote recoil in shallow water was investigated in Bühler and McIntyre (2003).

The possibility of exponentially fast straining of internal waves during ray tracing was pointed out by Jones (1967, 1969) and the 'glueing' of the wave field into the mean flow that is implied by the reduction in intrinsic group velocity is described in Badulin and Shrira (1993). The attendant nonlinear wave–mean interactions were investigated in Bühler and McIntyre (2005), which also introduced the conservation law for impulse plus pseudomomentum. The computation of the mean flow at the wavepacket location appeared in Bühler (2009).

Evidence for wave capture in numerical simulations of atmospheric flows is described in Plougonven and Snyder (2005); whether similar dynamics is active in the ocean is discussed in Polzin (2008). A recent study of how three-dimensional refraction affects the wave-induced global transport of angular momentum in the atmosphere is given in Hasha et al. (2008).

References

Andrews, D. G., and McIntyre, M. E. 1976a. Planetary waves in horizontal and vertical shear: asymptotic theory for equatorial waves in weak shear. *J. Atmos. Sci.*, **33**, 2049–2053.

Andrews, D. G., and McIntyre, M. E. 1976b. Planetary waves in horizontal and vertical shear: the generalized Eliassen–Palm relation and the mean zonal acceleration. *J. Atmos. Sci.*, **33**, 2031–2048.

Andrews, D. G., and McIntyre, M. E. 1978a. An exact theory of nonlinear waves on a Lagrangian-mean flow. *J. Fluid Mech.*, **89**, 609–646.

Andrews, D. G., and McIntyre, M. E. 1978b. Generalized Eliassen–Palm and Charney–Drazin theorems for waves on axisymmetric flows in compressible atmospheres. *J. Atmos. Sci.*, **35**, 175–185.

Andrews, D. G., and McIntyre, M. E. 1978c. On wave-action and its relatives. *J. Fluid Mech.*, **89**, 647–664.

Andrews, D. G., Holton, J. R., and Leovy, C. B. 1987. *Middle Atmosphere Dynamics.* Academic Press.

Arnold, V. I., and Khesin, B. A. 1998. *Topological Methods in Hydrodynamics.* Springer.

Badulin, S. I., and Shrira, V. I. 1993. On the irreversibility of internal waves dynamics due to wave trapping by mean flow inhomogeneities. Part 1. Local analysis. *J. Fluid Mech.*, **251**, 21–53.

Baines, P. G. 1995. *Topographic Effects in Stratified Flows.* Cambridge: Cambridge University Press.

Baldwin, M. P., Gray, L. J., Dunkerton, T. J., Hamilton, K., Haynes, P. H., Randel, W. J., Holton, J. R., Alexander, M. J., Hirota, I., Horinouchi, T., Jones, D. B. A., Kinnersley, J. S., Marquardt, C., Sato, K., and Takahashi, M. 2001. The quasi-biennial oscillation. *Revs. Geophys.*, **39**, 179–229.

Batchelor, G. K. 1967. *An Introduction to Fluid Dynamics.* Cambridge: Cambridge University Press.

Booker, J. R., and Bretherton, F. P. 1967. The critical layer for internal gravity waves in a shear flow. *J. Fluid Mech.*, **27**, 513–539.

Boyd, J. 1976. The noninteraction of waves with the zonally-averaged flow on a spherical earth and the interrelationships of eddy fluxes of energy, heat and momentum. *J. Atmos. Sci.*, **33**, 2285–2291.

Bretherton, F. P. 1969. On the mean motion induced by internal gravity waves. *J. Fluid Mech.*, **36**, 785–803.

Bretherton, F. P., and Garrett, C. J. R. 1968. Wavetrains in inhomogeneous moving media. *Proc. Roy. Soc. Lond.*, **A302**, 529–554.

Bühler, O. 2000. On the vorticity transport due to dissipating or breaking waves in shallow-water flow. *J. Fluid Mech.*, **407**, 235–263.

Bühler, O. 2006. *A Brief Introduction to Classical, Statistical, and Quantum Mechanics*. Courant Lecture Notes, vol. 13. American Mathematical Society.

Bühler, O. 2007. Impulsive fluid forcing and water strider locomotion. *J. Fluid Mech.*, **573**, 211–236.

Bühler, O. 2009. Wave–vortex interactions. In: Flor, J.B. (ed), *Fronts, Waves and Vortices in Geophysics*. Lectures Notes in Physics, no. 805. Springer.

Bühler, O., and Jacobson, T. E. 2001. Wave-driven currents and vortex dynamics on barred beaches. *J. Fluid Mech.*, **449**, 313–339.

Bühler, O., and McIntyre, M. E. 1998. On non-dissipative wave–mean interactions in the atmosphere or oceans. *J. Fluid Mech.*, **354**, 301–343.

Bühler, O., and McIntyre, M. E. 2003. Remote recoil: a new wave–mean interaction effect. *J. Fluid Mech.*, **492**, 207–230.

Bühler, O., and McIntyre, M. E. 2005. Wave capture and wave–vortex duality. *J. Fluid Mech.*, **534**, 67–95.

Centurioni, L. R. 2002. Dynamics of vortices on a uniformly shelving beach. *J. Fluid Mech.*, **472**, 211–228.

Church, J. C., and Thornton, E. B. 1993. Effects of breaking wave induced turbulence within a longshore current model. *Coastal Eng.*, **20**, 1–28.

Courant, R., and Hilbert, D. 1989. *Methods of Mathematical Physics, vol. 2*. Wiley-Interscience.

Craik, A.D.D. 1985. *Wave Interactions and Fluid Flows*. Cambridge University Press.

Drazin, P.G., and Su, C.H. 1975. A note on long-wave theory of airflow over a mountain. *J. Atmos. Sciences*, **32**, 437–439.

Dritschel, D. G., and McIntyre, M. E. 2008. Multiple jets as PV staircases: the Phillips effect and the resilience of eddy-transport barriers. *J. Atmos. Sci*, **65**, 855–874.

Dysthe, K. B. 2001. Refraction of gravity waves by weak current gradients. *Journal of Fluid Mechanics*, **442**(Sept.), 157–159.

Foias, C., Holm, D.D., and E.S., Titi. 2001. The Navier–Stokes–alpha model of fluid turbulence. *Physica D*, **152**, 505–519.

Grianik, N., Held, I. M., Smith, K. S., and Vallis, G. K. 2004. The effects of quadratic drag on the inverse cascade of two-dimensional turbulence. *Physics of Fluids*, **16**, 73–78.

Hasha, A. E., Bühler, O., and Scinocca, J.F. 2008. Gravity-wave refraction by three-dimensionally varying winds and the global transport of angular momentum. *J. Atmos.Sci.*, **65**, 2892–2906.

Hayes, W. D. 1970. Conservation of action and modal wave action. *Proc. Roy. Soc. Lond.*, **A320**, 187–208.

Haynes, P. H. 2003. Critical layers. In: Holton, J. R., Pyle, J. A., and Curry, J. A. (eds), *Encyclopedia of Atmospheric Sciences*. London, Academic/Elsevier.

Haynes, P. H., and Anglade, J. 1997. The vertical-scale cascade of atmospheric tracers due to large-scale differential advection. *J. Atmos. Sci.*, **54**, 1121–1136.

Haynes, P. H., Marks, C. J., McIntyre, M. E., Shepherd, T. G., and Shine, K. P. 1991. On the "downward control" of extratropical diabatic circulations by eddy-induced mean zonal forces. *J. Atmos. Sci.*, **48**, 651–678.

Hinch, E. J. 1991. *Perturbation Methods.* Cambridge University Press.

Hinds, A. K., Johnson, E. R., and McDonald, N. R. 2007. Vortex scattering by step topography. *Journal of Fluid Mechanics*, **571**, 495–505.

Hoskins, B. J., McIntyre, M. E., and Robertson, A. W. 1985. On the use and significance of isentropic potential-vorticity maps. *Q. J. Roy. Meteorol. Soc.*, **111**, 877–946.

Johnson, E. R., Hinds, A. K., and McDonald, N. R. 2005. Steadily translating vortices near step topography. *Physics of Fluids*, **17**, 6601.

Jones, W. L. 1967. Propagation of internal gravity waves in fluids with shear flow and rotation. *J. Fluid Mech.*, **30**, 439–448.

Jones, W. L. 1969. Ray tracing for internal gravity waves. *J. Geophys. Res.*, **74**, 2028–2033.

Keller, J. B. 1978. Rays, waves and asymptotics. *Bulletin of the American Mathematical Society*, **84**(5), 727–750.

Killworth, P. D., and McIntyre, M. E. 1985. Do Rossby-wave critical layers absorb, reflect, or over-reflect? *J. Fluid Mechanics*, **161**, 449–492.

Landau, L. D., and Lifshitz, E. M. 1959. *Fluid Mechanics.* 1st Eng. ed. Pergamon.

Landau, L. D., and Lifshitz, E. M. 1982. *Mechanics.* 3rd Eng. ed. Butterworth–Heinemann.

Leibovich, S. 1980. On wave–current interaction theories of Langmuir circulations. *J. Fluid Mech.*, **99**, 715–724.

Leibovich, S. 1983. The form and dynamics of Langmuir circulations. *Ann. Rev. Fluid Mech.*, **15**, 391–427.

Lighthill, J. 1978. *Waves in Fluids.* Cambridge University Press.

Lindzen, R. S. 1981. Turbulence and stress owing to gravity wave and tidal breakdown. *J. Geophys. Res.*, **86**, 9707–9714.

Lindzen, R. S., and Holton, J. R. 1968. A theory of the quasi-biennial oscillation. *J. Atmos. Sci.*, **25**, 1095–1107.

Longuet-Higgins, M. S. 1970a. Longshore currents generated by obliquely incident sea waves 1. *J. Geophys. Res.*, **75**, 6778–6789.

Longuet-Higgins, M. S. 1970b. Longshore currents generated by obliquely incident sea waves 2. *J. Geophys. Res.*, **75**, 6790–6801.

Maas, L. R., and Lam, F. P. 1995. Geometric focusing of internal waves. *J. Fluid Mech.*, **300**, 1–41.

Majda, A. J. 2003. *Introduction to PDEs and Waves for the Atmosphere and Ocean.* Courant Lecture Notes, vol. 9. American Mathematical Society.

Majda, A.J., and Wang, X. 2006. *Non-Linear Dynamics and Statistical Theories for Basic Geophysical Flows.* Cambridge University Press.

McIntyre, M. E. 1980a. An introduction to the generalized Lagrangian-mean description of wave, mean-flow interaction. *Pure Appl. Geophys.*, **118**, 152–176.

McIntyre, M. E. 1980b. Towards a Lagrangian-mean description of stratospheric circulations and chemical transports. *Phil. Trans. Roy. Soc. Lond.*, **A296**, 129–148.

McIntyre, M. E. 1981. On the 'wave momentum' myth. *J. Fluid Mech.*, **106**, 331–347.

McIntyre, M. E. 2008. Potential-vorticity inversion and the wave–turbulence jigsaw: some recent clarifications. *Adv. Geosci.*, **15**, 47–56.

McIntyre, M. E., and Norton, W. A. 1990. Dissipative wave–mean interactions and the transport of vorticity or potential vorticity. *J. Fluid Mech.*, **212**, 403–435.

McIntyre, M. E., and Weissman, M. A. 1978. On radiating instabilities and resonant overreflection. *J. Atmos. Sci.*, **35**, 1190–1198.

Morrison, P. J. 1998. Hamiltonian description of the ideal fluid. *Reviews of Modern Physics*, **70**, 467–521.

Nazarenko, S. V., Zabusky, N. J., and Scheidegger, T. 1995. Nonlinear sound–vortex interactions in an inviscid isentropic fluid: A two-fluid model. *Phys. Fluids*, **7**, 2407–2419.

Peregrine, D. H. 1998. Surf zone currents. *Theoretical and Computational Fluid Dynamics*, **10**, 295–310.

Peregrine, D. H. 1999. Large-scale vorticity generation by breakers in shallow and deep water. *Eur. J. Mech. B/Fluids*, **18**, 403–408.

Plougonven, R., and Snyder, C. 2005. Gravity waves excited by jets: propagation versus generation. *Geophys. Res. Lett.*, **32**, L18802.

Plumb, R. A. 1977. The interaction of two internal waves with the mean flow: implications for the theory of the quasi-biennial oscillation. *J. Atmos. Sci.*, **34**, 1847–1858.

Plumb, R. A., and McEwan, A. D. 1978. The instability of a forced standing wave in a viscous stratified fluid: a laboratory analogue of the quasi-biennial oscillation. *J. Atmos. Sci.*, **35**, 1827–1839.

Polzin, K. L. 2008. Mesoscale eddy–internal wave coupling. I Symmetry, wave capture and results from the mid-ocean dynamics experiment. *Journal of Physical Oceanography.*

Ruessink, B. G., Miles, J. R., Feddersen, F., Guza, R. T., and Elgar, S. 2001. Modeling the alongshore current on barred beaches. *Journal of Geophysical Research*, **106**(C10), 22451–22463.

Saffman, P. G. 1993. *Vortex Dynamics.* Cambridge: Cambridge University Press.

Salmon, R. 1998. *Lectures on Geophysical Fluid Dynamics.* Oxford University Press.

Shaw, T.A., and Shepherd, T.G. 2008. Wave-activity conservation laws for the three-dimensional anelastic and Boussinesq equations with a horizontally homogeneous background flow. *Journal of Fluid Mechanics*, **594**, 493–506.

Shepherd, T. G. 1990. Symmetries, conservation laws, and Hamiltonian structure in geophysical fluid dynamics. *Adv. Geophys.*, **32**, 287–338.

Shepherd, T. G., and Shaw, T. A. 2004. The angular momentum constraint on climate sensitivity and downward influence in the middle atmosphere. *J. Atmos. Sci.*, **61**, 2899–2908.

Spiegel, E. A., and Veronis, G. 1960. On the Boussinesq aproximation for a compressible fluid. *Astrop. J.*, **131**(Mar.), 442.

Thorpe, S. A. 2004. Langmuir circulation. *Ann. Rev. Fluid Mech.*, **36**, 55–79.

Vallis, G. K. 2006. *Atmospheric and Oceanic Fluid Dynamics: Fundamentals and Large-Scale Circulation.* Cambridge, U.K.: Cambridge University Press.

Vanneste, J., and Shepherd, T. G. 1999. On wave action and phase in the noncanonical Hamiltonian formulation. *Proc. Roy. Soc. Lond.*, **455**, 3–21.

Wallace, J. M., and Holton, J. R. 1968. A diagnostic numerical model of the quasibiennial oscillation. *J. Atmos. Sci.*, **25**, 280–292.

Whitham, G. B. 1974. *Linear and Nonlinear Waves.* New York: Wiley-Interscience.

Young, W. R. 2010. Dynamic enthalpy, conservative temperature, and the seawater Boussinesq approximation. *Journal of Physical Oceanography*, **40**, 394.

Index

Printed in the United States
by Baker & Taylor Publisher Services